Proceedings of the Sixth Annual Telecommunications Policy Research Conference

Proceedings of the Sixth Annual Telecommunications Policy Research Conference

Edited by
Herbert S. Dordick
University of Southern California
and the Massachusetts Institute of
Technology

Lexington Books
D.C. Heath and Company
Lexington, Massachusetts
Toronto

Library of Congress Cataloging in Publication Data

Conference on Telecommunications Policy Research, 6th Airlie House, 1978.

Proceedings of the sixth annual Telecommunications Policy Research Conference.

1. Telecommunication–Congresses. 2. Telecommunication–Data processing–Congresses. 3. Communications research–Congresses. 4. Television broadcasting–Social aspects–Congresses. I. Dordick, Herbert S., 1925-
HE7645.C66 1978 384 78-20380
ISBN 0-669-02732-4

Published simultaneously in Canada

Printed in the United States of America

International Standard Book Number: 0-669-02732-4

Library of Congress Catalog Card Number: 78-20380

Contents

Preface

The Airlie Idea

Each spring for the past six years, researchers and government officials concerned with telecommunications policy have met at Airlie House in Virginia to exchange findings and views. This series of meetings was started in 1972 by Bruce Owen, of Stanford University. At that time Owen was an official at the Office of Telecommunications Policy, and he felt the need for more academic input into government policymaking. As a result, he convened a group of about forty researchers and policymakers at Airlie House to discuss mutual interests. The meeting was so successful that the experiment was tried again the next year, and it has been repeated each year since then.

The Airlie House Conferences are unique. They are not, strictly speaking, a professional gathering or a briefing session. Rather, they contain elements of both. At each meeting there are exchanges of views between researchers. There are also state-of-the-art meetings for nonexperts as well as sessions in which government people can express their needs to the research community. Finally, there are sessions that combine all three of these functions.

Common to all these sessions is an interest in research that bears on public policy in communications. Most often, this research is in the field of economics but in recent years, and at the 1978 conference in particular, there have been sessions on technology, behavioral science, and the law. Some of the researchers come from the academic world, but many also come from think tanks and industry. Participants come from the Senate and House Communications Committees and the Executive Office of the President. There is also significant industry participation.

The Airlie House meetings render much needed services. Nowhere else can policy researchers meet and concentrate primarily on their major concerns; nowhere else can they get feedback from government officials who should be the recipients of their findings. Professional communications organizations generally neglect telecommunications policy research. Only at Airlie House can government officials who are interested in making better-informed policy choices about communications concentrate on new findings reported in the fields of interest to them and obtain background in areas where they may be deficient. The most important task of the conference planner, then, is to invite contributions that represent current and significantly important research, research that is of value to decision makers facing today's issues or that warns of issues to be faced in the near future. Airlie House is the site of new and often important research disclosures, for leaders in the field prefer to make their initial presentations before their peers to invite comment and debate. The Telecommunications Policy Conferences provide a unique and important focus for telecommunications policy research.

Communications policymaking is the full-time occupation of many groups in the United States. The web of government, industry, foundation, university, and citizen organizations that enter policy debates creates an awesome montage for the observer, especially if the observer is from foreign shores where national postal, telegraph, and telephone companies (PT&Ts) hold sway within highly structured societal environments.

Government bodies in the United States that are involved in communications policymaking include the Executive Office of the President, the Department of Commerce, the Federal Communications Commission, and committees of both Houses of Congress. Policy and regulatory activities of other agencies laterally affect communications; for example, the Federal Trade Commission is vitally concerned with advertising issues, and the Department of Justice is often the last resort on matters concerning industry structure. Policy is made by other government organizations that one might think have little to do with communications directly; the Department of State enters international matters, and the Internal Revenue Service establishes rules and regulations that have far-reaching effects on the communications industries.

Industry associations that look after the interests of their clients include the National Association of Broadcasters, the National Cable Television Association, and the Telecommunications Association of America. The National Association of Regulatory Utility Commissions looks after the interests of the states.

Corporations, too, spend a great deal of money influencing policy. Within our free and more or less competitive society, they are often very effective in lobbying whatever policies they advocate.

In recent years, as the science of policy analyses has grown, foundations including Ford and Markle have sponsored research and, some say, have unduly influenced the course of communications policy in the United States. Citizens' groups—Action for Children's Television and more recently the national PTA—have begun to take active roles in seeking to rid television of "gratuitous violence and sex."

Finally the universities are building facilities and staffs capable of performing policy research, but because of the way such research tasks are undertaken, they frequently find themselves defining policy options, even making policy choices.

The United States' experience in communications policymaking is unique, for it is based on a decentralized system that thrives on competition, controversy, and confrontation. It is this very environment that makes for a robust system that continually seeks new solutions to new problems.

Such a system, however, also creates problems that can be extremely frustrating to those who consider themselves champions of communications. There is no orderly process of development. There is no structured way in which new services and technologies are explored, evaluated, and compared. Tensions are created that often result in inappropriate use of technology, wasteful spending, and, in the international arena, poor bargains.

Some argue that we should centralize policymaking and create a more structured procedure for innovation. Others contend that such a course risks destruction of the potential for innovation inherent in the present system. Still others hold that in the coming periods of decreasing availability of investment dollars and resources, it may be necessary to become more efficient, and that if this means a more structured policymaking environment, so be it. This is the dilemma that a democracy faces.

In this environment the role of the university policy researcher becomes especially critical. While the actors in each of the government and corporate agencies participating in policy research frequently change, those at universities and research centers often do not. Their tenure can and indeed does create a substantial power base. By virtue of their sustained experience, constant contacts, and specialized expertise, academic policy researchers exert great influence on communications policy. This is a considerable responsibility. For this reason alone they must carefully examine the processes by which they perform policy research. Recognizing that policy research, like other research, is subjective, these policy researchers must be just as careful in the assumptions they make as in those they assign to the public. Whether they want to or not, academic researchers exert powerful influences on communications policy in the United States, and they ought not to consider themselves merely researchers who explicate and foreclose options.

The conference themes, the selection of session topics, the invitations to authors and discussants are the sole responsibility of the Organizing Committee. During the past ten years or so, telecommunications policy research has blossomed into a diverse and far-ranging area of scholarly inquiry. Its multidisciplinary character is illustrated by the nature of the topics now being explored: the social, behavioral, and political effects of television, the policy and institutional impacts of emerging technologies, regulated and alternative telecommunications marked structures, the international consequences of spectrum allocation, and speculations into the new information socety.

The 1978 Telecommunications Policy Research Conference

The 1978 Telecommunications Policy Research Conference was especially significant because appointees of the new communications policy administration attended. They had the opportunity to assess their new roles in the light of the large body of research completed and in process and to discuss their agenda with the research community.

As in previous years, the 1977 Organizing Committee selected candidates for the following year's organizing committee from the research community. The following individuals accepted the responsibility for planning and bringing about the 1978 conference (with their affiliation at the time of their selection):

Lawrence H. Day, The Diebold Group
Aimee Dorr, Harvard University
Henry Goldberg, Verner, Liipfert, Bernhard, and McPherson
Carol A.V. Keegan, Science Applications Inc.
Harvey J. Levin, Hofstra University
William A. Lucas, The Rand Corporation
Robert D. Willig, Bell Laboratories

At their first meeting the committee selected Herbert S. Dordick, of the University of Southern California, as chairman of the Organizing Committee.

The Organizing Committee's responsibility was to determine the focus of the 1978 conference, drawing on their understanding and appreciation of research in progress and on the needs of the policymaking community, which were determined from several meetings with government officials. The committee invited leaders of both the research and policymaking communities to serve as session chairpersons and, working together, invited papers from these communities.

The selection of topics and contributions was especially difficult this year because of the avowed intent of the committee to reflect the broad multidisciplinary mosaic that uniquely identifies all policy research but especially telecommunications policy research. The committee accepted three criteria by which to judge and select contributions. First, the research had to be of high quality and reportable. Second, the research had to be relevant to significant issues facing decision makers now or likely to be facing them in the near future. Third, analytical or descriptive research had to provide early warning of policy issues that are likely to emerge in the more distant future. While the conference itself provided several opportunities for informal discussions of timely topics, these proceedings contain only the papers deemed appropriate by the editor and the authors for publication and often represent the authors' rethinking of their initial versions presented at the conference.

Four themes reflecting present and forthcoming policy issues and current trends in policy research were selected.

1. Television: Its social effects and the policy issues they create
2. Computer-communications networks and the emerging information economy
3. Spectrum allocation
4. Economics of regulation and the economics of pricing of telecommunications services

Over three-hundred researchers and policy officials attended the 1978 Conference. Participants represented the Executive Office of the President, the Federal Communications Commission, the new National Telecommunications and Information Agency of the Department of Commerce, the Departments of Justice, Health, Education and Welfare, and State, the National Science

Foundation and several private foundations with interest in communications policy, several consumer interest groups, state and local regulators, the Corporation for Public Broadcasting, the Public Broadcast Service, and several major telecommunications and information industries. In addition, researchers and policy officials from Great Britain, Canada, France, Japan, and West Germany attended. As in previous years, the majority of the participants were from universities and other research-based organizations.

Acknowledgments

Funding for the 1978 conference was provided by the cosponsors: the Assistant Secretary for Communications and Information of the Department of Commerce, the Federal Communications Commission, the John and Mary R. Markle Foundation, the National Science Foundation and the Corporation for Public Broadcasting. The Organizing Committee is very grateful for this financial support. However, the committee as well as the research community are even more pleased with the hands-off policy of the cosponsors in the planning of the conference, a tradition of these conferences which is often lacking in similar endeavors.

The Center for Communications Policy Research of the Annenberg School of Communications at the University of Southern California provided much needed and appreciated administrative support. We are especially grateful to Rhonda Mange who was indispensible as administrative assistant. Without her tireless work and amazing patience the conference could not have been held nor these proceedings published.

To the session chairpersons who had the difficult task of soliciting papers and discussants, and prodding them into delivering their work, we owe much of the success of the conference. They were: Helen Franzwa-Loukas, U.S. Commission on Civil Rights; Charles L. Jackson, Subcommittee on Communications, U.S. House of Representatives; Elizabeth Baily, Civil Aeronautics Board; Bradley S. Greenberg, Michigan State University; Donald Dunn, Stanford University; Alfred Kahn, Civil Aeronautics Board; William Read, Harvard University; John Richardson, Department of Commerce; Philip A. Rubin, Corporation for Public Broadcasting; Walter Baer, The Rand Corporation; Tracy Weston, Federal Trade Commission; Emilie Griffin, Compton Advertising.

Cambridge, Massachusetts *Herbert S. Dordick*
1979

Part I
Television: Its Social Effects and the Policy Issues They Create

Introduction

Herbert S. Dordick

If one can identify a consumer movement in communications, it is certainly focused on television and its social effects. What television has done and is doing to our children is being fiercely debated by small but active groups of parents, researchers, and future parents. Congress has been shaken by demands for the regulation of advertising that children have been exposed to, in apparent contradiction to long-cherished traditional respect of the First Amendment. One solution that leaves the First Amendment intact is to seek ways to increase diversity and thereby choice for the viewer. This can be achieved through more active participation by citizens in the operation of television stations in making their voices heard and in the creation of new production and distribution organizations that work by diverse sets of rules. Another means to achieve diversity is through technology, and new technologies holding the promise of a television of abundance is emerging in the marketplace.

Atkin, Greenberg, and McDermott address racial differences in dependence on television as a general learning source, comparing responses of black and white youths. They also examine the impact of television portrayals of blacks on young white viewers. Their paper opens the policy argument about stereotyping—a matter that has not yet become a public issue, certainly not as compelling as violence, sex, or advertising, but could nonetheless become more important as the races and sexes achieve equality of portrayal or stereotyping on television.

Jeffries-Fox and Signorielli examine the question of the extent to which television's portrayals of occupations is reflected in children's or young people's conceptions about them. If, indeed television is a more potent source of occupational status knowledge than either personal contact or general community culture, then the power of television knows no bounds. Again we deal with the stereotyping power of television. The paper by Jeffries-Fox and Signorielli expands the potential policy question concerning stereotypes but with respect to jobs and professions. It is complementary to the paper by Atkins, Greenberg, and McDermott.

Reeves and Faber argue that most studies seek to determine the extent to which children assimilate the specific behaviors and attitudes shown on television into their lives. They tend to study individuals or single groups. They do not link television messages with actual behaviors and attitudes, which may result in premature policymaking. Reeves and Faber examine the impressions that children form of television characters in groups, concentrating especially on age and sex differences in these perceptions, the impact of television on children's perceptions of people in real life. Reeves and Faber suggest that policymaking might focus on conditioning processes, rather than only on codes, and regulations or commission rulings that may conflict with First Amendment issues since content is so important. They suggest, in part, education to help children become better consumers of television and specifically to assist them to understand the people on television.

Wartella adds to the study of stereotypes by examining children's impressions of television families and how this affects their impressions of real-life families. She finds that families on television present fairly prosocial portrayals of family characters. There is a limited range of family models. Television may have an impact on children's expectations of how people should behave in real life. To the extent that the lack of diversity restricts this range, policy should be addressed to this issue.

If diversity is the key to altering stereotypes or adding to the dimensions perceived by young people, then one means is to produce television programming with the specific objective of offering new visions of reality. FREESTYLE, a report by Williams on using television to combat sex-role stereotyping in children, does just that. FREESTYLE is a large project requiring the participation of a public television station, a county school office, a major provider of educational materials, a city college, and two university research organizations for formative and summative evaluations. Hollywood talent is used to capture both public television and possibly commercial television audiences. Williams describes the project's formative evaluation process, indeed, the entire role of formative evaluation and its importance to the production of educational materials with specific and behavioral and attitudinal objectives. The results to date appear positive; children are not brainwashed by television.

Regulatory or legal attempts to create diversity are founded on the First Amendment. Hammond argues for structural change in the media through increased minority ownership of broadcast outlets on the grounds that this might reduce the need to resort to regulatory sanctions and the attendant danger of government censorship.

Focusing on the portrayal of women as an example, Cantor discusses and disputes another alternative for increasing diversity that has been suggested by many, including the Federal Communications Commission (FCC)—that of employing more women as decision makers and creators. Pointing out that the FCC has jurisdiction only over the local stations and not the network organizations or the production companies, Cantor concludes that the orientation of mass communicators is shaped by the structure and culture of the organizations in which they work. To the extent that more women may influence this structure and culture, perhaps content will change. More careful research is needed, but an even greater need is citizen participation in media matters for the development of advocacy to raise the organizational level of women in the media workforce.

Hanks and Pickett examine the influence of the audience on the media, rather than the media on the audience, by examining the impact of the citizen movement on broadcasting. They report on an exploratory study of the activities and perceived influence of selected citizen groups on several television broadcasters in five major Eastern cities. They report that the immediate goals of the citizen groups were realized to varying degrees; employment for minorities and women as well as the reporting of events relevant to these groups increased. But little significant progress has been made toward achieving the long-range goals of social change, including increasing diversity of opinion and public debate on issues through citizen-initiated free speech messages.

Gutiérrez describes the Spanish International Network, a unique case of a foreign television network's operation in the United States. A variety of beneficial effects of this foreign network operating within U.S. borders are discussed, including making viable heretofore unprofitable UHF stations and serving the specific interests of a significant minority in the United States. However, several policy issues arise that require discussion. In the light of the potential for these imported networks to provide a significant amount of program diversity, their future should be carefully evaluated.

Richard Adler reminds us that concern about the effects of television advertising on children dates back to 1961, when the National Association of Broadcasters adopted guidelines governing the advertising of toys to children. The new Federal Trade Commission (FTC) inquiry is an excellent example of the use of empirical research in policymaking. In a review of this research Adler points out that while such research has helped identify areas of concern it is not likely to play a dominant role in determining the specific regulatory actions that will ultimately emerge from the FTC.

Adding to the concern about this television advertising is the more recent arousal of interest in food advertising to children. Griffin argues that in designing policy and specifying policy-relevant research some important and valuable lessons could be learned from the accumulated experience of advertising and marketing people and their collective wisdom. Too often research conferences have ignored these practitioners of the arts. Griffin suggests that advertising can be used in a prosocial way to communicate nutrition messages that teach young children good practices—perhaps a "Sesame Street" for nutrition education.

Jerome and Freese recognize that children's food choices can be influenced by television food advertising. But how, if at all, has this development altered the family's traditional role as a transmitter of dietary practices and nutritional ideology? They conclude that the family exerts a strong influence on the food habits of young children but that television advertising appears to significantly increase the individual's awareness of the availability of certain products. Jerome and Freese point out the need to determine just how and where forces in the larger society converge with those of the family to alter dietary patterns.

Manoff tackles the question of nutrition education head-on. He argues that nutrition education requires constructive, positive access to the public media. And this can only be achieved through the time bank and the inner fourth network, which would command a percentage of all network and station time for public uses including nutrition education. In short, rather than regulation of the airways, there is a need for a share of the air—a significant change in the structure of the broadcast system.

If diversity and increased viewer choice are promising solutions to the issues being discussed in part I, then technology appears to offer some hopeful possibilities. Performing its congressionally assigned role as innovators, the Public Broadcasting Service is experimenting with the concept of the public telecommunications center, a means by which new distribution channels can be made available to meet the many program service needs of an increasingly diverse audience. Dordick provides the backdrop for these demonstrations and describes several in process. Schaar offers a historical perspective to the development of the public telecommunication center via cable television (CATV) in Spokane, Washington. Smith describes how the public station and the Toledo city government are planning to cooperate through the uses of both Instructional Television Fixed Services (ITFS) and CATV to increase and improve public participation in local government. Smith appropriately raises several important policy matters that, in the near future, will require FCC and possibly congressional consideration. Technology, too, carries with it issues of public policy impinging on the First Amendment. Finally, Steen discusses the application of a variety of technologies including radio to reach the many audiences of San Diego.

1 Race and Social Role Learning from Television

Charles Atkin,
Bradley Greenberg, and
Steven McDermott

This chapter focuses on two aspects of television and racial socialization. The first issue concerns racial differences in dependence on television as a general learning source, comparing responses of black and white youths. The second issue involves the impact of television portrayals of blacks on young white viewers.

Responses of black and white youngsters can be compared along a number of dimensions. We examine program viewing, perceived reality of fictional portrayals, identification with television characters, perception of racial portrayals, learning of social role information, and parental influence on viewing.

As the television networks present more clearly defined stereotypes of black people, questions arise concerning the effects of these portrayals on white viewers. To what extent do white youth acquire knowledge and beliefs about blacks from watching television? What psychological and sociological factors mediate this learning process?

The effects on white viewers include (1) beliefs about traits associated with black people, such as strength and humor, (2) beliefs about dissimilarities between the attributes of white and black people, including dress, talk, behavior, and appearance, (3) beliefs about the incidence of blacks in various segments of the population, such as the proportion of police or women who are black, and (4) subjective general knowledge gained about blacks from watching television, as reported by the respondent.

These learning variables can first be related to the amount of exposure to black programming to assess the impact of sheer exposure. A number of antecedent and intervening variables may also play a role in determining effects. We devote the most attention to five factors: (1) contact with blacks: it is expected that whites with the least direct experience with black people will be most influenced by television inputs; (2) perception of television content: viewers who selectively perceive certain attributes of black characterizations (such as physical attractiveness or speech style) will develop beliefs that correspond to the images they derive from television; (3) motivation to learn: young people who say they watch TV in order to learn about the world are expected to be most influenced, compared with those who watch to relax, escape, or enjoy themselves; (4) character identification: it is anticipated that whites who identify with black role models on television will be most affected by these portrayals; and (5) perceived reality: if the viewer thinks that televised depictions are true to life, then televised inputs should be more directly applied to actual beliefs about real blacks.

The evidence on these problems is derived primarily from a survey of black and white youth conducted under Project Castle at Michigan State University. In addition, we rely on an earlier study of white and black youth by Greenberg (1972), along with several other key studies in the literature.

Method

The primary data base is an omnibus survey administered to 550 students in the fourth, sixth, and eighth grades. Questionnaires were administered in the classroom, requiring approximately forty-five minutes to complete. The schools, located in inner-city neighborhoods in Michigan (Detroit) and California (San Jose and South Berkeley), were selected to provide substantial proportions of whites (N = 316) and blacks (N = 234).

The secondary set of evidence comes from a similar survey by Greenberg (1972) involving 409 fourth and fifth graders. Schools were chosen for their relative distance from an urban setting, to yield white youngsters with varying degrees of direct contact with blacks. There were 104 white children from a rural public school, 101 from a suburban district, and 95 from the inner city. There were also 107 black children in the urban site.

In both studies the following variables were measured: television program exposure, character identification, perceived reality of television, and subjective learning from television. The primary survey also measured motivations for watching television; perceptions of traits and behavioral patterns of characters; and parental coviewing, guidance, and control. The measurement for these variables is described in more detail in the following portion of the methods section and in the presentation of results.

For the assessment of impact on beliefs and knowledge of white children, a number of viewing and response indexes were constructed from questionnaire items. The following measurement approach was used for the key variables.

Black program viewing. From a list of forty-seven prime-time and Saturday programs, respondents could mark frequency of watching six shows that featured black casts. Exposure level varied from every week to never. The programs were "Good Times," "The Jeffersons," "Sanford and Son," "That's My Mama," "Grady," and "Fat Albert."

Motivation to learn. Four items tapped the learning component of television-viewing motivations, with the common stem "Do you watch television to learn . . . " followed by specific reasons such as " . . . how different people behave" or "talk." Response alternatives were yes, maybe, and no.

Black character identification. Identification with 37 television characters was assessed with a scale from "want to be like" to "don't care" to "don't want to be like." Ten of the names on the list were black performers: J.J., Thelma, Florida, and James Evans, Lamont Sanford, Freddy Washington, Bill Cosby, George and Louise Jefferson, and Dr. Cal.

Perceived reality of television. Among the battery of items dealing with the perceived reality of television portrayals were five items dealing with black roles. Respondents were asked to agree or disagree with statements asserting a correspondence between televised and real blacks; for example, "Black men I see on television are like black men in real life."

Content perceptions. Four questions asked the respondents to compare the portrayal of whites and blacks on television. The first item asked "On television shows, do you think that black people are stronger than white people, about as strong as white people, or not as strong as white people." Three other items involved the attributes of "funnier," "prettier/more handsome," and "do more things." For the perceived traits index the four items were summed. For the perceived dissimilarity index the responses indicating that either blacks or whites had more of a trait were scored higher than responses indicating no difference perceived on an attribute. A third perceptual index, perceived frequency of blacks, was composed of items asking, "Are there a lot of black kids/men/women on television?"

Trait beliefs. Beliefs about the actual attributes of black people were measured in a fashion parallel to that of content perception items. For this set of questions respondents were asked whether "Black people in real life are stronger than white people, about as strong as white people, not as strong as white people." The other three items covered "funnier," "active," and "prettier or more handsome."

Dissimilarity beliefs. Nine items composed the index of beliefs about the dissimilarities between black and white people. One set of statements tapped whether respondents thought that "black people in real life sound the same as white people or sound different from white people," "use the same/different words," "say hello the same/different way," "dress differently/the same," and "behave differently/the same." In addition, the four trait belief items were recorded so that either blacks or whites possessed more of a trait was scored as dissimilar.

Frequency beliefs. To measure respondents' estimates of the proportion of blacks in the population, a technique was developed to assess how many out of every ten categories of people would be black. For instance, one item asked, "Out of every ten men in this country, how many do you think are black?" followed by the numbers 0 through 10. Three other questions dealt with policemen, doctors, and women.

Self-report learning. The extent to which the sample felt that they acquired knowledge about blacks from television was examined with a set of statements with the stem "From television, I learn most of the things I know about . . ." and the referents "black people," "how black people dress," "how black people behave," and "how black people talk." For each item, the responses were yes, not sure, and no.

Contact with blacks. Four items asked the respondent to report how often they "play with or go places with black kids your age," "see black kids in your neighborhood," "talk with black kids your age," and "see black adults." The frequency of contact was measured from "less than once a week" to "almost every day."

Parental coviewing. The list of forty-seven television programs was again presented to respondents, who were asked to indicate which shows they "usually" watched with their mothers or fathers. There were six programs with predominantly black casts.

Parental guidance. From a list of sixteen shows, respondents could identify whether "your parent ever suggests or tells you to watch any of these shows." "Good Times," "Sanford and Son," "Fat Albert," and "The Jeffersons" were included.

Parental control. Several items tapped the manner of parental restrictions over television viewing. The first question asked, "How late do your parents let you stay up to watch television on school days?" The alternatives ranged from 8:00 to 11:00 or later. The second asked, "Compared to most kids your age, do your parents make more rules about television or less rules about television, or is it about the same as your friends?" This was followed by, "How often do your parents know what programs you are watching on television?" and "How much do your parents care what you are watching on television?" with four-step response scales. The final sequence inquired, "When your parents see you watching a program that they don't think you should watch, what do they do about it?" Respondents described the frequency that parents would, "turn off the television set or change channels," "explain why you should stop watching," "suggest some other activity," "send you out of the room," and "let you watch anyway."

Results

The presentation of findings begins with analyses of black-white differences in response to television. This is followed by a description of the correlation results for the white sample.

Viewing. Black youth spend considerably more time than white youth watching television programming. Two earlier studies by Greenberg and Dervin (1972) found that poor black children in Cleveland watched seven hours per day, compared with six hours for poor white children and four hours for middle-class white children. Poor black Philadelphia adolescents watched six hours per day, while poor whites saw four and a half hours and middle-class whites viewed three and a half hours.

In the Greenberg (1972) survey black children regularly watched an average of 7.6 of the twelve nonblack shows listed in the questionnaire, while whites saw an average of 6.8. For thirteen programs featuring black characters, black children viewed 9.0 and whites 6.5.

The findings from the primary survey show that the six predominantly black programs were viewed "every week or most weeks" by 52 percent of the black youth and 30 percent of the white sample. For the forty-one nonblack shows blacks also watched more often, by a margin of 34 percent to 29 percent.

Perceived reality. Across several studies black children have been found to judge television to be true to life to a greater extent than white children (Greenberg and Reeves, 1976). The secondary survey showed that more black children thought that television, in general, was realistic, but there was no racial difference in specific perception of television blacks. In the primary survey a variety of role portrayals were evaluated by the respondents. The question asked whether the child agreed with each statement; the percentages in table 1-1 indicate how many children of each race selected the yes choice. Approximately 30 percent to 40 percent additional children indicated that they were not sure, so the proportions in table 1-1 underestimate the degree of belief in the reality of television.

Although the differences are not large, there is a consistent tendency for blacks to believe in the validity of televised characterizations.

Character identification. In the secondary survey respondents were asked to name people on television whom they would most want to be like.

Table 1-1
Reality of Television Role Portrayals
(in percent)

Statement	*Blacks*	*Whites*
Black men/women I see on TV are like black men/women in real life	29	25
Blacks on TV talk like blacks in real life	46	42
Blacks on TV behave like blacks in real life	26	15
Black teenages on TV are like black teenagers in real life	41	33
The jobs men/women do on TV are like the jobs men/women do in real life	60	52
Mothers/fathers on TV are like mothers/fathers in real life	38	25
Wives/husbands . . . like real life	26	17
Teenage boys/girls . . . like real life	41	31
Men/women . . . like real life	30	28
Families . . . like real life	22	21
Police officers/doctors/secretaries/nurses . . . like real life	44	37

Note: Percentages indicate how many children of each race agreed with the statement.

Three-fourths of the black children cited at least one black character, as did 43 percent of the white children. The primary survey presented a list of characters and asked whether the respondents wanted to be like each one. The black youth were more than three times as likely to express identification with the twelve black characters. The role models selected by the sample are shown in table 1-2.

For white television characters (data not shown) blacks and whites are equally likely to identify. Averaging across twenty-six white television characters shows that 17 percent of black youth and 16 percent of white youths say that they want to be like the character. Thus black children relate to a larger number of characters, because they identify with both black and white actors.

Motivations for viewing. Watching televisions serves many functions for viewers. Among these are opportunity to relax, to get excitement, to find diversion from in-house or out-of-home unpleasantries, to pass time, and to combat loneliness (Greenberg, 1974). But the most repeated finding is that the black child expects to learn something from his television experiences. This child anticipates that learning will occur primarily in ways that will be socially adaptive. The results of the primary survey illustrate this basic point: 38 percent of the whites and 50 percent of the blacks say that they watch television to learn how different people behave, talk, dress, and look (average of four items). The proportions saying that they watch television to learn what police, doctors, secretaries, and nurses are like (average of four items) are 39 percent for whites and 52 percent for blacks.

Perception of racial portrayals. Different children differentially perceive the stereotypes presented on the television screen. Even though they may consume the same basic content, their interpretations of the favorability of

Table 1-2
Role Models Selected by Young Television Viewers
(*in percent*)

Character	*Black Children*	*White Children*
Freddy Washington	57	27
Lamont Sanford	45	13
J.J. Evans	45	19
Bill Cosby	43	15
Thelma Evans	43	7
George Jefferson	41	8
Louise Jefferson	30	6
Roger Thomas	28	11
James Evans	29	4
Florida Evans	21	3
Average for ten black characters	37	11

Note: Percentages indicate the proportion of children of each race who said that they wanted to be like the character.

the portrayals vary considerably. The differences between the black and white youth are substantial; the differences are shown in table 1-3.

A large majority of the white sample rated black and white performers as equivalent on each trait. The white children perceived the blacks as possessing more of each trait except for physical attractiveness; the black children's perceptions were even more favorable toward black characters.

Subjective learning. According to the self-reports of the young blacks and whites in the primary sample, television is a major source of learning about the world. In particular, job roles and behavior patterns are reportedly learned from television. Blacks acquire much more from television than whites; they even learn more about their own race. Respondents were asked whether they learned most of their information about the subjects in table 1-4 from watching television. The proportion who said yes is reported; on the average; another 25 percent said "not sure," so the percentages in table 1-4 underestimates.

The secondary survey provides more extensive data on learning about blacks by white children. Approximately two-fifths of the overall white sample said that television was their main source of information about black people. The reliance on television was much greater as proximity to blacks decreased, as indicated by the comparison by locale shown in table 1-5.

As table 1-5 shows, rural children with little direct experience learn the most about blacks from television. The findings for urban respondents are quite similar to the findings from the primary study done in a similar setting.

Parental influence. Parents of black children take a more active role in their children's television viewing. First, black young people in our primary sample experience closer control over television exposure than white respondents.

Table 1-3
Perceptions of Racial Portrayals
(*in percent*)

Statement	*Whites*	*Blacks*
On TV shows, do you think that black people are		
Stronger than white people	15	41
About as strong as white people	74	51
Not as strong as white people	10	8
Prettier or more handsome than white people	3	47
About as pretty or handsome as white people	71	49
Not as pretty or handsome as white people	26	4
Funnier than white people	23	44
About as funny as white people	60	41
Not as funny as white people	17	15
Do more things than white people	20	36
Do about as many things as white people	63	38
Don't do as many things as white people	17	26

Table 1-4
Subjective Learning from Television
(in percent)

Topic	*Whites*	*Blacks*
Jobs that men and women have	35	47
How men and women make decisions	21	34
How men and women solve problems	29	42
How parents and children interact	44	57
How husbands and wives interact	33	45
How teenagers act	37	48
Police, doctors, secretaries, nurses	35	41
How black people dress, behave, talk	23	40

Note: Percentages indicate the proportion of children of each race who report that they obtain most of their information about the topic from television.

Table 1-5
Reliance on Television as Main Source of Information about Black People by Locale
(in percent)

Statement	*Urban*	*Suburban*	*Rural*
From TV, I learn most of what I know about			
How black people talk	24	38	68
How black people look	20	28	66
How black people dress	20	33	68

One-third said that their parents "make more rules about television" compared with most children their age, while one-fifth of the whites reported this. Half of each group said that their parents "almost always" knew what programs they were watching, and a higher proportion of blacks said their parents "care very much" about the shows they watch. In addition, the black youth are no more likely to be allowed to stay up late to watch television on school days, and they have just as much prohibition against watching programs that parents do not approve of.

Second, parents of black youngsters are more likely to provide guidance toward approved programs on television. For a set of four programs featuring black performers ("Good Times," "Sanford and Son," "The Jeffersons," and "Fat Albert"), an average of 40 percent of the black parents and 22 percent of the white parents suggested that their child should watch. For other prosocial programs (such as "Little House on the Prairie," "The Waltons," "Rhoda"), about one-fifth of both the white and black youth received guidance. Black parents are more likely than white parents to direct their child's viewing toward police shows such as "Police Women" and "The Rookies."

Third, black parents are somewhat more likely to sit down and watch television programs with their children. This is especially the case for black programs, where more than 80 percent of the black parents watch with their child. For thirty-seven of forty-six programs studied a higher proportion of black than white youth reported their parents watched with them. The difference averaged 10 percent to 15 percent higher for black youth.

Fourth, black and white parents provide equivalent amounts of interpretation of television depictions. About two-fifths of the black and white youth said that their parents tell them about the real and unreal elements in televised portrayals of occupations, family roles, racial roles, and sex roles.

Next we describe the relationships between the five viewing response variables and the four belief variables in the white youth sample. For each criterion variable the zero-order correlation coefficients and beta weights are presented in the upper portion of table 1-6. The bottom portion of the table presents contingent correlations between amount of exposure and each criterion, at high and low levels of the five antecedent and intervening variables.

Trait beliefs. There is no relationship between the amount of exposure to black programs and beliefs about the traits of real-life black people (table 1-6).

Table 1-6
Relationships between Black Program Viewing and Beliefs

Predictor Variable	*Trait Beliefs*		*Dissimilarity Beliefs*		*Frequency Beliefs*		*Self-report Learning*	
Black program viewing	−.02	−.05	+.09	+.06	+.17	+.13	+.20	+.12
Motivation to learn	.00	.00	+.10	+.04	+.21	+.18	+.28	+.24
Black character identification	+.12	+.06	+.01	−.02	+.03	.00	+.10	+.06
Perceived reality of TV	.00	+.01	−.07	−.10	+.06	+.02	+.21	+.19
Content perceptions	+.51	+.50	+.46	+.46	+.11	+.10	—	—
Black Program Viewing								
Fourth grade (144)	−.11		+.07		+.25		+.25	
Sixth and eighth grade (172)	+.07		+.10		+.11		+.19	
High contact (114)	+.04		+.11		+.09		+.23	
Low contact (202)	−.05		+.09		+.20		+.18	
High motivation (108)	−.03		+.12		+.19		+.02	
Low motivation (208)	−.01		+.04		+.10		+.24	
High reality (115)	−.02		+.08		+.03		+.11	
Low reality (201)	−.01		+.09		+.24		+.24	
High content perception	+.08		+.13		+.20			
Low content perception	−.09		+.04		+.14		—	

Note: In the upper part of the table the left column in each pair represents the zero-order correlation between the predictor variable and the belief variable. The right columns present the standardized regression coefficient (beta weight); $N = 316$. Correlations greater than 0.11 are significant at $p < 0.05$ level.

In the lower part of the table contingent zero-order correlations between viewing and each belief variable are presented at high and low levels of each test variable. N's are given in parentheses.

The content perception variable corresponds to the appropriate belief variable: perceived traits, dissimilarities, or frequencies.

The strongest predictor is the content perception that television blacks possess the four traits corresponding to the real-life attributes measured in the survey; that is, if the viewer perceives that black performers are active or attractive, there is a strong tendency to believe that actual blacks are active or attractive people, regardless of the quantity of exposure to television blacks. In addition, youth who identify with black characters in these programs have slightly more positive trait beliefs. Perceived reality of television blacks and learning-motivated viewing do not relate to this belief variable. Regression analysis indicates that these five predictors produce a multiple correlation of 0.57, primarily due to the content perception variable.

Of the four trait items only one is positively associated with exposure. There is a positive correlation of 0.10 between exposure and belief that blacks are funny. The correlations are slightly negative for strong (−0.09), active (−0.06), and attractive (−0.01).

The relationship between amount of exposure and beliefs is slightly positive under the condition that these traits are perceived to be linked to black characters; the correlation is negative if these traits are not perceived by the television viewers. Thus there is a mild interaction between exposure and perception, such that the impact of programming depends on the interpretation of the audience. The only other small interaction occurs for grade in school; older respondents exhibit a positive viewing-belief relationship, while younger respondents have the opposite association.

Dissimilarity beliefs. To a slight degree those who watch more black programs believe that black people differ from whites along various traits and behaviors (table 1-6). But the perception that television content presents dissimilar images is the major correlate of this belief pattern; if black characters are seen as stronger or weaker or as dressing differently, then viewers clearly tend to believe that this is the case in the real world. Motivation to learn about life from television is the only other positive predictor. The five predictors combine to produce a multiple correlation of 0.48.

Among those who perceive dissimilarities between television black and white performers, the contingent correlation is more positive than for respondents perceiving little racial difference in portrayals. The relationship is also stronger among the subset of viewers highly motivated by learning. Otherwise no other interactions are apparent.

Frequency beliefs. There is a mild tendency for heavy consumers of black programs to believe that black people are more frequently found in the actual population. Those who watch television in order to learn also estimate higher proportions of blacks in the real world. There is only a weak association between perception of many blacks on television and the parallel beliefs about reality. The multiple correlation is 0.27.

The most sizable interactions with amount of exposure are for grade and perceived reality. Fourth graders and those who perceive television blacks to be less real have the strongest correlation between viewing and frequency beliefs.

Self-report learning. Young people who watch the most black programming tend to report that television teaches them most of what they know about black people. Those who are motivated to learn and who perceive black portrayals as realistic are also more likely to say that they learn from television. The multiple correlation is 0.37 for this criterion variable.

Among the contingent correlations the relationship is stronger under conditions of low perceived reality and low learning motivation.

Discussion

This section discusses some of the more significant findings relating to black-white differences in dependence on television and the learning of white youth about blacks from television. Then policy implications of the research evidence are explored.

Young blacks watch somewhat more television programming than whites and are much more likely to view shows featuring black characters. The overall viewing difference cannot be explained solely by socioeconomic differences, since the two groups in the primary sample were equivalent on this dimension and other studies have held socioeconomic status constant. Although some of this heightened viewing behavior may be explained by deficits in the social or physical environment, a major factor seems to be the greater benefits that black youth derive from television. They rely on television as an important source of learning about the world. They also appear to enjoy viewing more than alternative activities; the black situation comedies are especially popular among young viewers. Finally, the family context encourages heavier viewing as parents suggest programs and exhibit higher coviewing.

Black youth display considerably greater dependence on television as a source of useful information to be applied to everyday life. This is reflected in higher levels of learning motivation for viewing, more extensive identification with role models on television, and greater self-reported learning from television. Part of this reliance may be due to a lack of experiential inputs, but part must be attributed to successful prior utilization of television-derived information in practice.

Blacks and whites interpret the televised portrayals of the races in a selective fashion. This can be most clearly seen in the perceptions of physical attractiveness of television blacks and television whites; each race almost unanimously rejects the notion that characters of the other race are prettier or more handsome. In general, however, black characters are perceived in a favorable light by both whites and blacks.

The effects of televised black portrayals on white viewers appears to be primarily indirect. Regarding beliefs about traits of black people, the independent contribution of sheer viewing is minimal. The only positive correlation occurs for funniness; since most black characters appear in comedy roles, this outcome is plausible. Consideration of motivation, identification, and perceived reality adds only slightly to the variance accounted for. More important are the perceived attributes of television performers; the viewer's interpretation of the content is closely related to parallel beliefs about the real world. While the perception-belief correlation may be partly due to selective distortion of the televised message (if real blacks are believed to be handsome, then television blacks are also seen to be handsome), it is likely that perception exerts much of the influence on the relationship. Thus it is the combination of exposure and interpretation that determines impact—consistent with traditional conceptions that what the child brings to television is just as crucial as what television does to the child.

In the case of beliefs about dissimilarities between real-life whites and blacks, there is marginal evidence that amount of viewing has a direct influence. Again it is the viewer's perception of distinctions between portrayals of whites and blacks that makes the most important contribution.

Television's largest impact occurs for beliefs about the frequency of blacks in society. The more one sees blacks on the television screen, the higher the estimates about prevalence of blacks in the real world.

Only one other factor seems to make a major difference in the nature of television effects. The amount of direct contact with black people interacts with television; rural children with little experience report learning far more than urban children who interact regularly with blacks. In the primary survey, however, the relationship between viewing and learning does not differ substantially according to level of contact; this is due to the restricted range of variation, since most of the white sample had fairly frequent interaction with black people.

The findings described in this chapter provide a tentative basis for addressing policy implications. Stereotyped portrayals on television do constitute a public interest issue, although the case is not as compelling as with violence, sex, or advertising. The research has yet to provide proof of demonstrable harm, and given the varying value judgments that may be applied to the stereotyping problem, it may be impossible to achieve widespread agreement when particular effects are documented. Nevertheless, the basic goal of diversity is at least moderately consensual. The finding that heavy viewers tend to believe that black people are funny may be traced to the predominance of comedy roles for television blacks; if diversity were encouraged, more frequent black characterizations in dramatic roles might alter this image among young white viewers.

The data showing that belief effects depend on the perceptions of the viewer indicate that the television's role may be to reinforce prior dispositions rather than cause dramatic changes. This suggests that parents and schools might play an important role in shaping the process of learning from television, since these sources directly influence the attitudes that the viewer brings to the set. These results can also be interpreted as minimizing the importance of focusing on the messages being sent, since the same portrayals are not perceived in a uniform and straightforward manner throughout the audience. Perhaps intervention programs aimed at teaching the child different television-processing skills may have more impact than changes in the content of television.

Those responsible for producing and disseminating programs should recognize the extent to which young people seek entertainment content to learn about the world. Clearly television is more than mere entertainment, as indicated by findings showing that the portrayals are widely judged to be realistic, that characters are used as role models, and that many viewers rely on television as their major source of information. Youngsters are a vulnerable audience because of their lack of cognitive sophistication and restricted experiences. The finding that reliance on television is greater under conditions of limited direct experience suggests that special care should be taken in presenting depictions of relatively remote subgroups of the population, such as racial minorities or less visible occupations. For most people in the audience portrayals of blacks or Mexican-Americans are likely to have a much bigger impact than portrayals of whites, simply because of the rarer opportunity for contact with the former groups.

The role of parents is particularly crucial in the case of stereotyping on television. Parents can have a major influence by suggesting or restricting exposure to programs, explaining and interpreting content, and inculcating underlying attitudes about social roles that the child will encounter on television. The parents can then influence the types of material that the child learns, according to their own values.

Researchers also bear a major responsibility in providing further empirical evidence on this problem. Little is known about the impact of stereotypes compared with other television fare. Researchers should identify the factors that stimulate youngsters to seek information from fictional content and how selective perception processes operate—whether experiences with family, peers, teachers or others are the primary determinants of predispositions or whether vicarious experience with prior television programs also play a major role. There is also a need for research on the effects of minority portrayals on the racial self-concept of young black viewers.

References

Greenberg, Bradley, and Dervin, Brenda. with the assistance of Joseph Dominick and John Bowes. 1972. *Use of the Mass Media by the Urban Poor.* New York: Praeger Publishers.

Greenberg, Bradley. 1974. "Ratifications of Television Viewing and Their Correlates for British Children." In *The Uses of Mass Communications,* ed. J.G. Blumler and E. Katz. Beverly Hills. Ca.: Sage.

Greenberg, Bradley. 1972. "Children's Reactions to TV Blacks." *Journalism Quarterly* 49: 5-14.

Greenberg, Bradley, and Reevers, B. 1976. "Children and the Perceived Reality of Television." *Journal of Social Issues* 32: 86-97.

2 Television and Children's Conceptions of Occupations

Suzanne Jeffries-Fox and
Nancy Signorielli

Television is undoubtedly one of the most pervasive sources of knowledge about occupations among young people today. Previous analyses of major characters who populate the world of prime-time dramatic television programs revealed that 57 percent of the characters were employed.[1] These characters were often portrayed as professionals (22.1 percent), managers (11.7 percent), or law enforcement personnel (11.7 percent); fields of employment included government, health, and education.

Researchers have also interviewed children about occupations. Beuf found that television viewing had an interesting relationship to career choice for preschoolers.[2] Moderate television viewers had a wider range of possible future jobs, while heavy television viewers were more likely to select stereotyped careers.

DeFleur and DeFleur investigated the relative influence of television on children's occupational knowledge. They selected three groups of occupations: those with which the ordinary child comes into direct contact, those that are rarely seen directly but frequently portrayed on television, and those that are neither encountered directly nor seen on television. When they asked children of various ages to identify, describe, and assess the relative prestige of these occupations, the influence of television was apparent.

> [It] seems safe to conclude that, within the limits of the present sample of children and occupations, television is a more potent source of occupational status knowledge than either personal contact or the general community culture.[3]

Television's portrayal of occupational roles is governed by dramatic considerations rather than educational ones. Consequently, only those jobs that serve a dramatic function in a story are portrayed. Moreover, these portrayals are often stereotyped in terms of the types of characters who are cast in these roles. Since television portrayals of people at work may be a major source, if not the only source, of knowledge about certain occupations among a large number of young people in our society, it is important to determine how occupations are portrayed and whether children's conceptions of these occupations reflect television images.

The research reported here addresses the question of the extent to which television's portrayals of occupations are reflected in young people's conceptions about them. The Cultural Indicators Project's eleven-year cumulative archives of systematic observations of network television's dramatic content and a study of young people's conceptions of social reality show that much of what young people know about occupations is quite consistent with the vivid images in television's dramatic content.

The research reported in this chapter is part of an ongoing research project called Cultural Indicators. This project began with the investigation of violence in network drama in 1967-1968 for the National Commission on the Causes and Prevention of Violence. It continued under the sponsorship of the Surgeon General's Scientific Advisory Committee on Television and Social Behavior, the National Institute of Mental Health, and the American Medical Association. Although violence-related findings and indicators have been published widely,[4] the approach was broadly based from the beginning to collect observations on the role and symbolic functions of several specific aspects of life, including occupations, presented in television drama.

The Cultural Indicators research paradigm consists of two interrelated parts: (1) message system analysis, monitoring the world of television drama, and (2) cultivation analysis, determining the conceptions of social reality that television tends to cultivate in different groups of viewers.

As part of Cultural Indicators' cultivation analysis, a three-year panel study of junior high school children was undertaken. In 1977, the third year of that study, 216 of the student respondents were interviewed to determine their conceptions about adult life and their sources of knowledge. This part of the research was funded by a grant from the ABC Television Network. Its sample is made up of students in the eighth, ninth, and tenth grades of a suburban-rural New Jersey school system. It is evenly divided among boys and girls who are predominantly white and come from a variety of social backgrounds.

This analysis focuses on the images and lessons related to the portrayal of six occupations in television drama. This report presents the results of the analysis of two data sets: (1) message system analysis data to isolate the image of doctors, psychiatrists, paramedics, judges, lawyers and police in prime-time network dramatic programming; and (2) cultivation analysis data to determine children's conceptions of these occupations. Since data collection for this part of Cultural Indicators Project has just recently been completed, the results reported in this chapter represent findings from the initial stages of the ongoing analysis.

The Image of Doctors, Psychiatrists, Paramedics, Judges, Lawyers, and Police in Prime-Time Drama

Message system analysis is designed to investigate the aggregate and collective premises defining life and its issues in representative samples of mass-produced symbolic material. Such analysis rests on the reliable determination of unambiguously perceived elements of communications. Its data base is not what any individual would select but what an entire national community absorbs. It does not attempt to interpret single or selected units of material or to draw conclusions about artistic merit or ability to "sell" products. The analysis is limited to functions implicit in the prevalence, rate, symbolic structures, and distribution of clear and common terms in the samples.[5]

The characters who populate the world of prime-time network television drama are portrayed in a large number of occupations. Since 1973 more than fifty distinct occupations have been isolated in these programs. The following discussion focuses on the portrayal of six of these occupations: doctors, psychiatrists, paramedics, judges, lawyers, and police. This information is then used to determine whether children's conceptions of these occupations reflect the images portrayed on television.

Table 2-1 presents the distribution of characters in these occupations by the type of role (major or minor) they portray. Major characters make up 20.5 percent of the character population. However, more of the characters in each of these occupations, except judges, portray major roles than is usually the case. Specifically, more than one-quarter of all doctors, lawyers, and police and more than one-third of all paramedics and psychiatrists portray roles that are essential to the story line.

Table 2-2 presents the distribution by sex for each of these six occupations. Since 1969 the world of prime-time dramatic television has been consistently dominated by males. Previous analyses have consistently found that the major character population is about three-quarters male and one-quarter female.[6] Table 2-2 reveals that male domination is especially important in the portrayal of these occupations. However, when compared with recent census figures (see table 2-3), these sex-related portrayals are realistic. They mirror the actual distribution of men and women in these jobs in our society. Because only a small number of women and major characters are cast in these occupations, the remainder of this discussion looks only at the image of all characters (major and minor, male and female) who portray these occupations.

Demographic and Descriptive Descriptions

Tables 2-4 and 2-5 present the distribution of demographic and descriptive content items for all human major and minor characters in prime-time network dramatic programs who are cast as doctors, psychiatrists, paramedics, judges, lawyers, and police. Table 2-4 presents the items coded for major and minor characters, and table 2-5 presents the items that were coded only for major characters.

More than 60 percent of these characters are found in action-adventure programs. This distribution is similar for paramedics and judges; an even larger

Table 2-1
Role of Characters Cast in Six Medical and Legal Occupations, 1969-1976

Role	*Doctor*		*Psychiatrist*		*Paramedic*		*Judge*		*Lawyer*		*Police*		*All Characters*	
	N	*%*	*N*	*%*	*N*	*%*	*N*	*%*	*N*	*%*	*N*	*%*	*N*	*%*
All characters	283	100.0	18	100.0	24	100.0	42	100.0	134	100.0	795	100.0	9371	100.0
Major	73	25.8	7	38.9	8	33.3	3	7.1	33	24.6	172	21.6	1924	20.5
Minor	210	74.2	11	61.1	16	66.7	39	92.9	101	75.4	623	78.4	7447	79.5

Table 2-2
Sex and Role of Characters Cast in Six Medical and Legal Occupations, 1969-1976

	Doctor		*Psychiatrist*		*Paramedic*		*Judge*		*Lawyer*		*Police*		*All Characters*	
Role	*N*	*%*	*N*	*%*	*N*	*%*	*N*	*%*	*N*	*%*	*N*	*%*	*N*	*%*
All characters	283	100.0	18	100.0	24	100.0	42	100.0	134	100.0	795	100.0	9371	100.0
Males	268	94.7	17	94.4	23	95.8	38	90.5	123	91.8	767	96.5	6892	73.5
Females	15	5.3	1	5.6	1	4.2	4	9.5	11	8.2	28	3.5	2479	26.5
Major characters	73	100.0	7	100.0	8	100.0	3	100.0	33	100.0	172	100.0	1924	100.0
Males	67	91.8	6	85.7	7	87.5	3	100.0	28	84.8	160	93.0	1392	72.3
Females	6	8.2	1	14.3	1	12.5	0	0.0	5	15.2	12	7.0	532	27.7
Minor characters	210	100.0	11	100.0	16	100.0	39	100.0	101	100.0	623	100.0	7447	100.0
Males	201	95.7	11	100.0	16	100.0	35	89.7	95	94.1	607	97.4	5500	73.9
Females	9	4.3	0	0.0	0	0.0	4	10.3	6	5.9	16	2.6	1947	26.1

Table 2-3
Percentage of Male and Female Doctors, Judges, Lawyers, and Police in the Television World and in the United States

	Television		*Census*	
Role	*Men*	*Women*	*Men*	*Women*
Doctors	94.7	5.3	91.0	9.0
Judges	90.5	9.5	94.9	5.1
Lawyers	91.8	8.2	95.2	4.8
Police	96.5	3.5	93.8	6.2

percentage of psychiatrists, lawyers, and especially police are found in these types of programs. Doctors are the only occupational group who seem to be evenly divided in action and nonaction programs. Most characters who portray these occupations are also found in programs whose tone has been classified as serious. Also these characters usually portray roles that are serious; only paramedics and psychiatrists portray fewer serious roles than the entire character population.

Settled adults make up the largest group of characters in prime-time dramatic programming, and most of the characters who portray these occupational roles are also so classified. However, when chronological ages are compared, characters classified as being in one of these occupations are somewhat older than the entire prime-time population of human characters. The only differences are that psychiatrists are categorized and judged to be younger than most characters (mean age = 30.2) and more than one-fifth of the judges are classified as old. The elderly status of judges is also revealed in their mean chronological age of 53.4 years. Doctors, paramedics, lawyers, and police are all about 40 years old.

Most of the characters included in this analysis cannot be classified as married or unmarried. More of the paramedics are married (30 percent) than any other group. Television drama is populated primarily by white characters; only 11.1 percent of the characters are classified as belonging to other racial groups. Psychiatrists, paramedics, judges, and lawyers are even more likely to be white, while the racial distribution of doctors and police is similar to that of the general character population.

About one-third of this group of characters is categorized as "good" or "neither good nor bad"; 12 percent are categorized as "bad"; and 17.1 percent cannot be classified on this item. However, the characters who portray these six occupations are much more likely to be categorized as "good," and all of these occupations have a smaller percentage of "bad" characters than the norm. The characters in these occupations are also more likely to be successful than the "average" character.

Characters in television drama are more likely to be victimized than to commit violence (violence-victim ratio = 1.26).[7] The distribution of characters on violence-related items reveals that less than one-fifth of these characters commit acts of violence while 23.9 percent are victimized. However, when we examine the distribution of characters cast in these occupations on items of violence, we find that except for the police fewer of these characters either suffer or commit violence and those who are involved in violence are more likely to be victimized than to commit violence. However, the police are 1.3 times more likely to commit violence than to be victimized.

Three content items focusing on employment satisfaction, happiness, and importance were recently added to the message system analysis of major characters. However, these results must be viewed very cautiously because the items do not meet acceptable reliability standards. We have included them in this discussion because they provide additional useful information about how these occupations are portrayed in television drama. In table 2-5 it can be seen that 28.5 percent of major characters are happy with their employment, 36.0 percent find it satisfying, and 52.7 percent are categorized as having their jobs be important for them. Except for one case, when the characters cast in these occupations are happy and satisfied with the work, they are also coded as considering their jobs important. The only exception is that the percentage of police coded as being happy with their jobs is the same as the entire population of major characters.

Table 2-6 presents the mean score for the major characters who portray these occupations on a number of five-point personality-trait scales.[8] Overall, on most of these scales, the characters who portray these occupations are coded more positively than "typical" major characters. Doctors, judges, and lawyers are coded somewhat more positively on these scales than police, paramedics, and psychiatrists. Police are coded as the most powerful and most violent occupational group. Lawyers are the most attractive and the smartest; judges and lawyers are the most fair and social. All these occupational groups except paramedics are coded as being taller than the average character. Surprisingly, psychiatrists are coded as a little less smart and somewhat less supportive than the typical television character. Finally, each of these occupations except the police is coded as being quite peaceful and supportive.

Young People's Conceptions about Doctors, Psychiatrists, Paramedics, Judges, Lawyers, and Police Officers

As part of a three-year panel study, young people were interviewed to identify their conceptions and sources of knowledge about these six occupations. Respondents were asked the following open-ended questions about each occupation.

Table 2-4
Descriptive and Demographic Items for Major and Minor Characters, 1969-1976

	Doctor		*Psychiatrist*		*Paramedic*		*Judge*		*Lawyer*		*Police*		*All Characters*	
Item	*N*	*%*	*N*	*%*	*N*	*%*	*N*	*%*	*N*	*%*	*N*	*%*	*N*	*%*
	(N = 283)		*(N = 18)*		*(N = 24)*		*(N = 42)*		*(N = 134)*		*(N = 795)*		*(N = 9,371)*	
Chronological age	X = 42.1		X = 30.2		X = 42.4		X = 53.4		X = 40.8		X = 37.7		X = 36.5	
Sex														
Male	268	94.7	17	94.4	23	95.8	38	90.5	123	91.8	767	96.5	6,892	73.5
Female	15	5.3	1	5.6	1	4.2	4	9.5	11	8.2	28	3.5	2,479	26.5
Status														
Major	73	25.8	7	38.9	8	33.3	3	7.1	33	24.6	172	21.6	1,924	20.5
Minor	210	74.2	11	61.1	16	66.7	39	92.9	101	75.4	623	78.4	7,447	79.5
Program type														
Action	150	53.0	16	88.9	14	58.3	27	64.3	96	71.6	752	94.6	5,826	62.2
Nonaction	133	47.0	2	11.1	10	41.7	15	35.7	38	28.4	43	5.4	3,545	37.8
Marital status														
Cannot code	160	75.5	12	85.7	13	65.00	27	84.4	68	73.1	654	82.3	4,279	64.2
Not married	19	9.0	2	14.3	1	5.0	1	3.1	8	8.6	79	9.9	1,284	19.3
Married	33	15.5	0	0.0	6	30.0	4	12.5	17	18.3	62	7.8	1,105	16.5
Social age														
Child-adolescent	0	0.0	0	0.0	0	0.0	0	0.0	0	0.0	0	0.0	588	6.3
Young adult	25	8.8	7	38.9	0	0.0	0	0.0	5	3.7	67	8.4	1,867	19.9
Settled adult	243	85.9	11	61.1	23	95.8	33	78.6	128	95.5	717	90.2	6,336	67.6
Elderly	8	2.8	0	0.0	0	0.0	9	21.4	1	0.7	2	0.3	339	3.6
Cannot code	7	2.5	0	0.0	1	4.2	0	0.0	0	0.0	9	1.1	241	2.6
Race														
White	253	89.4	17	94.4	23	95.8	41	97.6	131	97.8	691	86.9	8,280	88.4
Nonwhite	30	10.6	1	5.6	0	0.0	1	2.4	3	2.2	104	13.1	1,062	11.3
Cannot code	0	0.0	0	0.0	1	4.2	0	0.0	0	0.0	0	0.0	29	0.3

Character type														
"Good"	163	57.6	14	77.8	11	45.3	20	47.6	54	40.3	437	55.0	3,412	36.4
Mixed	70	24.7	2	11.1	9	37.5	17	40.5	62	46.3	225	28.3	3,306	35.3
"Bad"	7	2.5	0	0.0	1	4.2	0	0.0	7	5.2	14	1.8	1,049	11.2
Cannot code	43	15.2	2	11.1	3	12.5	5	11.9	11	8.2	119	15.0	1,604	17.1
Success														
Successful	111	39.2	5	27.8	1	4.2	13	31.0	42	31.3	235	29.6	2,143	22.9
Mixed	88	31.1	4	22.2	15	62.5	12	28.6	57	42.5	328	41.3	3,484	37.2
Unsuccessful	17	6.0	1	5.6	2	8.3	3	7.1	18	13.4	58	7.3	1,501	16.0
Cannot code	67	23.7	8	44.4	6	25.0	14	33.3	17	12.7	174	21.9	2,243	23.9
Violence														
Does not commit	268	94.7	17	94.4	24	100.0	40	95.2	124	92.5	501	63.0	7,583	80.9
Commits nonfatal	12	4.2	1	5.6	0	0.0	2	4.8	9	6.7	245	30.8	1,427	15.2
Commits fatal	3	1.1	0	0.0	0	0.0	0	0.0	1	0.7	49	6.2	361	3.9
Victimization														
Does not suffer	259	91.5	16	88.9	22	91.7	38	90.5	115	85.8	567	71.3	7,125	76.0
Suffers nonfatal	19	6.7	2	11.1	2	8.3	2	4.8	15	11.2	196	24.7	1,755	18.7
Suffers fatal	5	1.8	0	0.0	0	0.0	2	4.8	4	3.0	32	4.0	491	5.2
Violence-victim ratio	−1.26		−2.00		−0.00		−2.00		−1.90		+1.29		−1.26	

Table 2-5
Descriptive and Demographic Items for Major Characters

	Doctor		Psychiatrist		Paramedic		Judge		Lawyer		Police		All Characters	
Item	*N*	%	*N*	%	*N*	%	*N*	%	*N*	%	*N*	%	*N*	%
Total, 1973-1976	43	100.0	7	100.0	8	100.0	2	100.0	21	100.0	172	100.0	1,924	100.0
Role														
Comic	5	11.6	0	0.0	2	25.0	0	0.0	3	14.3	8	4.7	274	14.2
Mixed	5	11.6	4	57.1	2	25.0	0	0.0	2	9.5	12	7.0	367	19.1
Serious	33	76.7	3	42.9	4	50.0	2	100.0	16	76.2	152	88.4	1,283	66.7
Total, 1975-1976	24	100.0	4	100.0	5	100.0	1	100.0	8	100.0	111	100.0	764	100.0
Employment/happiness														
Cannot tell	2	8.3	0	0.0	1	20.0	0	0.0	1	12.5	14	12.6	88	11.5
Happy	14	58.3	4	100.0	3	60.0	1	100.0	4	50.0	32	28.8	218	28.5
Mixed	8	33.3	0	0.0	1	20.0	0	0.0	3	37.5	63	56.8	227	29.7
Unhappy	0	0.0	0	0.0	0	0.0	0	0.0	0	0.0	2	1.8	31	4.1
Not employed	–	–	–	–	–	–	–	–	–	–	–	–	200	26.2
Employment/satisfaction														
Cannot tell	2	8.3	0	0.0	1	20.0	0	0.0	1	12.5	12	10.8	93	12.2
Satisfied	20	83.3	4	100.0	2	40.0	1	100.0	6	75.0	56	50.5	275	36.0
Mixed	2	8.3	0	0.0	2	40.0	0	0.0	1	12.5	38	34.2	167	21.9
Unsatisfied	0	0.0	0	0.0	0	0.0	0	0.0	0	0.0	5	4.5	29	3.8
Not employed	–	–	–	–	–	–	–	–	–	–	–	–	200	26.2
Employment/importance														
Cannot tell	0	0.0	0	0.0	1	20.0	0	0.0	0	0.0	3	2.7	67	8.8
Important	24	100.0	4	100.0	4	80.0	1	100.0	7	87.5	96	86.5	403	52.7
Mixed	0	0.0	0	0.0	0	0.0	0	0.0	1	12.5	12	10.8	83	10.9
Unimportant	0	0.0	0	0.0	0	0.0	0	0.0	0	0.0	0	0.0	11	1.4
Not employed	–	–	–	–	–	–	–	–	–	–	–	–	200	26.2

Table 2-6
Means Score on Personality Trait Scales for Major Characters, 1973-1976

	Doctor		*Psychiatrist*		*Paramedic*		*Judge*		*Lawyer*		*Police*		*All Characters*	
	X	*s*	*X*	*s*	*X*	*s*	*X*	*s*	*X*	*s*	*X*	*s*	*X*	*s*
Personality Trait	*(N = 43)*		*(N = 7)*		*(N = 8)*		*(N = 2)*		*(N = 21)*		*(N = 172)*		*(N = 1,923)*	
Repulsive-attractive	4.0	0.79	3.9	0.69	3.1	0.35	4.0	–	4.2	0.54	3.7	0.70	3.8	0.88
Unfair-fair	3.9	1.08	4.0	0.82	3.8	0.89	4.5	0.71	4.1	1.06	3.8	0.89	3.4	1.21
Unsociable-sociable[a]	4.0	0.82	4.0	0.58	3.6	0.52	4.5	0.71	4.2	0.75	3.7	0.79	3.7	0.96
Cold-warm	4.0	0.67	3.3	0.49	3.6	1.19	4.0	–	4.0	0.92	3.6	0.85	3.6	1.02
Powerless-potent[a]	3.9	0.86	3.6	0.53	3.4	0.74	4.0	–	3.9	0.94	4.1	0.74	3.6	0.93
Short-tall	3.9	0.80	3.6	0.53	2.5	0.53	4.0	–	3.9	0.57	3.7	0.74	3.5	0.77
Stupid-smart[a]	4.4	0.63	3.7	0.49	4.2	0.46	4.0	–	4.7	0.48	4.0	0.67	3.8	0.80
Irrational-rational[a]	4.1	1.01	3.6	0.53	4.1	0.35	4.0	–	4.2	0.87	4.0	0.78	3.6	0.97
Unstable-stable[a]	4.2	0.91	3.7	0.49	3.6	0.92	4.0	–	4.2	0.98	4.0	0.78	3.5	1.01
Bungling-efficient	4.1	1.01	3.7	0.49	3.1	0.83	4.0	–	4.0	0.97	3.9	0.82	3.5	0.97
Feminine-masculine	4.1	1.00	3.7	1.38	3.8	0.89	4.0	–	3.9	1.18	4.2	0.86	3.5	0.68
Elderly-youthful	3.0	0.72	3.7	0.49	3.2	0.46	2.5	0.71	3.3	0.73	3.2	0.59	3.3	0.80
Unhappy-happy[a]	3.3	0.78	3.7	0.49	3.4	0.52	3.0	–	3.0	1.22	3.0	0.66	3.1	0.89
Rich-poor	3.4	0.59	3.0	–	3.6	0.74	3.0	–	3.6	0.50	3.0	0.13	3.2	0.72
Violent-peaceful	4.0	0.75	3.7	0.76	4.5	0.53	4.0	–	4.0	1.00	2.7	0.86	3.1	1.13
Unsupportive-supportive[a]	3.9	0.95	3.2	0.96	3.8	1.10	4.0	–	3.9	1.64	4.0	0.80	3.6	1.08

[a]Caution, reliability marginal

1. What does a _____ [occupation] do on the job?
2. What are the good parts about being a _____? the bad parts?
3. What kinds of people make good _____?
4. What kind of education and training does someone need to be a _____?
5. Which person [selected from a show card of forty portrait-style photographs] looks like a _____? Why?
6. What are the differences between a _____ and a _____ [related job already discussed]?
7. How did you find out what you know about _____?

Respondents' answers to these questions were coded into categories inspired by the respondents descriptions of these occupations. Sex, race, and age stereotypes for each occupation were derived from the picture selected in response to question 5.

Responses to these questions, especially to question 7, revealed that television is by far the most important source of knowledge about the six occupations among the young people in this sample. Table 2-7 indicates that for each occupation nearly all the respondents could recall having seen the occupation portrayed on a specific television program. Almost all respondents mentioned interpersonal sources as providing knowledge about doctors, and a majority of the respondents cited an interpersonal source for knowledge about police officers. However, interpersonal sources of information were much less pervasive than television for the four other occupations. The preliminary analyses of these data also indicate that for some of the occupations interpersonal discussions about the jobs may actually consist of talking about how the occupation is portrayed on television. Reading was not claimed as a source of information by most respondents for any of the six occupations.

Since television is such an important source of information about these occupations, it is useful to investigate the extent to which young people's conceptions about these jobs are consistent with television portrayals. We next discuss the five most frequently suggested categories that were coded from respondents' descriptions of each of the six occupations or its incumbents. This discussion then compares these conceptions to the images of the occupations young people see when they watch prime-time network television drama.

Respondents gave many descriptions of doctors and their work. The five categories mentioned most frequently were as follows.

1. It is a doctor's job to *help others.* (86 percent)
2. Doctors are *friendly*, understanding. (48 percent)
3. Doctors face *unpleasant tasks* as part of their work, for example, having to tell someone that they or a loved one will die soon, working around depressing people, facing blood and gore. (47 percent)
4. Doctors *save lives*. (44 percent)
5. Doctors make *good money.* (41 percent)

Table 2-7
Percentage Claiming Sources of Knowledge for Six Occupations (N = 216)

	Television			*Reading*			*Interpersonal*			
Occupation	*All Programs*	*Drama*	*Comedy*	*All Reading*	*Fiction*	*Non-Fiction*	*All Inter-personal*	*Met Someone with the Job*	*Related to Someone with Job; or Want Job in Future*	*Talking to Some-one about Job*
Doctor	96	94	69	35	11	25	90	73	7	44
Psychiatrist	90	52	77	25	9	18	43	6	–	40
Paramedic	92	92	–	7	–	7	24	13	–	18
Judge	89	58	68	14	–	14	32	14	–	27
Lawyer	90	87	13	21	5	18	43	24	6	39
Police	97	94	82	29	8	23	67	57	12	51

Note: If fewer than 5 percent of respondents claimed a source, it was dropped from the analysis and is not reported here.

Respondents' conceptions that a doctor's job is to help others is consistent with the television portrayal. In the message system analysis doctors were found to be relatively supportive of others. The concept that doctors are friendly is consistent with the finding that television doctors are coded as being relatively warm. The image of doctors' saving lives may be considered consistent with the high proportion of television doctors who were coded as being successful.

Other frequently mentioned concepts were not consistent with the television portrayal of doctors. Television doctors were coded as only slightly above average in affluence, although many respondent's mentioned that they make good money. Finally, even though respondents mentioned the unpleasant aspects of a doctor's job, television portrayals showed doctors as extremely happy and satisfied in their work.

The five categories most frequently mentioned in descriptions of psychiatrists and their work were as follows.

1. It is a psychiatrist's job to *help others* solve their problems. (89 percent)
2. Psychiatrists are *friendly*, understanding. (45 percent)
3. Psychiatrists show *patience* in listening to people's problems. (36 percent)
4. Psychiatrists have *power* over other people's lives. (32 percent)
5. Psychiatrists face *unpleasant tasks* as part of their work, for example, having to work with depressing or crazy people. (32 percent)

Overall the image of psychiatrists in television drama is somewhat inconsistent. They are shown as "good" people who are sociable, happy, and warm. Yet they are coded as less supportive than most characters. The conceptions that the job of psychiatrists is to help others solve problems and show patience are consequently ambiguous in relation to the television portrayal. However, the concept that they are friendly and understanding is consistent with the television findings.

The fact that television psychiatrists are happy in their work also seems inconsistent with the conception that they face unpleasant tasks. Television portrayals show psychiatrists as only average in their power and control over others, a portrayal that does not support the conception that psychiatrists have power over other people's lives.

The five categories frequently mentioned in descriptions of paramedics and their work were as follows.

1. It is a paramedic's job to *help others.* (80 percent)
2. Paramedics *save lives*. (47 percent)
3. Paramedics face *unpleasant tasks* as part of their job, for example, facing blood and gore. (35 percent)
4. Paramedics have to think and *act quickly*. (32 percent)
5. Paramedics are *physically strong*. (34 percent)

Television paramedics are coded as average in warmth and supportiveness and seem to be less successful in their work than most characters. The most frequently mentioned statements about this job—that paramedics help others and save lives—are not particularly consistent with the television image. Only 11 percent of respondents mentioned that paramedics sometimes fail to save lives.

Message system analysis found that while paramedics are coded as about average in happiness, they are not categorized as being happy with their job. This is consistent with the concept that paramedics must deal with unpleasant tasks. The conception of paramedics as thinking and acting quickly is consistent with their television portrayal as rational and smart.

The five categories most frequently mentioned in descriptions of judges and their work were as follows.

1. Judges must make *difficult decisions*. (56 percent)
2. Judges must be *fair* and free of prejudice. (39 percent)
3. Judges face possible *danger* from criminals seeking revenge. (26 percent)
4. Judges are knowledgeable and *experienced*. (26 percent)
5. It is a judge's job to *uphold the law*. (23 percent)

The stereotype of judges as making difficult decisions is consistent with television portrayals of judges as smart and rational. Judges also scored very high on fairness, which is also consistent with respondents' conceptions of this job. The conception that judges uphold the law seems to be consistent with aspects of their television image such as being regarded as relatively stable.

The portrayal of judges as considerably older than most television characters may be seen to be related to respondents' conception that judges are knowledgeable and experienced, that is, the "old and wise" stereotype. The notion that judges face danger from criminals seeking revenge is consistent with the fact that judges on television are as likely as policemen to be killed.

The five categories most frequently mentioned in descriptions of lawyers and their work were as follows.

1. It is a lawyer's job to *defend people* and their rights. (67 percent)
2. It is a lawyer's job to *help others*. (47 percent)
3. Lawyers feel sad if they *lose a case* when the client is innocent. (46 percent)
4. Lawyers make *good money*. (34 percent)
5. Lawyers are *intelligent*. (32 percent)

Respondents' conceptions of lawyers as defending people and helping others are consistent with the television portrayal that lawyers are supportive, fair, sociable, and warm. Television lawyers are also likely to have mixed success, which is consistent with the stereotype that they are sad about losing cases.

The high scores for television lawyers as rational and smart are consistent with a conception that they are intelligent. But while respondents often mentioned that lawyers make good money, television lawyers were coded as only slightly above average in affluence.

The five categories most frequently mentioned in descriptions of police officers and their work were as follows.

1. Police officers are in *physical danger*. (65 percent)
2. It is a police officer's job to *protect the public* and prevent crime. (57 percent)
3. It is a police officer's job to *arrest criminals*. (56 percent)
4. Police officers are *physically strong*. (46 percent)
5. It is a police officer's job to *help others*. (39 percent)

The stereotype that police officers are in physical danger is consistent with their television portrayal. In television drama police are likely to be victims of violence and are even more likely to commit violence. These findings are also consistent with the television portrayal that police are not satisfied with their work. The high score for television police officers on the personality-trait scale of power, together with the fact that almost all portrayals of police are found in action programs, is consistent with the stereotypes that the police protect the public and arrest criminals. Police officers are also portrayed as supportive, which is consistent with the stereotype that the police help others.

Respondents were asked to select one picture from a series of forty portrait-style photographs that represented people who looked as if they would be employed in each of the six occupations. These pictures had been coded reliably in the past for sex, race (white or black), and age (older and younger adults). Most respondents had no difficulty in selecting pictures; pictures of males were overwhelmingly chosen for all occupations. Psychiatrist was the only occupation for which respondents selected a relatively higher proportion of women (14 percent) than was found in the television portrayals.

The proportion of pictures of blacks that were selected to represent the occupations was higher than the proportion of blacks in the world of television. Doctors were the only occupation group for which respondents selected almost all pictures of whites. Respondents selected only pictures of older people as judges, which is consistent with television's portrayal of judges as elderly. They also selected fewer pictures of older people as paramedics.

Summary

In general, many of the respondent's open-ended responses revealed conceptions about the six occupations that were consistent with many aspects of the television portrayals of these jobs. This finding was especially true for the law and law enforcement jobs. Television portrayals do not reveal the degree of affluence often attributed to doctors and lawyers by these respondents. Television doctors and psychiatrists were coded as very happy in their work, yet many respondents mentioned unpleasant aspects of these jobs. Psychiatrists on television were shown as less powerful than average, a finding that is inconsistent with the conception that psychiatrists have power over others. Their portrayal as less supportive is also inconsistent with the conception of psychiatrists as patient and helpful. Paramedics were not portrayed as particularly masculine, which is not consistent with respondent's conception that they are physically strong; and television portrayal of paramedics as mixed in their success is not particularly consistent with the notion that paramedics save lives and help people.

On the other hand, most stereotypes about these occupations, particularly for the law-related jobs, were consistent with television portrayals. Each of the five most frequently mentioned conceptions about judges and police officers had support in findings from the message system analysis; respondents' stereotypes about lawyers were supported in all areas except money.

Selections of pictures to represent the incumbents of each of the six occupations were generally consistent with both television portrayals and with real-world data. Most respondents selected males, although they were more likely to select women as psychiatrists. Respondents were somewhat more likely to select blacks and young people as paramedics than one would predict from television, their almost exclusive source of knowledge about that job.

Although these findings suggest that young people's images of occupations are correlated with televisions's portrayals, they do not address questions of effects. Future analyses will investigate whether exposure to television images has a contribution independent of other learning sources to stereotypes about occupations and the world of work.

Notes

1. Nancy Signorielli (Tedesco), "Patterns in Prime Time," *Journal of Communication*, Spring 1974.

2. Ann Beuf, "Doctor, Lawyer, Household Drudge," *Journal of Communication*, Spring 1974.

3. Melvin L. DeFleur and Lois B. DeFleur, "The Relative Contribution of Television as a Learning Source for Children's Occupational Knowledge," *American Sociological Review* 32 (1967) 777-789.

4. George Gerbner, Larry Gross, Marilyn Jackson-Beeck, Suzanne Jeffries-Fox, and Nancy Signorielli, "Cultural Indicators: Violence Profile No. 9," *Journal of Communication*, Summer 1978.

5. A description of the analytical framework can be found in "Toward Cultural Indicators: The Analysis of Mass Mediated Public Message Systems"

by George Gerbner, in the *Analysis of Communication Content: Developments in Scientific Theories and Computer Techniques*, ed. George Gerbner et. al (New York: John Wiley and Sons, 1969); and George Gerbner, "Cultural Indicators: The Third Voice" in *Communication Technology and Social Theory*, ed. George Gerbner, Larry Gross, and William H. Melody, (New York: John Wiley and Sons, 1973).

6. George Gerbner, Larry Gross, Marilyn Jackson-Beeck, Suzanne Jeffries-Fox, and Nancy Signorielli, "Violence Profile No. 9: Trends in Network Television Drama and Viewer Conceptions of Social Reality, 1967-1977," (Annenberg School of Communications, University of Pennsylvania, March 1978); Signorielli, "Patterns in Prime Time."

7. Violent-victim ratios are calculated by dividing the more numerous of these roles by the less numerous within each group. A plus sign indicates that there are more violents than victims and a minus sign indicates that there are more victims than violents. A ratio of 0.00 means there were no victims or violents. A +0.00 ratio means that there were some violents but no victims; a −0.00 ratio means that there were victims but no violents.

8. The results for judges, paramedics, and psychiatrists should be viewed cautiously because of the small number of characters who portray these roles.

3

Children's Impressions of Television Characters

Byron Reeves and *Ronald Faber*

A dominant concern in research on children and television is the extent to to which children assimilate into their own lives the specific behaviors and attitudes shown on television. Researchers typically seek to measure aggressive responses to violence and demand for sugared cereal and candy; for those heavily exposed to Saturday morning advertising or the development of political cognitions for those who begin watching public affairs programs. Though there are certainly exceptions, the attitudes, behaviors, ideas, and themes of television have been more important in media research than the characters who express them.

When researchers have focused on the portrayals of television people, they have tended to study single groups of people, usually a minority group. Recent studies have concentrated on the portrayal of black people on television (Dominick and Greenberg, 1970; Fife, 1974), portrayals of women (Tuchman, Daniels, and Benet 1978; Miller and Reeves, 1976), and most recently the elderly (Aronoff, 1974; Peterson, 1973). Although these studies deal with people on television, two aspects of the ways in which the portrayals are studied should be noted. First, most of the studies on television people rely on content analyses. Research questions have typically asked about the frequency of portrayals, changes in the portrayals over time, or the types of programming the people appear in most often. While these studies provide necessary information about what content may have an impact on children, they do not link television messages with actual attitudes and behaviors. In this sense policy based on this type of data may be premature.

Second, people on television are usually studies in isolation from the other people on television. For example, the frequency of blacks in prime-time television has been traced for the past several years. While this strategy is valuable in determining changes in media content and the potential of the portrayals to have an impact, it also assumes that the single attribute of people that has been focused on, race in this example, is an important attribute relative to other attributes that might be apparent if additional characters were included. It is possible in these cases that researchers may isolate attributes that are important in the adult world but are not primary attributes used by children to understand television portrayals.

This report was completed as part of a project grant to Byron Reeves from the Graduate School, University of Wisconsin-Madison. Portions of this chapter appear in vol. 7 of the *Sage Annual Reviews of Mass Communication Research*, ed. Ellen Wartella (Beverly Hills, Ca.: Sage Publications)

Very seldom has research concentrated on more general perceptions of television people, especially children's impressions of how people or groups of people are different. Television provides children with possibly the most complete presentation of different types of people that they are likely to see during childhood. Focusing on the attributes that children use to differentiate television portrayals may tell us more about media impact on how children categorize and relate to people than the study of any one group of people or type of person. The ways in which children learn to differentiate people may be very important in determining which attributes of people are dominant in children's real-life perceptions and interpersonal relationships.

This chapter examines the impressions that children form of television characters, concentrating especially on age and sex differences in these perceptions. The general concern in this discussion is the possible impact of television on children's perceptions of people in real life. Before the television research is reviewed, it is first important to examine children's perceptions of people in general. The more general process of impression formation and person perception could both affect and be affected by children's perceptions of people on television. A final section of the discussion deals with the implications of this research area for different types of public policy.

The Person Perception Literature

Only a small portion of the literature on person perception is concerned with the way children form impressions of other people. Most of the research has studied adults and has dealt with questions about the accuracy of perceptions rather than their content or organization. The studies with children have generally asked them to write impressions of people they know or people shown in experimental films. This method of gathering data has frequently limited the age of the youngest respondents to six or seven. The primary research interest in these descriptions is on the content and organization of children's responses. Content refers to both the number of traits and the type of traits a child uses. Organization is concerned with the complexity and abstractness of impressions and with the ability to handle discrepant information.

Content of Impressions

Much of the research on children's impressions of others has come from a cognitive developmental perspective. These studies are concerned primarily with changes in children's impressions at different ages, and not surprisingly, they have consistently found an increase in the number of traits that children use at older ages (Maddock and Kenny, 1973; Peevers and Secord, 1973; Livesley and Bromley, 1973; Scarlett, Press, and Crockett, 1971). More interesting, Livesley and Bromley found that a major shift in the number of traits used occurs between the ages of seven and eight. The increase in traits mentioned between these two age groups was larger than the difference between eight- and fifteen-year-olds. This period of greatest change coincides with the approximate age when children change from preoperations to concrete operations in Piaget's stage theory of cognitive development (Piaget, 1953).

While age differences in the number of traits employed is the most consistent finding, several researchers have also reported sex differences. These studies primarily find that girls use more traits than boys (Livesley and Bromley, 1973; Beach and Wertheimer, 1961; Wegner, 1974); however, this sex difference may simply reflect girls' greater verbal skills at these ages.

Differences in the number of traits used also varies with the characteristics of the person being described. Scarlett, Press, and Crockett (1971), for example, found that first-, third- and sixth-grade boys use significantly more constructs to describe other boys than to describe girls. Livesley and Bromley (1973), using children of both sexes, also found more terms being used to describe males.

A greater number of constructs are also used when describing people the child likes than for disliked others (Supnick, 1964; Peevers and Secord, 1973; Livesley and Bromley, 1973). Third- and fifth-grade boys in the Scarlett, Press, and Crockett study used significantly more terms to describe boys they liked than boys they disliked. Among third graders there was no difference between liked and disliked females; however, for the fifth graders, disliked girls elicited more traits than liked girls. Supnick (1964), on the other hand, found that by high school, boys used more traits for liked girls than for disliked ones.

Two additional characteristics have been found to influence the number of traits employed in descriptions. Children use more traits to describe people their own age than to describe adults (Livesley and Bromley, 1967, 1973). Finally, more traits are used to describe well-known people than to describe those who are less well known (Yarrow and Campbell, 1963).

The type of traits that children employ has also been found to change with age. Younger children tend to focus primarily on physical appearance (Watts, 1944; Livesley and Bromley, 1967, 1973; Brierley, 1966; Pratt, 1975). Most of these studies found a noticeable decrease in descriptions of physical appearance around seven years old; this decrease again corresponds to the time when most children have moved into the concrete operational stage of development.

Sex differences in the type of traits that children use have also been found. Girls seem to emphasize personality terms such as bossy or friendly (Livesley and Bromley, 1967; Brierley, 1966) and interpersonal and social skill traits like nurturance (Yarrow and Campbell, 1963; Campbell and Radke-Yarrow, 1956; Supnick, 1964; Beach and Wertheimer, 1961). Boys, on the other hand, focus more on nonconforming and aggressive behaviors (Yarrow and Campbell, 1963; Dornbusch et al., 1965). Dornbusch and his colleagues found that of the twelve top categories of traits used by males and females, eleven were on both lists. However, the frequency of using any specific category differed by sex.

Organization of Impressions

Organization of impressions typically refers to the complexity, abstractness, or hierarchical integration of impressions. Unfortunately, researchers have used different categories to conceptualize organizational traits. These categories make it difficult to directly compare results across studies; however, there appears to be a definite developmental trend running through most findings. This trend can best be epitomized by Werner's orthogenetic principle: "Whatever development occurs, it proceeds from a state of relative globality and lack of differentiation to a state of increasing differentiation, articulation, and hierarchic integration" (Werner, 1957, p. 126). Using the categories that researchers have imposed, we find that as children grow older their impressions become more qualified (Leahy, 1976; Livesley and Bromley, 1973), more situationally specific (Peevers and Secord, 1973), more complex (Yarrow and Campbell, 1963; Livesley and Bromley, 1973); and there is a greater use of abstract traits (Scarlett, Press, and Crockett, 1971; Livesley and Bromley, 1973) and internal and dispositional traits (Pratt, 1975).

The final attribute of impression formation to be discussed is the way that children integrate conflicting information about people. Children are increasingly able to use qualifiers and recognize that behaviors are situationally determined. Thus it should not be surprising that older children more frequently mention both positive and negative characteristics of people (Peevers and Secord, 1973; Gollin, 1958; Livesley and Bromley, 1973). Gollin (1958) looked at the ability to handle conflicting information about people by showing children aged ten to seventeen a five-scene film in which the main character was portrayed in both positive and negative ways. Children were then asked to write their impressions of this character. Responses were coded for the presence of inferences and concepts. Inferences were scored if the subject tried to explain or ascribe a motive for any action in the film, and concepts were coded when the respondent attempted to account for the diversity of the actor's behavior. The use of inferences increased most dramatically between ten and thirteen years old and continued to increase between thirteen and sixteen. Concepts were relatively rare among the younger children but were used by about half of the sixteen-year-olds. In the two oldest age groups (thirteen and sixteen) females used significantly more inferences and concepts than males.

In summary, researchers have found that impressions become more detailed as children grow older. With increasing age children also become more aware of the temporal and situational factors affecting the behaviors of others. Older children are more frequently able to deal with conflicting behaviors, and their impressions of others change from being univalent to divalent.

Sex differences are less clear. It appears that girls use more traits in describing people than boys and also emphasize personality and psychological traits more than boys. Girls also appear to comment on and account for conflicting behaviors more. Several researchers have hypothesized that females are both quicker and more likely to stereotype people than males (Sarbin, 1954; Kohn and Fiedler, 1961; Shapiro and Tagivri, 1959); however, all these studies have been conducted with adult subjects. Whether girls are also more likely to stereotype people than boys is still an open question.

Studies on Perceptions of Television Characters

Compared with the literature on children's perceptions of people in real life, the number of studies that focus on children's perceptions of television characters as a special group of people is very small. Few of the studies just reviewed can be compared with similar media studies where television characters have been the object of perception, because most of the media research has centered around only one or two questions. The purpose of most of the television research has been to identify character attributes that are important in children's attraction to various characters. In this sense researchers have begun with characters that children identify with and then asked questions about the attributes of those characters that predict the attraction. Only a couple of recent studies have been concerned with the attributes that children use to differentiate and describe characters independent of other criterion variables usch as identification.

Reeves and Greenberg (1977) found four primary dimensions that third-, fifth-, and seventh-grade children used to differentiate a sample of fourteen prime-time and Saturday morning entertainment characters. In order of importance, the attributes were humor, strength, attractiveness, and activity. Figures 3-1 and 3-2 show the sample of characters plotted along four spatial dimensions which represent the four attributes used to differentiate the characters.[1]

Dimension 1, humor, shows that children perceived characters such as Fred Flintstone, Gilligan, Fred Sanford, and Archie Bunker to be the funniest characters while Marshal Dillon, Steve Austin, John-Boy Walton, and Laura were the characters perceived to be the least humorous.[2] Further analysis of this attribute also revealed that humor was not necessarily a desirable attribute. The characters perceived as being the most humorous were also rated as being unsupported by their television peers—which probably makes them more like the classic boob than desirable role models.

The second two attributes (shown as dimension 2, figure 3-1, and dimension 3, figure 3-2) represented stereotypic evaluations of television characters. Characters rated highest on dimension 2, physical strength, were Steve Austin, Marshal Dillon, Reed, and Hawkeye. Characters highest on dimension 3, physical attractiveness, were Samantha, Mary Tyler Moore, Steve Austin, and Hawkeye. The final dimension, activity, differentiated the characters on the basis of the number of things they did in their television roles. Characters highest on that attribute were Gilligan, John-boy Walton, Laura, and Chico.

Figure 3-1. Dimensions 1 and 2 of a Four-Dimensional Space Representing the Perceived Similarities among Fourteen Television Characters

A replication of this study that was recently completed attempted to empirically derive the same four attributes using a different sample of children (grades six and eight) and a different sample of television characters (Reeves and Lometti, 1978). Among other new characters in this sample were three additional women to more evenly balance the ratio of television men and women. Essentially the same four dimensions emerged. The one important difference was that dimension 2 from the original study, physical strength, more clearly differentiated television characters in the replication in terms of sex. In fact, dimension 2 in the second study was the only attribute that clearly separated the male and female television characters in the sample. Another difference in the replication was that the attribute smart was also negatively related to humor.

Figure 3-2. Dimensions 3 and 4 of a Four-Dimensional Space Representing the Perceived Similarities among Fourteen Television Characters

The most striking finding in the two studies, however, was that there were no age or sex differences in the number or content of the four attributes used to differentiate the characters. This finding was unexpected and was especially divergent from the expectations about boys' and girls' use of the second two dimensions, strength and attractiveness. It was expected that children would stereotypically orient to just one of these physical descriptors, males depending on the attribute strength and females on physical attractiveness. In this study, however, the boys were just as likely to differentiate the characters using the attribute attractiveness as were the girls, and the girls were just as likely to use the attribute strength as were the boys.

When compared with the psychological literature on person perception, the absence of sex and age differences is conspicuous. Older children develop more complex perceptions of others that depend on increasingly inferential qualities rather than merely on physical descriptions. Sex differences included the use of more attributes by girls and the emphasis on personality and psychological traits. In terms of children's perceptions of television characters, however, none of these findings was confirmed. The location of the characters along the four dimensions was almost identical for children from grades three to eight, for children of both sexes.

Unfortunately, there appear to be at least two possible explanations for these differences, one theoretical and one methodological. At first, it seemed appropriate to make predictions about children's perceptions of television characters from theory and research concerned with their perceptions of actual people. The expectation was that television characters represent people who should be evaluated and categorized similar to people in real life. The lack of correspondence between the two literatures, however, suggests a counter-hypothesis. While the perceptual skills of children are developing to the extent where cognition is abstract, inferential, and generalized, children may also be learning that television, unlike real life, requires only the most simple evaluations. Consequently, while older children are making complex personality inferences about peers, for example, they may have learned early that television characters need only be evaluated on the basis of whether they are funny or serious. A similar rationale could be applied to the lack of sex differences in the perception of television characters. The sex-typed use of certain attributes to describe others may apply only to real people and not to the special case of television people.

Further, there is some indication that the actual attributes that emerge from descriptions of real people are different from those used to describe television characters. The major attributes that Livesely and Bromley (1975) report in their studies include attributes such as nice, kind, good, temperamental, helpful, bossy, generous, and smart. The primary attribute in the television studies, humor, does not even rank in the top ten attributes.

It is important, however, to note differences in the methods typically employed to study person perception among children. There has been little uniformity in the ways in which the actual attributes and the organization of the attributes have been studied, and it is relatively easy to speculate on differences in study results that could be attributed solely to research procedures. The methods used most often in the psychological literature have been either children's free descriptions of people, semantic differential scales, or the use of adjective checklists where children indicate which attributes they feel apply to various people.[3] Critics of these methods have argued that research results could be due as much to different abilities to use language as to developmental shifts in person perception. This criticism seems especially applicable to the finding that children use increasingly more attributes to describe others as they develop.

The two studies on children's perception of television characters relied on a multidimensional scaling procedure which does not require that children describe characters or rate them on attributes supplied by the researcher. Children rate only the *similarity* of all possible pairs of characters in each study using as the basis for their judgments whatever attributes they choose to think about. These data are then analyzed to provide a geometric representation of the psychological distances between each of the characters (the spaces in figures 3-1 and 3-2 were calculated using this method). The dimensions that emerge in the spaces are then assumed to represent fundamental psychological processes by which the children differentiate the people in the space. Other data are then used to determine how each of the dimensions should be conceptualized. Researchers have criticized this method because it is a mathematical procedure that attempts to obtain maximum parsimony (the fewest number of dimensions) from the data. In this sense the fewer number of dimensions that emerge using this method could be attributed to the scaling procedure used rather than simple perceptions. Problems have also been pointed out concerning children's ability to make relative comparisons using the words "alike" and "different" (Palermo, 1973).

Only one study in the psychological literature attempted to use multidimensional scaling to study children's perceptions of peers (Olshan, 1971). Similar to the television studies, the author found that three dimensions adequately represented the data and, further, that no age differences existed in the number of dimensions used to differentiate peers. Although this study did not find an age difference in the complexity of perception, there were differences in the actual attributes that emerged. The primary attributes that third-, sixth-, and ninth-grade children used to describe peers were good/bad and many friends/few friends. A smart/dumb dimension was found for third and sixth graders, adult/child for third graders, and male/female for sixth graders. In addition to the similarity with the television studies in terms of perceptual complexity, this study also suggests that television people and real people may be evaluated using different attributes even when the method for determining the perceptions (multidimensional scaling) is held constant.

One study that is currently under way is attempting to separate differences in the attributes ascribed to people and television characters that can be attributed to methods from actual differences in the perception of the people (Wartella and Alexander, 1979). Until the methodological issue is resolved, it will be difficult to argue with certainty that television people are evaluated differently from real people; however, the lack of differences in the studies using multidimensional scaling at least suggests the possibility that more uniformity exists in the perception of television people than researchers have expected.

Organization of Impressions about Television Characters

Similar to the focus of the psychological literature on person perception and children, the focus of media researchers has been on the content of children's perceptions of television characters. Most researchers, however, point out that perceptions of other people have both content and organizational aspects (Livesley and Bromley, 1973). Questions that could be asked about the organization of children's impressions of television characters deal with how children relate the various content statements into meaningful unified impressions.

Only one recent study has attempted to study the organization of children's perceptions of television characters (Wartella and Alexander, 1978). Children in grades two, five, and eight were asked to describe television characters "so that someone would know what he was like and why he was like that." The study found two age-related differences in the organization of children's perceptions of television people: (1) greater use of internal descriptors (such as motivational and personality characteristics) and (2) increased use of causal descriptions (such as describing reasons and/or motivations for characters behaviors). No sex differences in these trends were reported.

Unlike the studies on the content of perceptions of television characters, these studies support the expectations of age-related differences in children's organization of impressions, as suggested by past research in the general area of person perception. While the television studies on content of perception used methods different from those typically employed in psychological studies, the single study on organization of television impressions used a method much more consistent with the previous literature. Despite methodological problems, however, the limited number of attempts to study the perception of television characters suggests that there are differences in the actual attributes that children use to describe real people and television people, although the perceptions seem to be organized in a similar manner.

Why Study Children's Perception of Television Characters

Few of the psychological studies on children's perceptions of other people suggest other variables that may depend on the content or organization of children's impressions of others. The actual perceptions of others and the process of organization and development seem to be themselves the criterion concerns. This is not the case with research on the effects of television. The primary purpose of studying perceptions of television characters has been to study the attributes of characters that are related to children's desires to model their behavior. Consequently, most of the research in this area has actually taken as a starting point children's desires to identify with characters and then asked which attributes of the characters are related to that attraction.

Generally, age and sex differences in the attributes that predict children's attraction to television characters have been much greater than differences in the attributes they use to describe the characters. This was very apparent in the studies that identified humor, strength, attractiveness, and activity as the primary attributes that differentiated television characters (Reeves and Greenberg, 1977; Reeves and Lometti, 1978). The strength and activity dimensions were excellent predictors of whether boys of all ages wanted to "be like" or "do things like" the characters in the sample. The attractiveness dimension was an equally good predictor of the same measures for girls. Furthermore, the primary attribute that differentiated the characters, humor, was not a predictor of identification for either sex at any age.

Sex differences in the salience of the attributes in choosing role models are especially interesting in relation to the lack of sex differences in the use of the attributes to discriminate among the same characters in the same study. Boys were just as likely as girls to differentiate characters on the basis of physical attractiveness, yet the attractiveness of characters was useful only to the girls in choosing television role models. Similarly, girls were just as likely to differentiate characters on the basis of strength, although strength was important only to the boys in identifying with characters.

The key suggestion from this finding is that cognitive structures for boys and girls are the same while the use of the structures is completely different. This is different from saying that only males use the attribute of strength and that strength is therefore a predictor of television effects only for males. Males use both strength and attractiveness, but only one attribute is applicable to their modeling decisions. The counterpart process existed for girls' use of attractiveness. While others have suggested that little data exist to support stereotypic differences in the ways males and females describe and orient to other people (Maccoby and Jacklin, 1974), these results constitute a conspicuous exception.

The television studies that have focused only on the attributes that predict attraction to television characters have essentially confirmed the same stereotypic evaluations using several different methods. Miller and Reeves (1976) reported similar stereotypic rationales for model choices. After third- through sixth-grade children had indicated their favorite television characters in open-ended personal interviews, boys said they chose the characters primarily because they were strong or tough, and girls reported choosing models because they were good looking or well dressed. Using a multidimensional measure of identification in a survey of third through sixth graders, Reeves and Miller (1978) found the same pattern of prediction for the attributes strength and attractiveness. Similar findings for these stereotypic attributes were also found in two other studies (Greenberg et al., 1975; Miller, 1976).

Although the attributes strength and attractiveness have been consistent predictors of modeling, the most powerful predictor, and the one for which large sex differences have been found, has been the sex of television character. Several studies have demonstrated that children are most favorable toward same-sex characters (Miller and Reeves, 1976; Reeves and Miller, 1978; Miller, 1976). There are, however, important differences in the willingness of boys and girls to nominate opposite-sex models. Two studies report that no boys chose opposite-sex television characters as people they would like to "be like" while approximately one-third of both samples of girls chose male characters as primary role models (Miller and Reeves, 1976; Miller, 1976). Further, boys were more likely to name any television model than girls (Miller and Reeves, 1976).

In most of the studies on character attributes related to identification, television models have been determined before children are asked to describe their primary attributes. When children do describe the characters, they are usually asked to name the features of the character that made them choose the character as a model. This strategy could be criticized because by definition it would ignore other attributes of television characters that may not predict identification but could be important for other reasons. For example, the primary attribute that differentiated television characters in two past studies, humor, was not found to be related to any identification measures, yet it should probably not be assumed that there are no consequences of that particular perception. While researchers have concentrated almost exclusively on impressions of characters that predict identification, other criterion variables may further explain how television people affect children.

A first question deals primarily with the content of the attributes that children associate with television characters and involves the generalization of television perceptions to person perception in general. Television represents a primary source for meeting new people, especially for younger children, and typically the variance in the types of people is nowhere greater than on television. As noted in several studies in the area, the attributes that children use to describe television characters obviously depend on the number and types of people there are to observe. Humor would not be an important attribute of television characters, for example, if there were only serious characters on television. The variance in television portrayals could then influence the attributes that children use to differentiate not only television people but also people in real life. People in real life may be evaluated primarily on the basis of humor even though children might not have naturally settled on that attribute solely on the basis of the people they meet in real life.

This question becomes more important if the attributes that relate to television and real people are in fact found to be different. Comparing the results of the television studies with those in the general area of person perception at least suggests that these attributes may be different. Not all of the four primary attributes found to differentiate television characters appear in similar lists involving children's perceptions of real people. A study in progress is attempting to find out whether there are unique sets of television attributes and real person attributes and whether high television viewers are using primarily television attributes to evaluate real people (Reeves and Faber, 1979, in progress).

A second important contribution of this research may be a better understanding of the effects of attitudes and behaviors expressed on television.

It is probably an error to assume that effects of television on children can be totally attributed to either television people or their actions. Typically, however, we study the effects of people and actions as if they were independent and additive. We assume that what we know about the behaviors on television (say, number of violent acts, altruistic behavior, information about occupations) can be added to what we know about the characters who express them (that they are funny, attractive, or strong) to gain a full understanding of the impact of the medium. It is highly probable, however, that the two interact to produce effects. The same act of violence, for example, may have an impact if expressed by a strong, active male but no impact or negative impact if attempted by an attractive female. Studying these interactions could be very important in understanding how children relate television content to their own lives.

Other questions concerning the development of children's perceptions of television characters remain. The validity of studying impressions of television portrayals at least assumes that children can recognize the consistency of a role both within and between the same program. When, for example, do children realize that a character remains constant throughout a program or that the same individual appears at the same time every week? Are there differences in the impressions held of one-time characters versus characters who are developed, although with little variance, across different episodes? When do children begin combining and organizing scattered character attributes into unified impressions? How do children combine attributes, and what happens when they receive information about characters that runs counter to past impressions or forces them to combine both positive and negative information? Is there some minimum level of development that these impressions must reach before they begin to determine how and whether children assimilate television behavior into their own lives? Are there developmental changes in children's orientation to television characters versus television actions and events? Finally, what effect does children's understanding of the nature and purposes of television in general have on their impressions of television people? Will children form and apply impressions differently if they believe that television is a "magic window" versus a source of fantasy drama?

Can Present Research Help?

Despite the need for additional study, there are at least some policy implications of this research area that can be linked to present data. Before actual suggestions are made, however, it is probably important to consider the parts of the effects process to which this area of research is most relevant. The first objects of social policy that we most often consider are the sources of media messages, the networks. While some areas of research can be effectively linked to the regulation of content, we should probably not consider communication sources the only outlet for policy solutions any more than we should consider their messages the only required piece of evidence for documenting television effects.

Children's perceptions of television characters represent a perceptual process that intervenes between exposure to messages and their impact. The perceptions should help explain how television messages affect children and especially why the same messages, or the same television characters, can have different effects on different children. The policy considerations that come from the research on children's impressions of television characters could therefore perhaps most appropriately focus on children's perceptions by educating them about television and the potential biases in the people and messages that it presents.

There are several reasons why policy that focuses on conditional processes of perception may be more appropriate than policy that results only in regulations, laws, codes, or commission rulings. First, the translation of policy into research requires that values at some point be imposed on the data during translation. There would be great difficulty in finding total agreement among those responsible for policy that it is wrong for children to perceive television characters or real people to be primarily funny or serious. By dealing with individual perceptions rather than changes in the content, different policy solutions could result from the same research and could better reflect the concerns of more specific groups of people. Similarly, content would not have to be changed for those who have no problem with the programming as it currently exists.

Second, the translation of media research into regulatory policy also requires that complex legal and economic concerns be considered. Dealing with perceptions of content rather than changes in its amount and form would bypass the especially difficult constitutional problems of regulating media. Finally, by changing media content, we have no assurances that we would not be replacing one set of incidental effects with another. It may be impossible to eliminate from television drama potentially negative consequences of primarily positive messages.

Two major suggestions will be proposed for helping children to become better consumers of television programming and specifically for helping them to understand the people on television. In terms of children's organization of impressions about television characters, children could benefit from attempts to develop the process by which they assign and account for attributes of people that are in conflict. One of the major differences between younger children's and older children's organization of impressions of other people is an awareness that temporal and situational factors affect how people act. Thus older children are able to account for conflicting behaviors in their impressions of others while younger children usually fail to do so. Even in free descriptions of people, younger children usually use only "good" or "bad" traits for an individual while older children use both. One way of aiding younger children to see situational constraints is by encouraging them to engage in role taking. The goal of role taking is to understand all the factors that influence a person's actions.

Parents and teachers can help and encourage children to engage in role taking. When parents watch television with children, they can increase their children's understanding of television characters' behavior by asking how the child would act or feel in the same situation. Teachers might also encourage role taking by having children imagine or act out situations in books they are reading or by role playing the actions of historical figures. When children ignore some constraint on the character, this should be pointed out. If children can learn to role take in these situations, it is likely that this skill will also carry over in their formation of impressions of other people.

Role-taking abilities are often considered a combination of maturational development and environmental interaction. If this is true, it would not be possible to teach very young children the highest levels of role taking. However, providing role-taking experiences could help children operate at their highest maturationally determined level.

The second suggestion deals with the actual attributes that children use to differentiate people on television and the possible generalization of those attributes to people in real life. Almost by definition, the number and types of people that fill a child's environment, including people on television, determines the attributes that children rely on to deal with others and possibly to make inferences about themselves. It may then be important to monitor the extent to which television people provide children with the sample of people that they use as the basis of their perceptions of others. Part of a child's television education could focus on the extent to which the variance in television people actually reflects the variance of people in real life. Further, it may help children judge whether people in real life deserve descriptions that differ from or are at least supplemental to those suggested by television.

Notes

1. These spaces are the result of a multidimensional scaling analysis of the perceived similarities among all possible pairs of the fourteen television characters. Children were first asked to rate the similarities of the character pairs on a five-point scale. The resulting matrix of character-by-character similarity ratings was then input into the INDSCAL multidimensional scaling program which yielded the four-dimensional solution shown in figures 3-1 and 3-2. Children were then asked to rate each of the fourteen characters on nine unidimensional univariate measures (funny, strength, attractiveness, reality, age, support received from television peers, activity, and goodness). Sex was the ninth

unmeasured attribute. The unidimensional measures were then related to the multidimensional spaces using multiple regression procedures to conceptualize each of the dimensions in the spaces. (See Reeves and Greenberg, 1977 for further details of the analysis procedure.)

2. The unidimensional ratings did not include polar adjectives (such as funny versus serious) but rather asked whether the attribute was applicable (funny versus not funny). Polar adjectives should probably not be assumed for children in the absence of evidence that the adjectives do in fact represent different ends of same continuum.

3. For a detailed review of the methods researchers have used to measure both the content and organization of children's impressions of others see Livesley and Bromley (1973, pp. 39-45).

References

Alexander, A., and E. Wartella, "Children's Impressions of Television Characters as Real People." Paper to be presented to the Mass Communications Division, International Communications Association, Philadelphia, 1979.

Aronoff, C. 1974. Old age in prime time. *Journal of Communication* 24:86-87.

Beach, L., and Wertheimer, M. 1961. A free-response approach to the study of personal cognition. *Journal of Abnormal and Social Psychology* 62:367-374.

Brierley, D. 1966. Children's use of personality constructs. *Bulletin of the British Psychological Society* 19:72.

Campbell, J., and Radke-Yarrow, M. 1956. Interpersonal perception and behavior in children. *American Psychologist* 11:416 (abstract).

Dominick, J., and Greenberg, B. 1970. Three seasons of blacks on television. *Journal of Advertising Research* 10:21-27.

Dornbusch, S., Hastorf, A., Richardson, S., Muzzy, R., and Vreeland, R. 1965. The perceiver and the perceived: Their relative influence on the categories of interpersonal cognition. *Journal of Personality and Social Psychology* 1:434-440.

Fife, M. 1974. Black image in American TV: The first two decades. *Black Scholar* 6:7-15.

Gollin, E. 1958. Organizational characteristics of social judgment: A developmental investigation. *Journal of Personality* 26:139-154.

Greenberg, B., Heald, G., Wakshlag, J., and Reeves, B. 1975. TV character attributes, identification, and children's modeling tendencies. Paper presented to the International Communication Association, Portland, Oregon, April 1975.

Kohn, A., and Fiedler, F. 1961. Age and sex differences in the perception of persons. *Sociometry* 24: 157-164.

Leahy, R. 1976. Developmental trends in qualified inferences and descriptions of self and other. *Developmental Psychology* 12:546-547.

Livesley, W., and Bromley, D. 1967. Studies in the developmental psychology of person perception. *Bulletin of the British Psychological Society* 20:67.

Livesley, W., and Bromley, D. 1973. *Person Perception in Childhood and Adolescense*. New York: John Wiley and Sons.

Maccoby, E., and Jacklin, C. 1974. *The Psychology of Sex Differences*. Stanford, Ca.: Stanford University Press.

Maddock, R., and Kenny, C. 1973. Impression formation as a function of age, sex, and race. *Journal of Social Psychology* 89:233-243.

Miller, M.M. 1976. Factors affecting children's choices of televised sex role models. Department of Communication, Michigan State University.

Miller, M.M., and Reeves, B. 1976. Dramatic TV content and children's sex-role stereotypes. *Journal of Broadcasting* 20:35-50.

Olshan, K.M. "The Multi-Dimensional Structure of Person Perception in Children." Ph.D. dissertation, Rutgers University, 1971.

Palermo, D. 1973. More about less: A study of language comprehension. *Journal of Learning and Verbal Behavior* 12:211-221.

Peevers, B., and Secord, P. 1973. Developmental changes in attribution of descriptive concepts to persons. *Journal of Personality and Social Psychology* 27:120-128.

Petersen, M. 1973. The visibility and image of old people on television. *Journalism Quarterly* 50:569-573.

Piaget, J. 1953. *The Origin of Intelligence in the Child*. London: Routledge.

Pratt, M. 1975. A developmental study of person perception and attributions of social causality: Learning the what and why of others. Doctoral dissertation. Harvard University.

Reeves, B., and Faber, R. 1979. The distinguishing attributes children use in their impressions of real people and television characters. In preparation.

Reeves, B., and Greenberg, B. 1977. Children's perceptions of television characters. *Human Communication Research* 3:113-127.

Reeves, B., and Lometti, G. 1978. The dimensional structure of children's perception of television characters: A replication. Paper presented to the International Communication Association, Chicago.

Reeves, B., and Miller, M. 1978. A multidimensional measure of children's identification with television characters. *Journal of Broadcasting* 22:71-86.

Sarbin, T. 1954. Role theory. In *Handbook of Social Psychology*, ed. G. Lindzey Cambridge, Mass.: Addison-Wesley.

Scarlett, H., Press, A., and Crockett, W. 1971. Children's descriptions of peers: A Wernerian developmental analysis. *Child Development* 42:439-453.

Shapiro, D., and Tagivri, R. 1959. Sex differences in inferring personality traits. *Journal of Psychology* 47:127-136.

Supnick, J. 1964. Senior Honors Thesis, Clark University, Worcester, Mass.

Tuchman, G., Daniels, A.K., and Benet, J. (eds.) 1978. *Hearth and Home: Images of Women in the Mass Media*. New York: Oxford University Press.

Wartella, E., and Alexander, A. 1978. Children's organization of impressions of television characters. Paper presented to the International Communication Association, Chicago.

Watts, A. 1944. *The Language and Mental Development of Children*, London: Harrap and Co.

Wegner, D. 1974. The development and articulation of attributes in person perception. Doctoral dissertation. Department of Psychology, Michigan State University.

Werner, H. 1957. The concept of development from a comparative and organismic point of view. In *The Concept of Development*, ed. D.B. Harris Minneapolis: University of Minnesota Press, 125-148.

Yarrow, M., and Campbell, J. 1963. Person perception in children. *Merrill-Palmer Quarterly* 9:57-72.

4 Children's Impressions of Television Families

Ellen Wartella

Research on the effects of television on children has tended to focus on children's learning of social behaviors from television (Surgeon General's Report, 1972; Stein and Friederich, 1975). Research has frequently, examined such social learning without reference to the characters who portray these behaviors. Similarly, far too little attention has been devoted to children's understanding of social roles and interpretations and evaluations of television characters apart from the social behaviors that such television characters portray (see, for example Leifer, Gordon, and Graves, 1974). Since behaviors are performed by characters, one might expect that children's evaluations of televised behaviors may be modified by their interpretations and evaluations of the characters themselves. For example, aggressive actions by "funny" characters may be interpreted differently from violent actions by "sinister-villain" type characters. Since social roles are elaborated by specific characters engaging in a variety of social actions, research focusing on children's interpretations and evaluations or impressions of characters on television should lead to better understanding of children's social learning from the medium. Social role learning from television should include children's learning about the range of acceptable behaviors, attitudes, and norms appropriate for various roles.

Indeed, the depiction of social roles, norms, attitudes, and behaviors of both a violent and a nonviolent sort is standard fare in American television. Of particular interest here is children's learning about appropriate family roles and behaviors from family-oriented television programs. In recent years family relationships have been depicted frequently in both situation comedy formats and family dramas. Child viewers may acquire much socially important information about how families—particularly fathers, mothers, and children—are supposed to behave by watching such programming. To this end, the research study reported in this chapter was designed to examine the types of social behaviors portrayed by various television families and child viewers' impressions of these television family members.

A general assumption that underlies the research is that children's social learning from complex naturalistic stimuli such as television programs is, in part, a function of their age-related cognitive capabilities. Support for this assumption

The research reported here was supported through a grant from the John and Mary R. Markle Foundation to Daniel B. Wackman and Andrew Collins of the University of Minnesota, project entitled "Children's Social Learning from Family-Oriented Television Programs."

comes from a variety of studies examining children's comprehension of television content (for example, Collins, 1975; Flapan, 1968; Leifer and Roberts, 1972; and Ward, Wackman, and Wartella, 1977). These cognitive capabilities may affect both understanding of a given televised episode and children's impressions of characters across a series. Learning from television, then, is assumed to be influenced by the interaction of cognitive capabilities and the stimulus characteristics of television programs.

A program of research carried out by Collins and his associates (Collins, 1975) has provided evidence that both age-related information-processing skills and bases for judgment of social acts affect children's understanding and evaluation of the social content of television programs. Similarly, research on impression formation, or how children develop understanding of other persons, has found that children's impressions of others (both real-life and television characters) are influenced by the child's general cognitive abilities to organize and relate information about their world (Gollin, 1958; Livesly and Bromley, 1973; and Wartella and Alexander, 1978). Thus we expect that children's social learning about television families should follow a developmental pattern.

To assess both the range of behaviors portrayed by television families and children's impressions of such characters, two types of data were collected in the current study. Content analysis of ten family-oriented television shows was employed to describe the range of behaviors of fathers, mothers, and children on television. Second, children in the second, fifth, and eighth grades were surveyed about their impressions of several of these television families.

Content Analysis

A 40 percent sample of all the late afternoon and evening television series that involved family settings broadcast in the Twin Cities area during the 1975-1976 season were taped for content analysis. These included three dramatic series—"Little House on the Prairie," "The Waltons," and "Swiss Family Robinson"—and seven situation comedies—"The Brady Bunch," "The Partridge Family," "All in the Family," "Good Times," "Happy Days," "The Jeffersons," and "Phyllis." Of these, eight series continued from prior years, and two—"Swiss Family Robinson" and "Phyllis"—were new in the fall of 1975. For each series, three shows were taped during September, and two each during the following four months. Thus eleven shows from each series were taped.

Behaviors protrayed were coded either prosocial or antisocial. Conceptually, prosocial behaviors represent a willingness to work with others in attaining goals, using socially acceptable methods. Antisocial behaviors, on the other hand, generally reflect an unwillingness to work with others or the use of aggressive methods to attain goals. Robert Liebert's research provided the basis for the measurement of pro- and antisocial behaviors. In his research five major categories of prosocial behavior and one category of antisocial behavior (aggression) are distinguished. In this study the same five prosocial behavior categories were retained, but for each of the categories antisocial behaviors at the opposite end of the same continuum were identified. The major categories of behaviors and specific behaviors included within each category are shown in table 4-1.

The ten major categories and sixteen specific behaviors represent a relatively complete catalogue of both pro- and antisocial behaviors. However, there are many other social behaviors that are not clearly either pro- or antisocial such as conversational statements, joking, laughter, and many nonverbal cues. These neutral behaviors, which in this research accounted for nearly two-thirds of the television characters' behavior, were not coded. In the following analysis, therefore, data reported indicate the relative frequency of pro- and antisocial behavior when a behavior was either clearly prosocial or antisocial.

About 150 behaviors were coded per half-hour show, or one behavior every ten seconds. Coders viewed each show three times, stopping the show as often as necessary to fill in details regarding each behavior and verifying initial coding. Although coders had a relatively low agreement in their identification of behaviors to be coded (about 60 percent), when there was agreement that a specific behavior was to be coded, there was high agreement on the behavioral category (86 percent). After all the coding was completed, the coders always

Table 4-1
Prosocial and Antisocial Behaviors Identified in Content Analysis

	Prosocial Behavior		*Antisocial Behavior*
1.	Altruism	1.	Refusing altruism
a.	Sharing	a.	Refusing to share
b.	Helping	b.	Ignoring request, refusing to help
c.	Cooperating	c.	Refusing to cooperate
2.	Expressing remorse	2.	Refusing to express remorse
a.	Admitting mistake, apologizing	a.	Refusing to apologize or admit mistake
b.	Attempting to repair physical damage	b.	Refusing to repair physical damage
3.	Concern for others	3.	Lack of concern for others
a.	Affection	a.	Rejecting affection
b.	Sympathy	b.	Rejecting sympathy
c.	Explaining self (feelings, intentions, interpretations)	c.	Refusing to listen
d.	Showing respect	d.	Selfishness
4.	Control of agression	4.	Agression
a.	Control of verbal agression	a.	Verbal agression
b.	Control of physical agression	b.	Ordering, demanding
		c.	Physical agression
		d.	Nonverbal agression
5.	Resisting temptation	5.	Succumbing to temptation

had high agreement on the total number of behaviors coded (within 10 percent) and on the distribution of pro- and antisocial behaviors coded (mean correlation of 0.91 for the major behavior categories).

Survey

The data were collected in the second wave of a panel survey conducted in October 1975 and May 1976 in St. Paul, Minnesota. The panel survey was designed to examine developmental changes in children's interpretations of television series and characters and the subject's own social behavior. After ascertaining the frequency with which children viewed each of the ten "target" family-oriented programs, the children were subsequently interviewed for their impressions of father, mother, and child characters on each of two of these programs.

An open-ended question, similar to that used by Livesley and Bromley (1973) was employed to measure children's impressions of the television characters. It asked the children to indicate the kind of person the character was: "Describe [character] so that someone would know what he was like and why he was like that." Interviewers recorded verbatim the children's responses to the task. Coding of the answers utilized a complicated coding scheme that indicated whether the child was making an interpretation in terms of appearance or identified characteristics, behavioral actions, or personality traits or motivations as well as whether the assertions about the character were positive (mention of prosocial characteristics), negative (mention of antisocial characteristics), or neutral. After the impression description task, the children were asked how much they liked each character and how realistic each character appeared to them.

To operationalize the developmental aspects of the research design, subjects from three age groups spanning six years of the cognitive development continuum were sampled—second, fifth, and eighth grades. In October 1975 a random sample of 510 children was taken from lists of students at five grade schools and two junior high schools in the Mounds View school district in suburban St. Paul. In the fall of 1975, 420 children were interviewed. The data reported here were collected in May 1976 in follow-up interviews with 388 of the children, including 127 second graders, 138 fifth graders, and 128 eighth graders. Approximately equal numbers of boys and girls in each group were interviewed.

Content Analysis Results

Data for only nine of the ten series that were taped are reported in this section. "Phyllis" has been deleted from consideration because, as the series progressed, her daughter assumed a very minor role and was not even written into a number of shows. Thus this series did not really qualify as a family series.

For purposes of analysis the series were grouped into two categories, based on expectations that the father characters in each group would vary in their multidimensionality of pro- and antisocial behaviors. (In the survey children were asked to respond to two series, one of which was chosen from the Group I father series and one from the group II father series. The intention in doing this was to insure that children would be responding to characters—fathers—who differed considerably in their social behavior.) Five series are included in Group I, the unidimensional, strongly prosocial category: the three family dramas ("Little House," "Waltons," and "Swiss Family Robinson") and two situation comedies ("Brady Bunch" and "Partridge Family"). Group II, multidimensional father programs, are those in which the father was expected to be depicted in less strongly positive terms with substantial doses of antisocial as well as prosocial behavior. Four series, all sitcoms, are included in this category ("All in the Family," "Good Times," "Happy Days," and "The Jeffersons"). The classification was based on general impressions of the series.

As table 4-2 indicates, each program generally included more prosocial than antisocial behavior. However, as expected, the unidimensional father series were more heavily prosocial, with almost three-fourths of the coded behaviors being prosocial, compared with less than three-fifths in the multidimensional father series.

Three major categories of social behaviors coded on each of these shows dominate, accounting for about 90 percent of the behaviors performed in the series. Almost half of the behaviors depicted involved showing concern for others in one way or another, and one-seventh were altruistic behaviors: helping, sharing, or cooperating. On the other hand, one-fourth of the behaviors were antisocial, aggressive behaviors of which 75 percent were verbal aggressive behaviors including verbal attacks (insulting, jeering, threatening) and demands (insistent demands for actions, commands involving yelling or hostility). The relatively low incidence of physical aggression (14 percent) is to be expected given the family nature of these programs (data not shown; reported in Wackman, Collins, and Wartella, 1976).

The major behavioral focus of these series, then, involves two prosocial behaviors—helping behaviors and explanations of feelings, intentions, and actions—and one antisocial behavior—verbal aggression. These three behaviors account for over 54 percent of all the coded behaviors.

A second major interest in the analysis is the character performing the behavior. Major differences were expected in the behavior of fathers in the two groups of series. But possible differences in the behavior of mother and child characters were also of interest.

In all but two series involving both parent families—"Waltons" and "Happy Days"—the father was a dominant character in terms of the number of behaviors coded, but the mother was a dominant character in only three series—"Brady Bunch," "Good Times," and, of course, "Partridge Family," which had no father character. One of the child characters played an equally dominant role with one or both parents in three series—"Little House," "Partridge

Table 4-2
Prosocial and Antisocial Behaviors by Television Series
(in percent; numbers in parentheses)

	Group I					Group II				
Behavior	*"Little House"*	*"The Waltons"*	*"Swiss Family"*	*"Brady Bunch"*	*"Partridge Family"*	*"All in Family"*	*"Happy Days"*	*"Good Times"*	*"Jeffersons"*	*Total*
Prosocial	79 (1,807)	76 (1,716)	66 (1,168)	66 (922)	62 (677)	57 (1,084)	60 (907)	60 (1,194)	54 (974)	65 (10,449)
Antisocial	21 (481)	24 (545)	34 (608)	34 (484)	38 (421)	43 (803)	40 (597)	40 (781)	46 (815)	35 (5,535)
Total	100 (2,288)	100 (2,261)	100 (1,776)	100 (1,406)	100 (1,098)	100 (1,887)	100 (1,504)	100 (1,975)	100 (1,789)	100 (15,984)

Family," and "Good Times"–and adolescent characters played the dominant role in two series–John-Boy in "Waltons," and Fonzie and Richie in "Happy Days.[1] Thus in most series the father character was a dominant character, either standing alone in his domination of the series–"All in the Family," "Jeffersons," and "Swiss Family Robinson"–or sharing domination with his wife or one of his children.

What is the portrayal of the characters? As table 4-3 indicates expectations about the portrayal of unidimensional and multidimensional fathers are solidly confirmed. Group I fathers' behavior is heavily positive by a ratio of four to one pro- to antisocial behavior, whereas Group II fathers are portrayed relatively more negatively with a 60/40 ratio of antisocial to prosocial behaviors. Only one of the Group II fathers performed more pro- than antisocial behaviors. On the other hand, mothers are uniformly portrayed in a positive manner, with all but two mothers performing at least 70 percent prosocial behaviors. Similarly, child characters are portrayed in a positive manner, although generally they are not as prosocial in their behavior as positive fathers or mothers. Interestingly, male and female child characters both exhibit two-thirds prosocial and one-third antisocial behaviors, indicating little difference between the portrayal of child characters of the opposite sexes in these series. Only two of the child characters exhibited more antisocial behavior than prosocial behavior–Fonzie in "Happy Days," and Peter in "Brady Bunch"–and two others exhibited about equal anti- and prosocial behavior–Mary Ellen in "Waltons" and Thelma in "Good Times."

Four shows have been selected for further analysis based on the number of children who discussed characters in each of the original ten programs possible for them to discuss. Children were questioned only about characters on two television series, one from each group of programs. From Group I shows "Partridge Family" and "Little House on the Prairie" were selected for further analysis and were discussed by 195 and 54 children respectively. From Group II shows further analysis was conducted on "All in the Family" and "Happy Days," discussed by 54 and 244 children respectively.

Table 4-3
Prosocial and Antisocial Character Behaviors across Television Series
(in percent; numbers in parentheses)

Behavior	*Group I Father*	*Group II Father*	*Mother*	*Male Child*	*Female Child*	*Total*
Prosocial	79 (1,026)	40 (762)	75 (2,249)	65 (1,845)	68 (1,041)	65 (6,923)
Antisocial	21 (270)	60 (1,146)	25 (738)	35 (1,109)	32 (498)	35 (3,761)
Total behaviors coded for each character	12 (1,296)	18 (1,908)	28 (2,987)	28 (2,954)	28 (1,539)	100 (10,684)

Table 4-4 presents data on the percentage of prosocial and antisocial behaviors coded in each of the eight major behavior categories for fathers, mothers, and child characters on the four selected programs. These data suggest that nearly all the characters who appear regularly in these series are presented in a relatively positive light, performing substantially more prosocial than antisocial behaviors. For instance, for both "Little House on the Prairie" and "Partridge Family," showing concern for others represents the majority of behaviors coded. This is overwhelmingly the case for "Little House," where clearly three-quarters of the behaviors coded for the father and mother character and over half of the behaviors coded for the child character, Laura, fall in this category. Thus one would expect these characters to present a relatively uniformly positive role model for child viewers.

On the other hand, several family characters portrayed on "All in the Family" and "Happy Days" are more differentiated in terms of their prosocial and antisocial behaviors. For instance, over half of the codable behaviors for Archie Bunker are aggressive behaviors, as are two-third of the behaviors for Howard Cunningham. Similarly, Fonzie represents a highly mixed character type; most of his codable behaviors are accounted for by aggression (42 percent) and showing concern for others (31 percent). These characters thus present mixed types and afford an opportunity to examine children's abilities to integrate mixed behavioral patterns of a character in developing an impression. In addition, these characters are frequently portrayed in humorous contexts, therefore it may be that the humor modifies the impact of these more antisocial behavior portrayals on children's interpretations of the characters. The interpretations that children make of these television characters can be seen from the data collected in the survey.

Children's Impressions of Family Characters

Children were asked to discuss three characters on each of two television shows of interest. The procedure began by asking the children which character on the show was his or her favorite character. Children chose a child character 85 percent of the time, and when this was the case, the respondent was asked a series of questions about the child character first. Next the respondent was asked the same questions about the father character, followed by identical questions about the mother character.

Children's free descriptions of the characters were subsequently content analyzed into three major categories: identity/appearance characteristics, behaviors, and personality traits/motivations. Up to five assertions about each character were coded for analysis.

In general, children's free descriptions of television characters confirmed age-related findings of previous research on children's impressions of others.

Table 4-4
Prosocial and Antisocial Category Behaviors Coded for Father, Mother, and Child Characters on Four Series
(in percent)

"Little House on the Prairie"

	Father	*Mother*	*Laura*
Altruism	16.1	17.9	13.4
Antialtruism	1.7	1.4	1.8
Apology	1.0	1.0	1.8
Antiapology	0	0	0
Concern	74.3	71.5	57.9
Anticoncern	0.2	0	2.6
Control of aggression	0.5	0.5	0.8
Aggression	4.8	4.3	18.7
	(N = 411)	(N = 200)	(N = 369)

"Partridge Family"

	Mother	*Keith*	*Laurie*
Altruism	11.9	13.8	15.5
Antialtruism	5.6	3.2	4.7
Apology	1.6	1.1	3.1
Antiapology	0.4	0.5	0
Concern	48.8	41.0	48.1
Anticoncern	0.8	2.7	0
Control of aggression	2.0	3.7	2.3
Aggression	29.0	34.0	26.4
	(N = 252)	(N = 188)	(N = 129)

"Happy Days"

	Father	*Mother*	*Richie*	*Fonzie*
Altruism	15.5	23.9	31.5	8.6
Antialtruism	3.1	0.9	2.2	3.9
Apology	1.9	2.7	1.4	2.0
Antiapology	0	0	0	0.5
Concern	33.5	61.1	36.9	31.3
Anticoncern	3.9	0.9	2.9	5.1
Control of aggression	3.1	0	1.8	1.6
Aggression	36.0	8.0	11.5	42.2
	(N = 156)	(N = 110)	(N = 246)	(N = 410)

"All in the Family"

	Father	*Mother*	*Michael*	*Gloria*
Altruism	5.0	18.8	13.9	12.0
Antialtruism	1.7	25.0	0.3	1.4
Apology	0.9	0.8	3.6	2.5
Antiapology	0.3	0.3	0.6	0.4
Concern	26.9	55.6	43.3	42.6
Anticoncern	3.2	3.0	2.7	3.5
Control of aggression	0.3	5.7	2.7	1.4
Aggression	57.2	12.8	28.8	34.5
	(N = 635)	(N = 363	(N = 323)	(N = 279)

Children in the two youngest age groups, second and fifth grades, are more likely to describe characters in terms of appearance/identity characteristics or surface attributes than are eighth graders. Similarly, children in these two age groups are less likely than eighth graders to use trait/motivation assertions in their descriptions, although this relationship holds most strongly for the mother and child characters. Fifth and eighth graders are more likely to describe characters in terms of behavioral actions than are children in the youngest age group.

Overall analysis of the children's impressions of all television characters supports previous research findings; as children grow older they tend to mention more internally based attributes (such as personality traits). However, in contrast to descriptions that children give for real people (Livesley and Bromley, 1973), the actions of television characters form the preponderance of children's impressions at each grade, with 85 percent of all children mentioning at least one behavior for each character discussed. In this sense descriptions of television characters diverge from descriptions of real-life people. (Data not presented; see Wartella and Alexander, 1978.)

Children's impressions of each character were also examined for assertions of antisocial, prosocial, and neutral characteristics. Each behavior or trait/motivation assertion about a character was classified into each of these three categories based on the general conceptual framework used in the content analysis. Most of the assertions about characters were neutral statements, for example, "He talks to Fonzie a lot," "She's trying to find a job," "They got married," "He's different," "He has troubles." Children also gave many pro- and antisocial assertions. The mean numbers of prosocial, neutral, and antisocial assertions (each with a possible range of 1 to 5) for each of the characters on the four family shows are presented in table 4-5.

Examination of this table indicates that the characters on "Little House" and "Partridge Family" are generally perceived by all children in prosocial terms. This is consistent with the content analysis of these shows. Indeed, almost no antisocial assertions were made for the mother and father on "Little House." The children, Laura and Danny Partridge, by contrast, are seen and described with mixed assertions, neither strongly prosocial nor antisocial. Further, where age differences in the use of the three types of assertions occur (such as for the mothers on both shows and the father on "Little House"), older children, particularly eighth graders are more likely to describe these characters in prosocial terms.

By contrast, children's descriptions of the family characters on "Happy Days" and "All in the Family" include more antisocial assertions than descriptions of characters in the previous two programs. While both fathers were shown to be similarly mixed in terms of their performance of pro- and antisocial behaviors in the content analysis, child viewers describe these two fathers somewhat differently. Children at each grade level make higher mean number of antisocial assertions about Archie Bunker than about Howard Cunningham. Moreover, older fifth- and eighth-grade children made more prosocial assertions

Table 4-5
Mean Frequencies of Prosocial, Neutral, and Antisocial Assertions about Characters, by Grade Level

Character		*Prosocial* ($\bar{X}$)	*Neutral* ($\bar{X}$)	*Antisocial* ($\bar{X}$)
"Little House on the Prairie"				
Mother				
Second grade	(n = 15)	0.91*	0.43	0
Fifth grade	(n = 19)	1.48	0.62	0.10
Eighth grade	(n = 8)	2.4	0.80	0
Father				
Second grade	(n = 23)	1.30*	0.43	0
Fifth grade	(n = 21)	2.05	0.33	0.10
Eighth grade	(n = 10)	3.10	0.40	0
Laura				
Second grade	(n = 23)	0.73	0.73*	0.40
Fifth grade	(n = 21)	1.00	1.00	0.21
Eighth grade	(n = 10)	1.38	1.88	0.38
"Partridge Family"				
Mother				
Second grade	(n = 61)	0.90**	0.48	0.15**
Fifth grade	(n = 74)	1.62	0.43	0.03
Eighth grade	(n = 60)	1.73	0.48	0.03
Danny				
Second grade	(n = 31)	0.16	1.39	0.29
Fifth grade	(n = 44)	0.27	1.61	0.34
Eighth grade	(n = 39)	0.26	1.51	0.41
"Happy Days"				
Mother				
Second grade	(n = 66)	0.74	0.65	0.03**
Fifth grade	(n = 92)	0.97	0.96	0.09
Eighth grade	(n = 86)	1.02	0.77	0.24
Father				
Second grade	(n = 66)	0.83**	0.48	0.32
Fifth grade	(n = 92)	1.00	0.67	0.29
Eighth grade	(n = 86)	1.45	0.81	0.14
Fonzie[a]				
Second grade	(n = 62)	0.89	1.16	0.02**
Fifth grade	(n = 92)	0.96	1.35	0.06
Eighth grade	(n = 68)	0.99	1.27	0.19
"All in the Family"				
Mother				
Second grade	(n = 23)	0.87	0.74	0.17
Fifth grade	(n = 17)	0.72	1.12	0.24
Eighth grade	(n = 14)	0.71	0.71	0.57

Table 4-5 (Continued)

Character		*Prosocial* ($\bar{X}$)	*Neutral* ($\bar{X}$)	*Antisocial* ($\bar{X}$)
Father				
Second grade	(n = 23)	0.17	0.70	1.48
Fifth grade	(n = 17)	0.18	0.65	1.71
Eighth grade	(n = 14)	0.07	0.71	1.92
Gloria				
Second grade	(n = 21)	0.67	0.71*	0.19
Fifth grade	(n = 11)	0.91	0.46	0.09
Eighth grade	(n = 10)	0.50	1.50	0.10

[a]Too few children chose to describe Richie to make any analysis of this character meaningful

*F-statistic, $p = 0.05$

**F-statistic, $p = 0.01$

about this latter character than did the younger children. In contrast to the fathers on these shows, neither mother is described in strongly prosocial or antisocial terms, although eighth graders did make slightly more antisocial assertions about Marion Cunningham than did younger children. Last, the two child characters on these shows are described with more neutral and positive assertions than antisocial assertions. Surprisingly, this is the case for Fonzie even though his behavior was shown to be relatively mixed in the content analysis, with high amounts of prosocial and antisocial behavior.

Might children's assertions about these characters be modified by their liking for each character? Table 4-6 presents age group means of the children's liking for these television family characters based on the question: "How much do you like [character]?" Responses were obtained on a four-point scale: not at all (1), not too much (2), a little (3), and a lot (4).

These data indicate almost no age-related differences in children's liking of each of the television characters discussed. Rather, there seem to be strong differences by character type, with mothers consistently less well liked than either father or child characters (this held for both males and females). Also, while the father on "Little House" is perceived by the children as more prosocial than the father on either "Happy Days" or "All in the Family," there is relatively little difference in liking of this father over the latter two fathers. As might be expected, Fonzie is almost uniformly liked a lot by children in each age group (overall mean of 3.95).

Last, children's perceptions of the realism of the characters were assessed by a single question: "Do you think [child/father/mother character] acts like [kids/father/mothers] in real life?" Three response alternatives were provided: yes, sometimes, and no. These data are presented in table 4-7.

Table 4-6
Liking of Television Program Characters, by Grade Level

"Little House on the Prairie"			
	Father (X)	*Mother (X)*	*Laura (X)*
Second grade	3.64	2.58	3.53
Fifth grade	3.74	3.73	3.70
Eighth grade	3.60	2.50	3.88
"Partridge Family"			
		Mother (X)	*Danny (X)*
Second grade		2.60	3.83
Fifth grade		2.58	3.84
Eighth grade		2.28	3.67
"Happy Days"			
	Father (X)	*Mother (X)*	*Fonzie (X)*
Second grade	3.23	2.32	3.94
Fifth grade	3.36	2.38	3.96
Eighth grade	3.60	2.47	3.93
"All in the Family"			
	Father (X)	*Mother (X)*	*Gloria (X)*
Second grade	3.30	2.81	3.65
Fifth grade	3.37	2.56	3.46
Eighth grade	3.40	2.43	3.55

Children in the oldest age group perceive all the characters on the unidimensional, prosocial programs as less realistic than do children in the younger two age groups. However, in judging the reality of characters on the multidimensional shows, "Happy Days" and "All in the Family," there are relatively minor differences among the three age groups. Moreover, second and fifth graders judge mothers and fathers on these two shows to be less realistic than their counterparts on the unidimensional, prosocial programs, but eighth graders do not make this distinction for mother and father characters.

All children distinguish between child characters on the two types of shows, with the children on the multidimensional shows being viewed as more realistic. Also, Fonzie of "Happy Days" is viewed as most realistic, with over 50 percent of all children reporting him as real. On the other hand, over 40 percent of all children view Laura and Danny Partridge as not real.

Summary

Social role depictions of families on television present fairly prosocial portrayals of family characters. Mothers particularly were found to be the most uniformly

Table 4-7
Perceived Realism of Television Characters, by Grade Level
(*in percent*)

	Father			*Mother*			*Child*		
"Little House on the Prairie"	*2nd*	*5th*	*8th*	*2nd*	*5th*	*8th*	*2nd*	*5th*	*8th*
Not real	0	22	10	5	14	10	46	39	50
Sometimes real	37	13	80	21	23	70	15	28	38
Real	63	65	10	74	63	20	39	33	12
N =	(22)	(23)	(10)	(19)	(22)	(10)	(13)	(18)	(8)
"Partridge Family"				*Mother*			*Child*		
				2nd	*5th*	*8th*	*2nd*	*5th*	*8th*
Not real				9	14	16	41	50	45
Sometimes real				32	20	40	24	25	24
Real				59	65	44	35	25	31
N =				(57)	(75)	(57)	(29)	(44)	(38)
"Happy Days"	*Father*			*Mother*			*Child*		
	2nd	*5th*	*8th*	*2nd*	*5th*	*8th*	*2nd*	*5th*	*8th*
Not real	15	17	17	16	25	16	15	11	9
Sometimes real	42	36	44	17	37	51	22	38	49
Real	42	39	38	30	38	33	62	51	42
N =	(59)	(90)	(82)	(56)	(89)	(80)	(62)	(81)	(64)
"All in the Family"	*Father*			*Mother*			*Child*		
	2nd	*5th*	*8th*	*2nd*	*5th*	*8th*	*2nd*	*5th*	*8th*
Not real	50	53	57	33	47	36	20	46	20
Sometimes real	27	31	36	33	21	28	35	9	50
Real	23	16	7	33	32	36	45	45	30
N =	(22)	(19)	(14)	(21)	(19)	(14)	(20)	(11)	(10)

and consistently prosocial in their behavior. On the other hand, fathers on television programs, typically the most dominant character in the program, fell into one of two categories: strongly unidimensional, prosocial fathers, such as Charles Ingalls of "Little House on the Prairie"; or more mixed, multidimensional fathers such as Howard Cunningham of "Happy Days," who present both prosocial and aggressive behaviors. Child characters similarly portray mixed behavior patterns on most family-oriented programs. The major categories of prosocial behaviors depicted on the programs are showing concern for others and altruistic behaviors; verbal aggression, such as yelling and jeering, accounts for the preponderance of antisocial behaviors. It would appear, then, that children are presented with a fairly limited range of family models from television, with mothers on television presenting the most restricted range of behaviors followed by child and father characters.

Examination of children's assertions of prosocial, neutral, and antisocial characteristics in several selected families on television indicates that children in all three age groups fairly accurately perceive characters' behavior patterns. However, when describing characters with mixed behavior patterns, eighth graders are more likely than the younger children to note the character's antisocial tendencies. Both father and child characters are uniformly well liked by the children regardless of their behavior patterns, while mother characters are less well liked overall. Seemingly, "liking" of characters does not modify children's perceptions of their behavior nor vice versa. However, perceptions of a character's realism does seem to be affected by the character's behavior portrayals for at least two character types. Younger children report that the unidimensional, prosocial mothers and fathers are more realistic than the multidimensional fathers and mothers. Similarly, the child characters on the multidimentional shows are viewed as more realistic than the other child characters. Perhaps the children's abilities to identify with child characters increase this perception of realism.

A fairly benign interpretation of these results can be offered. Types of behaviors depicted on family-centered television programs and children's interpretations of these characters are of prosocial. Perhaps the only jarring finding in the research is the relatively stereotyped portrayals of mothers. They are accurately perceived in prosocial terms by the children and are uniformly less well liked by all age groups than are father or child characters. This latter finding may reflect primarily the less dominant role of television mothers on the series discussed here.

If the range of characters depicted on television is restricted, what impact may this have on the child viewer? It seems likely that television may have an impact on children's expectations of how people *should* behave in real life. Consequently, where the range of character portrayals is limited, subsequent expectations of real-life roles for child viewers may also be limited. This would seem to be at the center of the issue of diversity in program portrayals. Future research should address the issue of how television character portrayals might build the child's expectations of how people should behave.

References

Collins, W.A. 1975. The developing child as viewer. *Journal of Communication* 25:35-44.

Flapan, D. 1968. *Children's Understanding of Social Interaction*. New York: Teachers' College Press, 1968.

Gollin, E.S. 1958. Organizational characteristics of social judgment: A developmental investigation. *Journal of Personality* 26:139-154.

Leifer, A.D., Gordon, N.J., and Graves, S.B. 1974. Children's television: more than mere entertainment. *Harvard Educational Review* 44:213-245.

Leifer, A.D., and Roberts, D.F. 1972. Children's responses to television violence. In *Television and Social Behavior*, ed. J.P. Murray, E.A. Rubinstein and G.A. Comstock, Washington, D.C.: U.S. Government Printing Office, 1972.

Livesley, W.J., and Bromley, D.B. 1973. *Person Perception in Childhood and Adolescence*. New York: John Wiley and Sons.

Stein, A.H., and Friederich, L.K. 1975. *Impact of Television on Children and Youth.* In *Review of Child Development Research*, vol. 5, ed. E.M. Heatherington. Chicago: University of Chicago Press.

Surgeon General's Scientific Advisory Committee on Television and Social Behavior, "Television and growing up: The impact of television violence." Report to the Surgeon General, United States Public Health Service, Washington, D.C.: U.S. Government Printing Office, 1972.

Wackman, D., Collins, W.A., and Wartella, E. 1976. Children's social learning from family-oriented television programs. Report to the John and Mary R. Markle Foundation, University of Minnesota, September 1976.

Ward, S., Wackman, D., and Wartella, E. 1977. *How Children Learn to Buy*. Beverly Hills, Ca., Sage Publications.

Wartella, E., and Alexander, A. 1978. Children's organization of impressions of television characters. Paper presented to the International Communication Association Convention, Chicago, April 1978.

5 FREESTYLE: Using Television to Combat Sex-Role Stereotyping in Children

Frederick Williams

FREESTYLE is an attempt to present counterstereotypic sex-role information on television. Although designed for an educational market and mainly for public broadcasting distribution, FREESTYLE is also meant to compete with commercial television; this is one reason why it is being produced in Hollywood (by public television station KCET). The series, with associated nonbroadcast materials for use in the schools and homes, began in October 1978. It has been developed through the TV Career Awareness Project funded under a grant from the National Institute of Education (NIE). The idea for FREESTYLE grew out of concerns by personnel in the office of Education and Work in the NIE that sex-role stereotyping, even in preadolescent children, may bias activities and school work so that later careers become improbable. For example, girls may avoid mathematics because they feel that society does not expect them to do well in the subject. There was also a concern that the great imbalance of positive male and female portrayals on commercial television was adding to the sex-role stereotyping by young children. Thus it appeared to productive to reach children in this age range through television in an attempt to present counterstereotypic portrayals that would combat sex-role stereotypes prevailing so often in the commercial medium. This is presumably one reason why the project eventually found its way to southern California and the heart of the talent field used in commercial television. The contract that developed FREESTYLE materials was awarded to a consortium comprising the following organizations:

Public television, KCET: Production and management of the project

Los Angeles County Schools office: Curriculum development, teacher inservice training

Science Research Associates: Nonbroadcast materials

East Los Angeles College: Outreach activities.

This report was funded under contract from the National Institute of Education of the Department of Health, Education, and Welfare. The views expressed do not necessarily reflect those of the National Institute of Education or any other agency of the U.S. government.

Institute for Social Research at the University of Michigan: Summative evaluation

Annenberg School of Communications at the University of Southern California: Formative evaluation

This chapter reports on certain of the formative evaluation activities carried out by the Annenberg School. Essentially the role of the formative evaluator is to assist in all phases of program development, whereas the summative evaluator engages in production validation after the finished product is out in the field. In the broadest perspective, formative activities were divided into four areas.

Baseline studies. What is the status of sex-role stereotyping in nine- to twelve-year-old children, particularly those tied to career attitudes? What types of television most appeal to children of this age range? What are some of the likely production strategies for presenting counterstereotypic objectives?

Ad hoc studies. What are teacher evaluations and production suggestions for early script materials? How do children respond to early examples of production materials? Do children comprehend intended objectives and sample production materials?

National site evaluations of pilots. Given pilot broadcast and nonbroadcast materials prepared for the project, what is the relative contrast in students' liking and comprehension of these materials? What additional effects can be attributed to use of the nonbroadcast materials in schools and classrooms? How successful was the communication of the anticipated objectives? Is the project favored by teachers, administrators, and parents?

Follow-up analysis of studies. What does theoretically oriented analysis of the research data indicate in terms of sex-role stereotyping in children? What insights can be gained about how children perceived counterstereotyping in the television materials?

As of Spring 1978 the third phase of the research—the national site evaluation of pilot materials—was completed and production of the final series was in progress. The next section of this chapter presents a summary of the results of the pilot evaluation which were the basis for determining many aspects of the series. Following this summary are some broad generalizations drawn from the research to date.

Objectives of the National Site Evaluations of Pilot Materials

The prototype materials were developed to focus on and were evaluated primarily in terms of the following objectives.

Children will practice behaviors such as independence and interdependence, aggressiveness, and yieldingness, depending on the most appropriate for the situation rather than on sex-role stereotyped behaviors. (This inclusive objective was not tested for in this pilot phase.)

Children will experience a greater variety of work-related activities not previously considered appropriate for their sex-roles (nontraditional jobs).

Children will experience situations that depict the limitations that sex-role stereotyping has placed on both male and female choices and activities (limitations).

Girls will experience situations that encourage them to take reasonable risks and not give up because of occasional failure (reasonable risk).

Children will experience situations that counteract the prevailing myth that females will be taken care of by males and will not need to be employed (Cinderella myth).

Children will experience situations that encourage the development of positive attitudes toward girls in leadership roles (female leaders).

Girls will experience situations designed to help them attribute their success to ability, not solely to luck (ability).

Girls will experience situations designed to evoke and reward appropriate degrees of independence, initiative, and self-direction (independence; not included in pilot materials).

Boys will experience situations that evoke and reward nurturant behavior (male nurturance).

In the preliminary phase of the FREESTYLE project three half-hour television pilots, a comic book, a school guide, and a home guide were developed as prototype materials.

The three television pilots, each divided into two fifteen-minute segments, varied among three production strategies, each known to appeal to children in this age range: (1) "The Mike Farrell Show," a mixture of fast-paced dramatic, song, dance, and cartoon segments; (2) "Felipe and Fran," a format with more continuity than the first and with a host and hostess who appeared frequently; 'Grease Monkey," a half-hour drama in two fifteen-minute segments about a young girl who aspires to fix cars.

The comic book, produced in English and Spanish versions, was twelve pages long and was printed in black and white. Because prior research had shown differences among children in preferences for realistic versus cartoon art styles, the book was produced in these two alternative art forms.

The home guide and the school guide presented activities for parents and teachers respectively to reinforce the project's objectives.

The project schedule provided an initial phase for the development of the phototype materials and their testing at selected national sites, as a basis for undertaking final planning for the full television series and associated non-broadcast materials. There were four overall questions in the field evaluation.

1. *Television format*. What are the relative contrasts in students' liking, comprehension of, and preference for the three pilots? What are the recommendations for selection of a television format?

2. *Nonbroadcast materials.* What are the effects of combining nonbroadcast materials with the television materials? What are the differences in the two art forms of the comic book? What are recommendations for the development of nonbroadcast materials for the final series?

3. *Comprehensive of objectives.* What is the relative comprehension of the different objectives as presented in different formats? What are recommendations for the treatment of objectives in the series?

4. *Utilization.* To what degree was the series responded to favorably by teachers, administrators and parents? What were their relative preferences among the pilots? What are recommendations for potential utlization of the series?

Method

Data were gathered in a quasi-experimental design in four national test sites (Paterson, New Jersey; Austin and Lockhart, Texas; in and near Lexington, Kentucky; in and near Lincoln, Nebraska) and with an ad hoc group of parents and children at a fifth site (Pittsburgh, Pennsylvania). Most of the data gathering took place in October 1977; it involved 1,639 students, 59 teachers, 11 administrators, and 83 parents. At each formal test site there three schools were typically involved in the project, and in each of these schools one fourth, one fifth, and one sixth grade received the "full treatment" condition, which included seeing the television pilots, receiving the comic book, and undergoing in-class activities as suggested by the school guide and as conducted by teachers who underwent in-service training.

In the same schools one fourth-grade class and one sixth-grade class judged comparable with the full treatment classes received the "partial treatment" condition. These classes saw only the television pilots, and their teachers did not undergo in-service training. All children at all test sites saw the three pilots, but the orders differed for the four sites. They were exposed to the pilots fifteen-minute segments approximately every other day.

Data included: (1) a student attitude questionnaire on sex-role biases, (2) student-liking ratings on each of the fifteen-minute segments, (3) student comprehension of other facts in the segments, (4) student preferences if they were to see more pilots, (5) student ratings of likability and comprehension of the comic book and (6) teacher questionnaires, (7) parent questionnaires, and (8) administrator questionnaires. Selected students, teachers, and parents were interviewed by project field consultants. Results of these interviews generally corroborated interpretations of the quantitative findings.

Results

Television Format

Figure 5-1 summarizes the comparison of the three pilots. These overall reactions were substantially similar across sites, sexes, ages, and ethnicities; thus primary interpretations can be based on this figure.

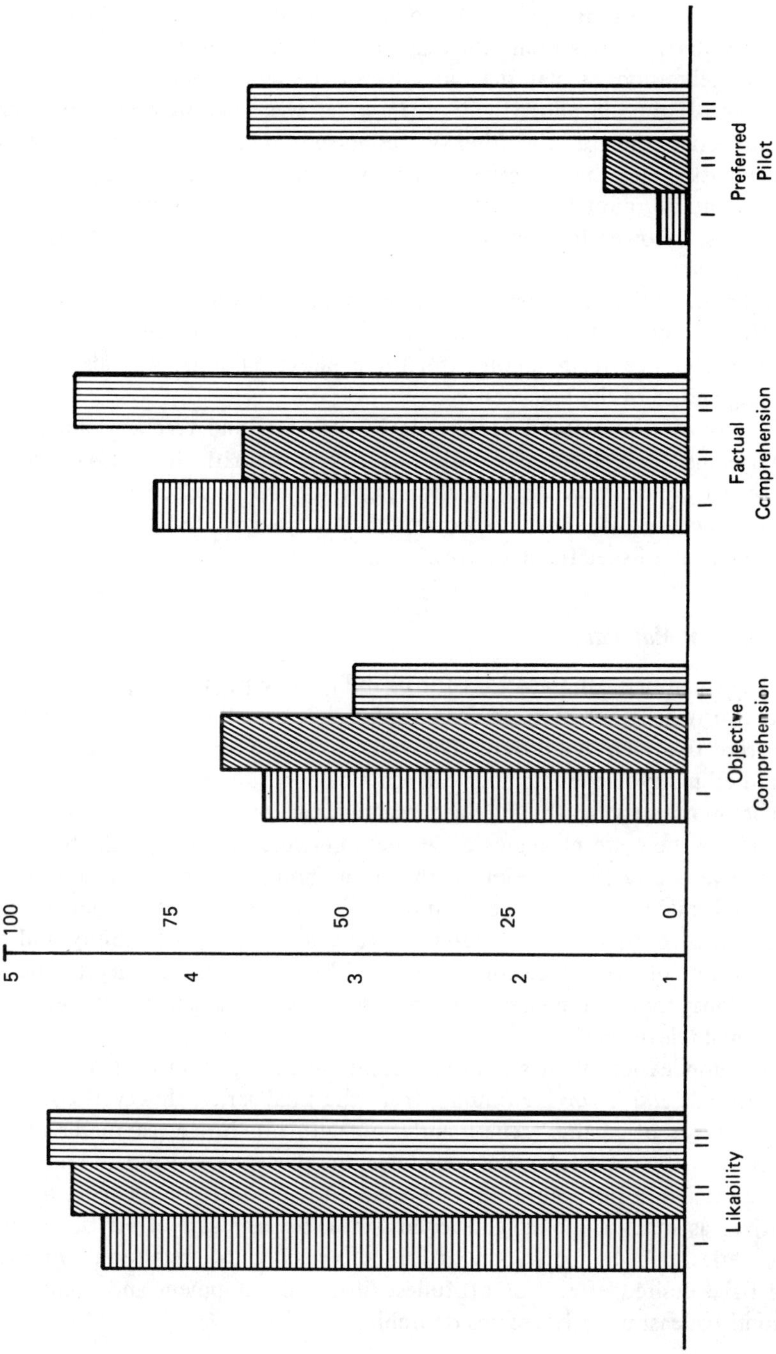

Figure 5-1. Comparison of Pilots

All three pilots, rated by children on a likability scale of 1 to 5 (highest), were evaluated most favorably. Although there are greater-than-chance differences ($p < 0.01$) among them, the magnitude is psychologically trivial.

Comprehension of material reflective of the series objectives was higher in pilots I and II than in pilot III. Pilot III, which was an entire drama, focused on fewer objectives than did the other two pilots and used a more indirect approach. This evidently made the objectives less comprehensible to the students.

By contrast, pilot III had the highest average factual comprehension of all three pilots, followed by pilot I. Pilot II had lower factual comprehension than the other two.

When the children were asked which pilot they wanted to "see more of," pilot III was by far the preferred choice. Twenty-two percent of the children had no preference from among the three pilots. Pilot III was also the overwhelming choice of the teachers.

Because of the marked preference by students and teachers and the high likability and factual comprehension ratings, a dramatic format such as used in pilot III would have distinct advantages for the final series. The dramatic format, however, will have to be modified or augmented to achieve higher comprehension of specific objectives.

Nonbroadcast Materials

Figure 5-2 compares the averages for likability, comprehension, and preference measures originally summarized in figure 5-1, but now they are compared by full and partial treatment conditions. These patterns have generality across sites, sex, age, and ethnicity. Hence findings can again be best summarized in a single figure.

The most striking finding in these comparisons is that on all measures on all pilots there are no major differences between full and partial treatments. This indicates that the addition of the comic book, the activities promoted by the school guide, and in-service training of teachers failed to augment any of the effects of the television materials as gauged in terms of likability and comprehension of or preference for the pilots. This finding is contrary to the usual expectations that complementary in-class activities augment the effects of educational television.

One implication of these findings could be the decision to drop nonbroadcast materials and in-service training from the final series. However, two factors argue in favor of further experimentation with non-broadcast materials. First because of different production schedules, the present television and non-broadcast materials are not closely parallel and thus do not complement the series objectives as much as possible. Second, since the aggregate effect of seeing the entire series will be partly on attitudes, and because classroom interaction could have desired effects on attitudes, further development and evaluation of the nonbroadcast materials seems desirable.

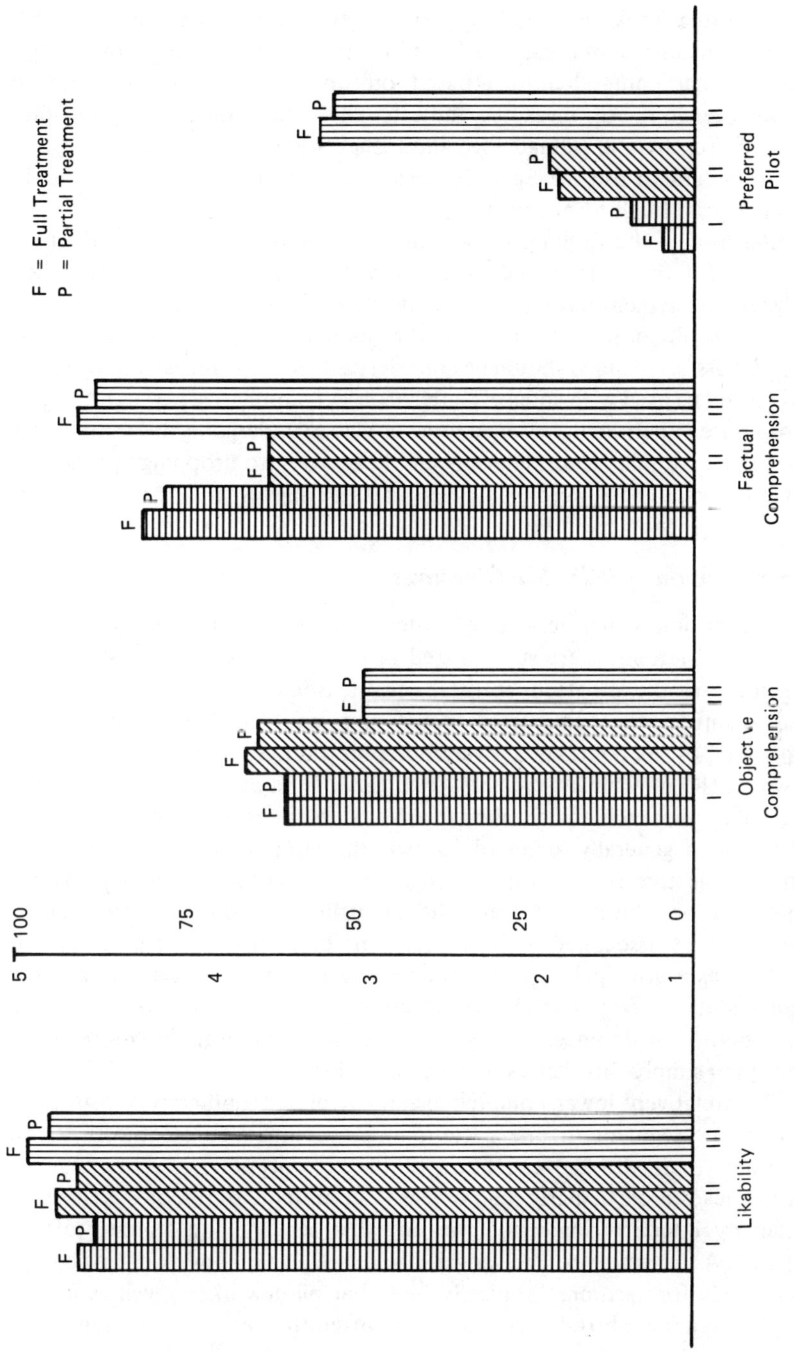

Figure 5-2. Comparison of Full and Partial Treatments

Figure 5-3 summarizes the comparison of the realistic with cartoon versions of the comic book. Results show no greater than a chance difference between the two versions on measures of likability, readership, and objective comprehension. Factual comprehension (based on one question) tended to favor the cartoon version. All measures indicate that the comic book was favorably received. Teachers in bilingual (Spanish-English) classes in Texas and New Jersey were reluctant to use the Spanish versions of the comic book and school guide, preferring the English versions instead.

Because of the findings of lack of effect, development of the nonbroadcast materials for the series should proceed with caution. The materials should be designed to augment more explicitly the series objectives, to have separate objectives, or to do both. Development that emphasizes promotion of attitudinally related class discussions should be considered. Either the realistic or cartoon comic book could be used with equal effectiveness. The summative research design should incorporate experimental comparisons of classes undergoing different uses of nonbroadcast materials. Consideration should be given to dropping Spanish language comic books. (There were insufficient data to suggest this for the home guide.)

Comprehension of Individual Objectives

Comprehension scores reflecting students' likelihood to identify the objectives in the television pilots are summarized in figure 5-4. A number of factors could contribute to the variations in these results. Some variation could be the result of different media (films or cartoons) or it could be the result of story line, characters, or the contexts of overall presentation. Some could reflect the degree to which the questions were easy or difficult for a particular segment. However, to combat the problem of bias, questions on the instructional objectives were written in a generally standard format; the child was requested to select the correct objective from a list of distractors composing other objectives of the project or objectives that were the opposite of those objectives. Distractors that could be associated with the segment even though the producers did not intend it as a principal objective were avoided. Given this attempt at a standard format and the fact that all the questions were pretested for bias, it is quite likely that the differences among objectives as shown in figure 5-4 reflect factors other than simply differences between questions.

The relatively lower comprehension for pilot III objectives, with the exception of one segment, is one pattern in the comparison of objectives. Another pattern is that with the exception of the stereotyping song, musical numbers did not lead to particularly high objective comprehension scores. (This is corroborated by interview material indicating that student and teachers often could not follow the words to the songs). There are also a number of low comprehension scores for cartoons; ironically in earlier pilot work, as well as in interview data in the national study, cartoons were often the best liked medium.

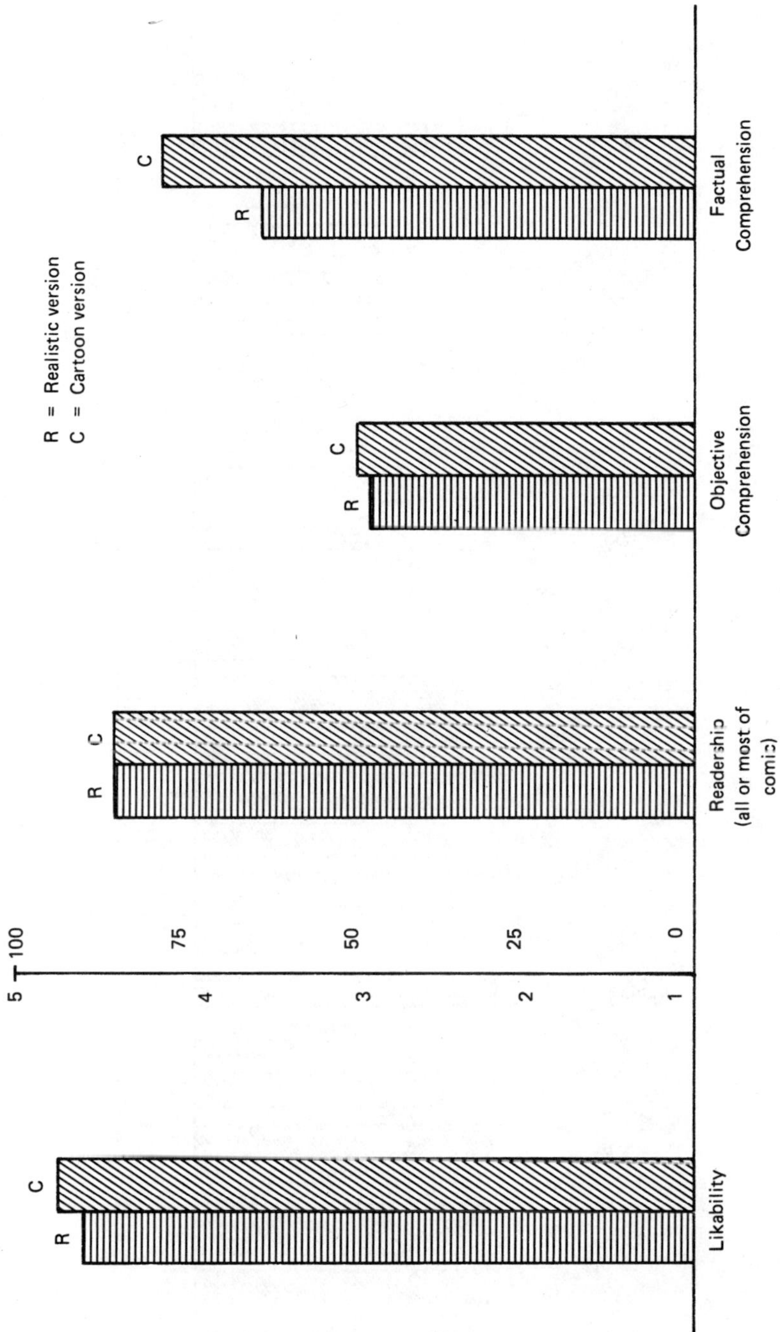

Figure 5-3. Comparison of the Realistic and Cartoon Versions of the Comic Book

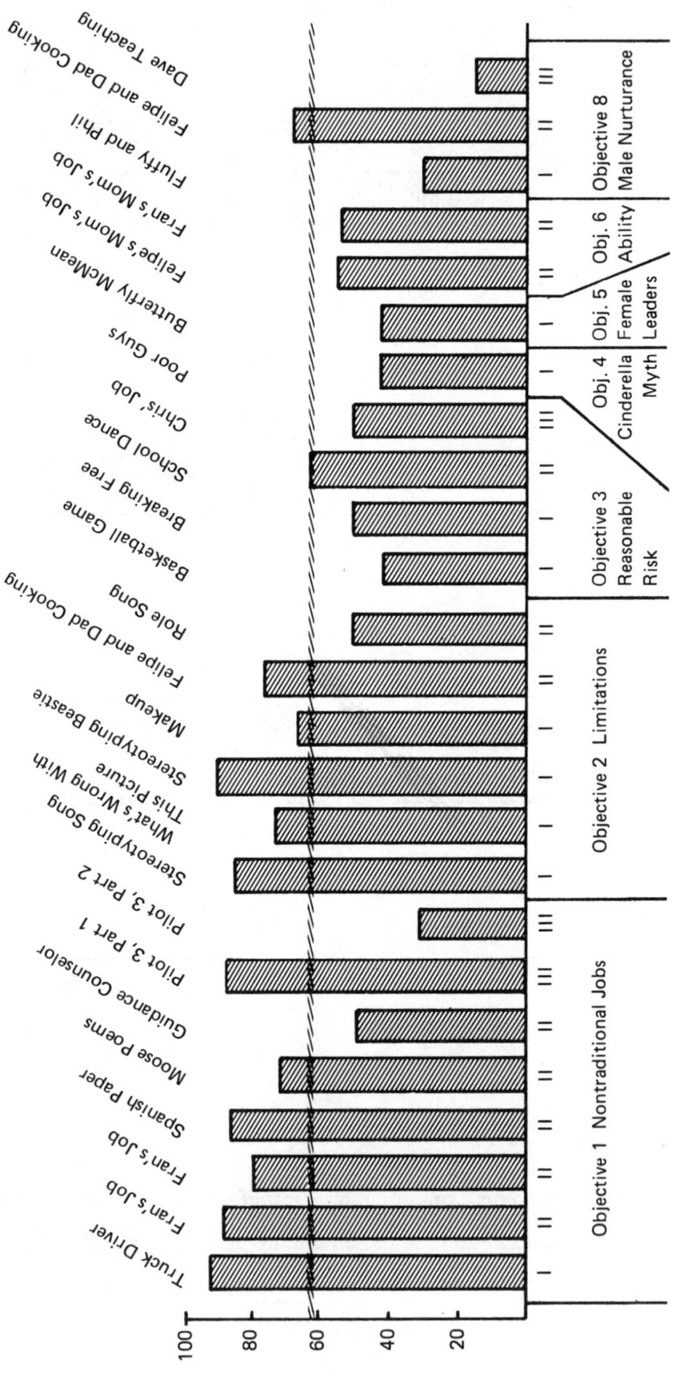

Figure 5-4. Comprehension of Objectives of Television Pilots

On a more subjective level, particularly the nature of each objective, and the segments relative to the development of questions for each, certain segments are markedly more explicit than others. The explicit segments seem consistently to have the highest objective comprehension scores. Of course, some of the objectives may lend themselves to a more explicit portrayal than others. This appears to be the case for nontraditional roles and to some extent for limitations of stereotypes, compared with female leadership, the Cinderella myth, or male nurturance.

Explicitness in the portrayal of an objective, rather than the length of a segment or the production technique, seems to determine the degree to which comprehension is achieved. Children seem to find it difficult to understand the explicit message of songs and musical numbers. If a dramatic context similar to pilot III, which the children and teachers preferred, is to be used for the final series, then strategies must be undertaken to make the objectives more explicit. Objectives should be written to lend themselves to explicit portrayal. Production personnel should strive for explicitness in the eventual media portrayal.

In retrospect, the initial set of objectives was probably insufficiently explicit and the production staff perhaps did not keep objectives clearly enough in mind, especially when producing dramatic segments. It is especially important for a child to understand the message if it is assumed that objectives such as those in the pilot materials are only the precursor to attitudinal and behavioral change objectives.

Utilization

Although the number of administrators was only eleven, the average ratings of content appropriateness, community acceptance, anticipated teacher acceptance, and encouragement for teacher use were 3.9 or higher on a scale of 1 to 5, where 5 was the most desirable rating (figure 5-5).

As shown in figure 5-6, teachers' ratings of content appropriateness averaged 4.1 or higher across all four sites. Recommendation for series use was also high on the average at all sites. Teachers showed a marked preference for pilot III.

In comments on questionnaires, as well in some interviews, both teachers and administrators commented that although they liked the series, the crowded curriculum, especially the educational television (ETV) curriculum, would make it difficult to use the series. The more the series could be incorporated into the existing curricula, the more likely that it would be used.

Content-oriented objections to the series were usually tied to a disagreement with the project goals, a judgment that the series was "too old" for the children, or both. Only a small minority of teachers or administrators expressed such objections.

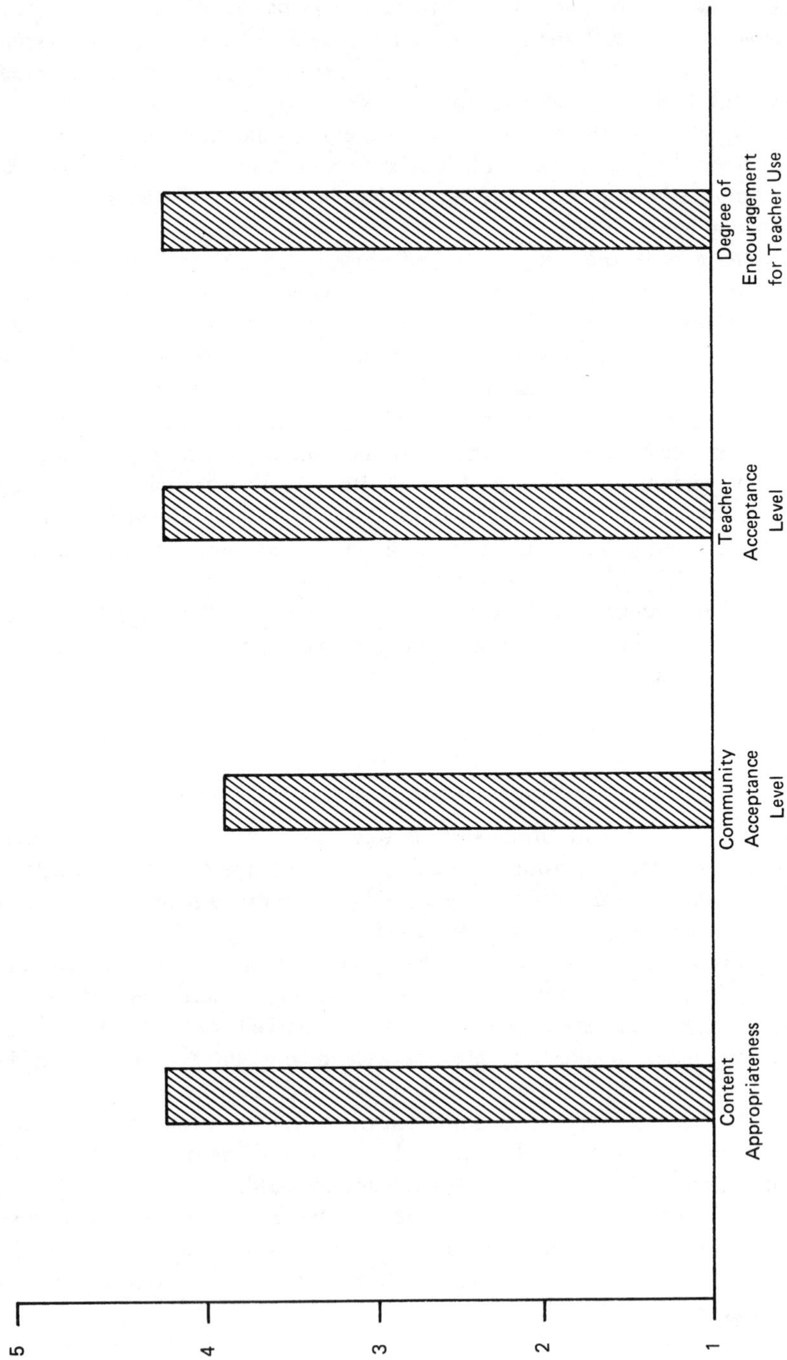

Figure 5-5. Mean Reactions of Administrators to Package Utilization (N = 11)

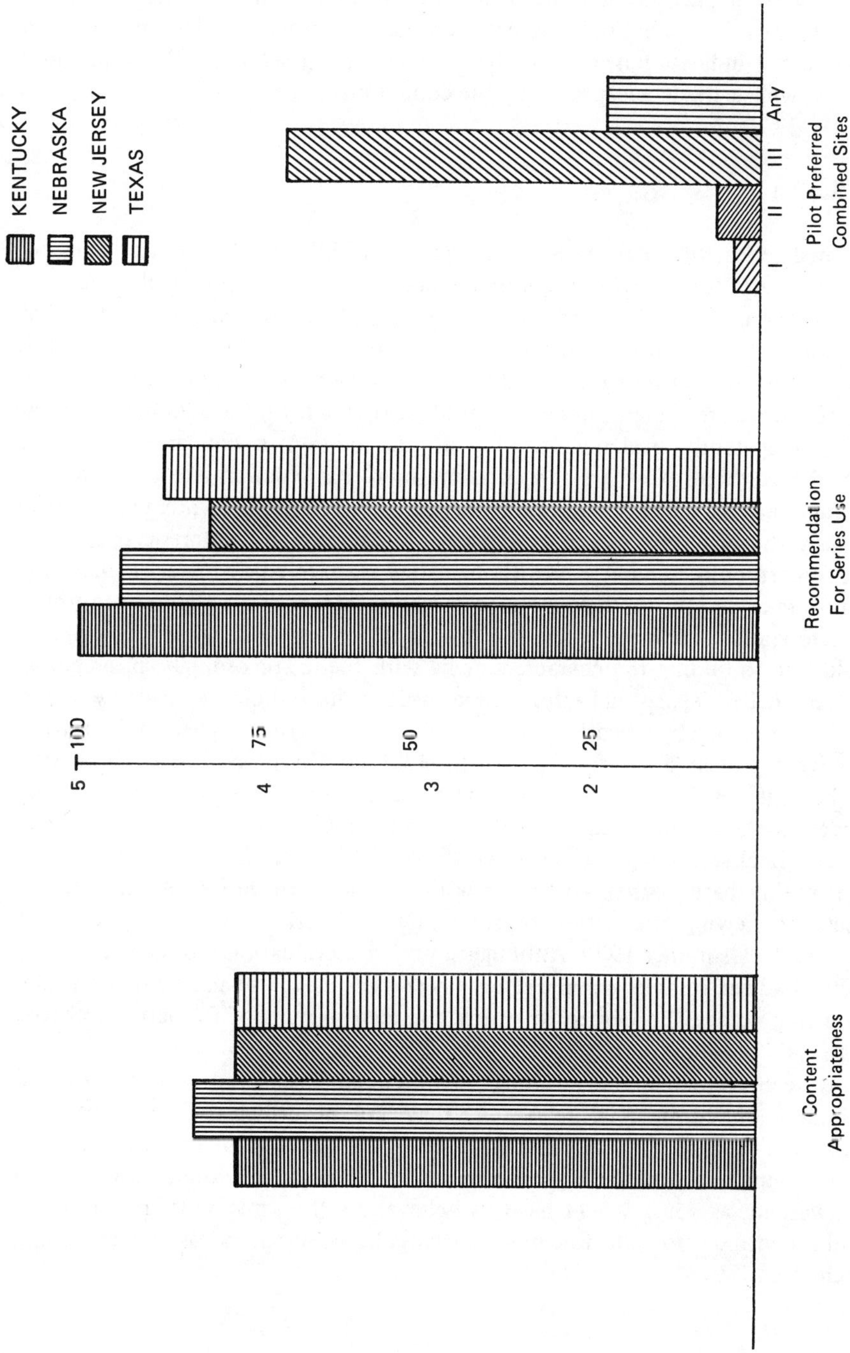

Figure 5-6. Reactions of Teachers to Package Utilization, by Site

Data from parents were more difficult to obtain, since in most cases the researchers had to obtain volunteer viewers. However, the data summarized in figure 5-7 indicate largely favorable responses to agreement with the program goals, as well as likely acceptance by the community.

Conclusion

The final series, which comprises twelve new half-hour programs, uses an essentially dramatic format with a continuing cast of children as principal characters. Certain characters are featured in each program. The dramas are designed so that they can be seen in almost any order; each program contains a story that is not intended to be continued across different programs. In addition to the more formal research results just described, another reason for using a dramatic format was to gain depth of characterization of the children in the series. It is clear from the research of others as well as my own that children like children. The programs are intended to be markedly realistic. Children do not tend to take seriously the television materials that have buffoons, counterstereotypic characters, or cartoonlike characters. Although they seem to enjoy materials that they do not take seriously, they have the clearest perception of antistereotypical characterization qualities when the drama is realistic and when the children would like to be like the characters or be with them. Accordingly, in preparing for the final series, special efforts were made to have dramatic characters with a high potential identification by the children. Most of the performers for the roles were pretested for their appeal to children. The process of identification has become an interesting theoretical research topic and is currently being pursued in evaluations of earlier portions of the final series.

The final series began in October 1978, and its first showings will continue on a weekly basis with versions for both in-school and home viewing. Overall effects of viewing this series, or summative evaluation, are scheduled to be completed by summer 1979. Although summative evaluation is designed to offer practical advice on whether the series meets its goals, it will also be theoretically interesting to see the degree to which eighteen months of formative activities led to valid production suggestions.

It seems obvious that although children are inundated with stereotyped characters on commercial television, they are nevertheless able to perceive accurately the specific antistereotypic qualities included in this experimental programming and occasionally on commercial television. Children have not been brainwashed, as some would have us believe. At the same time, however, we should continue to call for more socially balanced broadcast programming for children.

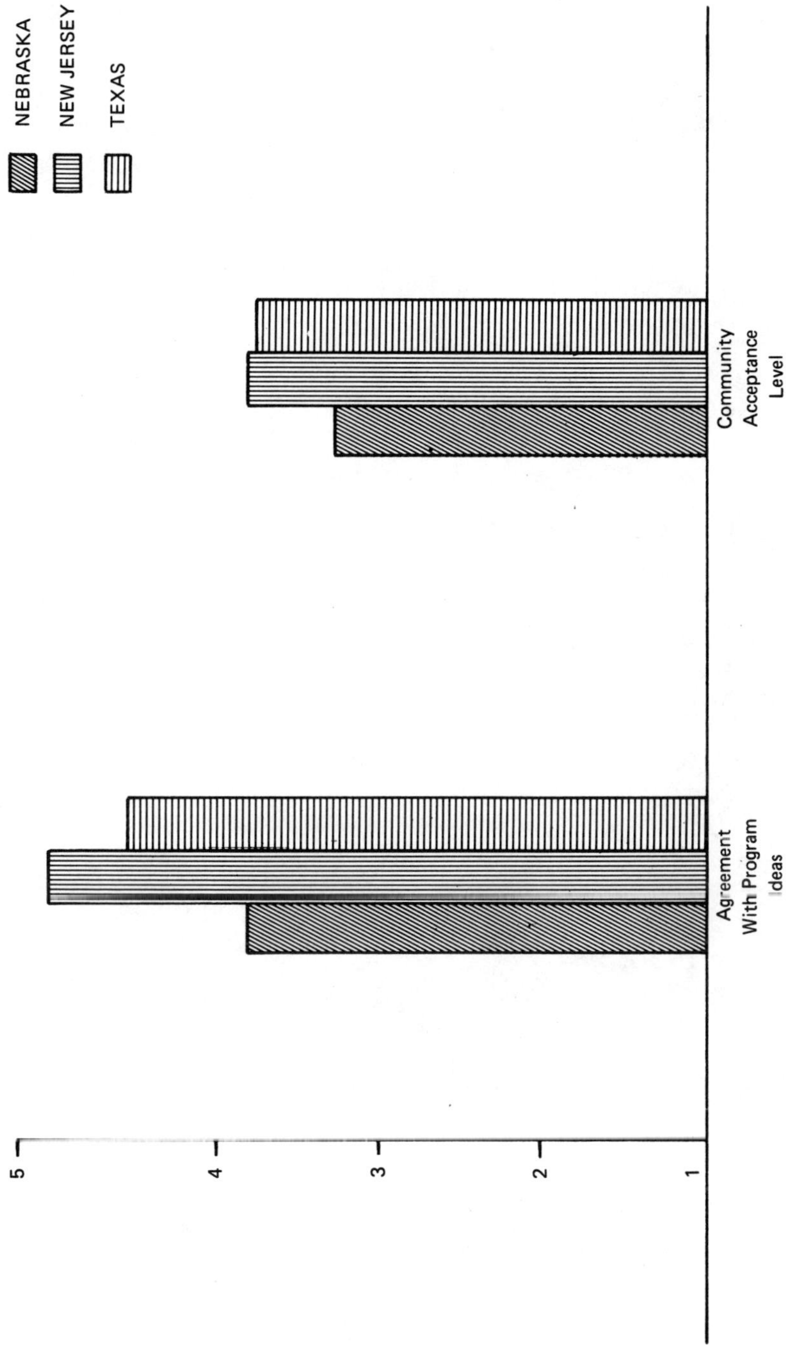

Figure 5-7. Reactions of Parents to Package Utilization, by Site

6 Meeting Minority Concerns: Structural versus Regulatory Approaches to Broadcasting

Allen S. Hammond IV

Lack of Diversity in Portrayals

> Although we are finally more frequently represented on television, we must now concern ourselves with how we are being depicted. . . . I'm all for being able to laugh at myself to a point. However, every time I see my people on screen, I don't want the show's essence to be merely crime or comedy. . . . I wonder why the shows that feature 99 percent black casts are comedies.[1]

It has increasingly been alleged that television's present programming fare perpetuates many negative images of minority Americans. For Example, television's black-oriented programs are said to be "white oriented," lacking in truth and in the diversity of black life-styles presented,[2] negative in their impact on the attitudes of both black and white Americans, and indicative of the continued insensitivity of the television industry to the needs and aspirations of black Americans.[3]

The criticism concerning the lack of diversity in the portrayal of blacks is extensive. In 1968 the National Advisory Commission on Civil Disorders stated in its report:

> If what the white American . . . sees on television conditions his expectation of what is ordinary and normal in the larger society, he will neither understand nor accept the black American. By failing to portray the Negro as a matter of routine and in the context of the total society, [the media] have . . . contributed to the black-white schism in this country.[4]

Some black critics have claimed that the images that television broadcasts produce largely unrecognized psychological and behavioral effects of unfounded feelings of superiority in whites and feelings of inferiority and self-hatred in blacks.[5]

Lack of Economic Representation

Although minorities constitute 18.5 percent of the nation's population and account for over 15 percent of the total population within thirty-one of the top fifty television markets, they own less than 1 percent of the nation's approximately nine thousand broadcast outlets.[6] As of February 1978 there were 987 television stations and 8,483 radio stations in operation in the United States. Of these outlets, only six television stations and approximately 69 radio outlets are owned by minorities. In sharp contrast, is that minorities make up approximately 20 percent of the broadcast consumer market.[7] Minorities are also substantially underrepresented in the workforces of local television stations and are excluded from decision-making positions at these stations.[8] As a result of their appalling absence as owners and staff within broadcasting, minorities exercise little direct control over their portrayal.

Consequent Failure to Achieve Economic Prosperity Attendant with Adequate Representation

The current absence of minorities in broadcasting has serious negative implications for the economic growth of minority business. Minorities, by virtue of their near total lack of ownership of broadcast facilities, are excluded from participation in an industry that has experienced annual revenue increases of more than 20 percent in recent years.[9] This exclusion, which continues even though minorities make up approximately 20 percent of the broadcast consumer market, denies minorities access to monies that could be flowing back into their communities to stimulate further economic growth. Moreover, given the economic underdevelopment that many minority communities are now experiencing, the exodus of purchasing power with little return is especially alarming.

The Need for a Minority Ownership Trust That Addresses the Problems

Minorities may be said to experience two fundamental problems in broadcasting. The first is the present lack of programming responsive to minority interests and needs (cultural diversity). The second is the present inability of minorities to secure a return of dollars for their investment in advertiser-supported broadcasting (as citizens and consumers) that is commensurate with that of the majority and could be used to stimulate further economic growth in minority communities. (economic diversity). Any proposed solution must address both problems.

Federal Regulation of Programming

Most minority efforts to assure programming responsive to minorities have focused on the use of the regulatory sanctions of the Federal Communications Commission through petitions-to-deny and citizen-broadcaster agreements.

Current regulatory sanctions take two basic forms; regulations aimed at assuring that licensees allot an established percentage of time to certain types of programming and, regulations that seek to assure balance in the licensee's presentation of information and opinion based notions of fairness, equity, and journalistic integrity. The serious enforcement of both types of regulations creates problems. It injects the determinations of a government tribunal into the licensee's exercise of discretion regarding how and what to program.[10] Commission reliance on licensee discretion, on the other hand, often renders the regulations ineffective to protect the public's stake in broadcasting. Such a result flies in the face of the FCC's congressional mandate to regulate in the public interest, convenience, and necessity.

Aside from the obvious tensions and dangers to the First Amendment inherent in the direct or indirect government regulation of broadcasting, the regulatory approach alone does not adequately assure that minority needs are addressed. While the regulations can be used to partially alleviate the present lack of cultural diversity, they do so by encouraging government intervention into programming decisions. This approach is limited because it requires a balancing of interests that often favors ineffective regulation to avoid possible censorship. Moreover, the program regulations do nothing to increase the economic self-sufficiency of minorities; instead they further entrench the economic status quo. As a result, the minority voice is insecure, and minority economic development (based on the use of the fair return on the dollars invested in broadcasting by minorities) is unattainable.

"Market" Regulation of Programming

Given the nature of the problems that minorities experience in broadcasting, direct ownership of broadcast facilities appears to be the most feasible alternative. Such a solution would not only assure the continued existence of a growing minority voice but also provide the opportunity for minorities to derive the economic benefits inherent in a thriving industry. These profits could then be used to further expand minority ownership of communications and noncommunications enterprises, thereby stimulating new economic growth.

The legal and regulatory policy justifications for such an alternative are already well documented.[11] As the D.C. Court of Appeals has stated:

> It is upon ownership that public policy places primary reliance with respect to diversification of content, and that historically has proven to be significantly influential with respect to editorial comment and the presentation of news.[12]

It is not unlikely that a minority licensee who is attuned to the concerns of the city of license would refuse to carry programming that the minority community considered degrading and would also require greater diversity in the portrayals of minorities.

Some may argue that the premise of minority broadcaster responsiveness to minority needs is conjecture. However, the history of minority involvement in broadcasting to date supports the premise. All black-owned outlets direct a large portion of their programming to minorities. Moreover, in only one instance has a petition to deny been brought against a minority-owned outlet by minorities. Although this record is not conclusive, it provides a strong indication that minority broadcasters have been responsive to minority citizens in the past.

Conclusion

Minority ownership of broadcast outlets would allow minorities to directly realize to a much larger degree the cultural diversity that the present broadcast industry has failed to adequately provide. Furthermore, it allows opportunities for further stimulating minority economic growth. While it would be folly to propose that minority ownership should be used as a substitute for federal regulation of broadcasting, it is fair to say that it will reduce the need to resort to regulatory sanctions and thereby reduce the danger of government censorship. For these reasons, substantially increasing minority ownership cannot help be in the public's interest.

Notes

1. Jon Mills, "Blackness in Televisionland," *Essence Magazine* 6 (1965): 15.

2. Beck, "Up from Amos 'n' Andy? Minority Images on TV," *Access* 2 (1975): 4; McPherson, "An Open Letter to CBS Studios and Quinn Martin Productions," *Essence Magazine* 6 (1975): 13. U.S. Commission on Civil Rights, Window Dressing on the Set (Washington, D.C.: U.S. Government Printing Office, 1977). See also, A. Hammond, "Toms, Coons, and Good Times: The Evidence of Black Stereotypes in Prime-Time Television" (master's thesis Annenberg School of Communications, University of Pennsylvania, 1977; available at NTIA (National Telecommunications and Information Agency) offices, Washington, D.C.).

3. Douglass, "Black Television: Avenues of Power," *Black Scholar* 5 (1973): 21-36; Nicholas, "Blacks in Television," *Black World*, April 1976, pp. 40-41.

4. *Report of the National Advisory Commission on Civil Disorders* (New York: Bantam Books, 1968), p. 383.

5. Hatcher, "Mass Media and the Black Community" *Black Scholar* 5 (1973): 4; Johnson, "Telecommunications Technology and the Socialization

of Black Americans: Issues, Concerns, and Possibilities" (Master's thesis, Washington University, 1974), p. 158.

6. See petition for Establishment of Policy to Promote Broadcast Ownership by Minorities, and Related Implementing Policies, RM 3055 filed before the Federal Communications Commission by the former Office of Telecommunications Policy and the Office of Telecommunications (Commerce), January 31, 1978.

7. *Broadcasting Magazine*, April 3, 1978, p. 112.

8. U.S. Commission on Civil Rights, *Window Dressing on the Set.*

9. Hickey, "Will The Network System Survive?" *TV Guide Magazine*, May 6, 1978, p. 40; Noll, Peck, McGowan, *Economic Aspects of Television Regulation* (Washington, D.C.: Brookings Institution, 1972).

10. See Bazelon, "FCC Regulation of the Telecommunications Press," *Duke Law Journal* 75 (1975): 213; Robinson, "The FCC and the First Amendment," *Minnesota Law Review* 52 (1967): 67.

11. Minority ownership can contribute to the ascertainment of minority needs and problems and provide added diversity of viewpoint (Minority Ownership Petition, pp. 5-9).

12. *TV-9 Inc.* v *FCC*, 495F. 2d 929, at 938, (D.C. Cir. 1973), cert. denied, 419 US 986 (1974).

7 Will More Women Make a Difference? A Sociological Perspective on Employment and Portrayal in Television

Muriel G. Cantor

Public interest groups, broadcasters, and several government agencies have been concerned with the way women are portrayed on television and with their failure to achieve the occupational success of men in the broadcast industry. This concern has generated a debate on how to change the portrayal of women on the screen. Some believe that the portrayal of women will not change unless more women are employed in the creative and decision-making aspects of the television industry; others, while advocating the employment of more women, also suggest that direct intervention in programming may be necessary. The issue of direct intervention is not addressed in this chapter.[1] Rather this chapter discusses and disputes the assumption that simply employing women as decision-makers and creators will necessarily change the way women are portrayed on the screen. The first section presents the argument and historical background of this issue. The second part is devoted to the slim research findings and factual information relating to the problem. The conclusion integrates and summarizes the two sections.

The most recent and the most complete report on the issues of portrayal and employment of women and minorities has come from the U.S. Commission on Civil Rights (referred to throughout as the Commission).[2] In its study *Window Dressing on the Sets* (1977) the Commission maintains that television drama does not reflect the sexual makeup of the United States: white males are overrepresented; sex stereotyping is prevalent; females are younger than the males shown on television; females are more often unemployed or family bound than the men; and they are more frequently seen in comic roles. The study also reports that women are rarely seen as newsmakers or news reporters on the three national networks, and civil rights issues relating to women receive minimal coverage on television network news. The Commission investigated employment as well as portrayal and found that women are unlikely to hold positions of importance at the stations. White males hold the overwhelming majority of decision-making positions. In contrast, women are likely to be found within the ranks of clerical workers.

These are only a small portion of the findings from the Commission's report. Those who have been concerned about the position of women in the broadcast industry, such as the United Church of Christ, the National

Organization for Women (NOW), and the American Association of University Women, will not be surprised by these findings. Since 1972 these groups and others have been documenting through content analyses that women are not represented on the screen in numbers equal to males and that they are not likely to hold important jobs at local stations. The Screen Actors Guild (1974, 1977) has also been concerned about the lack of employment for all actors and has done special studies to show how limited the work opportunities are for women actors in particular. It is news that a government agency issues a report on this topic, synthesizes old research, verifies some of the findings through original research, and disseminates the information to a wider audience.

The report performs a most important service by making recommendations to other parts of the government so that positive steps might be taken to change the position of women. In particular, the recommendations are addressed to the FCC and the broadcasting industries. The Commission suggests that the FCC conduct an inquiry on the portrayal of women in both commercial and public television because the problem is industrywide. Although it is not stated explicitly, it is implied that the local stations have minor responsibility for most television content. In the section on portrayal three of the four recommendations are directed to changes in the operations of production companies and the commercial television networks who are responsible for television series where stereotyping is most prevalent. It is noted that these production companies and program suppliers are exempt from the Equal Employment Opportunity (EEO) regulations imposed by law on the local stations. These recommendations also suggest that scripts should be solicited from women writers and that television dramatic series portraying women in a variety of roles should be developed.

Before the report was published, a draft was sent to the FCC for comments. As a result, the published report contains a direct response from the FCC, which states in part:

> Programming is, of course, the essence of broadcast service in the public interest . . . we agree that the lack of adequate role models may have an adverse effect on minorities and women . . . . Increasing numbers of minorities and women are now actively seeking employment in higher-status jobs in broadcasting and are attending professional schools to prepare themselves for such jobs. Further, the period of improved programming for minorities and women has followed and in part coincided with this agency's efforts to encourage and promote improved job opportunities for minorities and women. In our view, the promotion of equal opportunities for such persons is a *better* [emphasis added] solution to the programming problem than the direct intervention advocated in the draft report.

Thus the FCC suggests that there is a direct connection between the portrayal of women on the screen and the shortage of women employed in decision-making positions in the industry. (Positions of importance are defined to include managerial occupations, creative production jobs, and on-camera work). The FCC is not alone in this belief.

The relationship between employment and portrayal is complicated by legal and policy issues as well as what appears to be a denial of the realities of how content is produced. At this time there are no legal mechanisms to encourage the employment of more women by producers and directors who create the dramatic series or by the networks who disseminate these series. The FCC has regulatory authority only over the local stations, who depend on the program suppliers and the networks for most of the programming broadcast in the United States. However, even if the Communications Act of 1934 were rewritten to include the networks (no one has suggested that the act also include the production companies), it is unlikely that increased employment of women would necessarily change the way women are portrayed.[3] Although it may appear that the simplest solution to the problem without resorting to direct intervention is through the hiring of more women, there is little evidence of a direct connection between the sex of the communicators and the content they produce. Furthermore, the evidence is weak that sex or any other social or psychological characteristic will make a difference in how women are portrayed.

Background

Because there has been little research into the questions raised in this chapter, all statements are conjecture. This issue has only recently been considered, and the direct research into the question is limited. It has become salient because of the involvement of public interest groups. Only after this became a political and social issue did social scientists become aware of the problem. While doing research on television producers in the late 1960s (Cantor, 1971), I found nothing on the relationship of ascribed characteristics of the communicator to the content of the media. Also, there was little discussion of how social and personality characteristics might influence the final product.[4] Several content analyses on the demographic characteristics of characters in dramatic series were often cited in mass communication literature to show that women were given a smaller number of parts and were not likely to be portrayed in occupational roles (Head, 1954; DeFleur and DeFleur, 1967). However, except for these, the literature before 1968 contained little on employment, portrayal, or the creative process in television.

In the late sixties there was a move to change the media. Television has been attacked for presenting women throughout all programmings as second-class citizens, sex objects rather than human beings, and both the network and local television stations were faulted for neglecting women's rights issues and other issues of public importance relating to women. As part of license renewals these public interest groups monitored television to show that women were treated unfairly (see Franzwa, 1978, for a review of some of these). These content analyses, though methodologically questionable, have been since replicated by several different social scientists, and the results can be considered both reliable and valid.

Public interest groups made the following arguments. Television failed to provide full and equal representation of women, and therefore is an issue of public importance. Because broadcasters did not afford reasonable opportunity for a contrasting point of view, they were not applying the Fairness Doctrine. The same social action groups also pressured for equal employment for women in the industry. As soon as the FCC included women in its EEO rules in 1971 and local stations were required to report each year the sex and ethnic composition of their workforces, the documentation was available verifying that women were indeed a minority of the workforce in local stations. In license challenges and other community action the issue of employment was separated from content. The NOW brief (1972) for the petition to deny the license of WRC-TV made no claim that employing more women at the station would change their portrayal. The content was presented as unfair to women, and the lack of women in positions of importance was presented as discriminatory. Although this pressure resulted in the employment of more women at the local stations, the increased number has had little effect on content.

Because this issue has only recently been investigated by social scientists, there is little hard evidence about the connection between the sex of the communicator and the content produced. The next section summarizes the available findings.

The Evidence

Mass communication research has focused on the audience and the effects of communication; the communicators and creators are generally neglected. The gatekeeper studies of decision making in news and entertainment provide one model of the kind of research that might be done in an organizational context to find out whether employing women would make a difference in the portrayal of women. These studies point out the complex relationship between the communicators and the programs created. In a review of some of these studies McQuail (1969) notes that the most significant point of emerge from such work is that the orientation of the mass communicators is shaped by the structure and culture of the organization in which they work. In reviewing some of the same studies and some newer ones as well, Paul Hirsch (1978) makes a similar point. Hirsch says that occupational, craft, and organizational norms concerning news and story categories explain the variety in story selection more than personal bias.

There are few studies on how content is "manufactured" or created in the production companies. In my studies of television producers (Cantor, 1971, 1972) I found few women in important positions at the studios. All directors, producers, and most writers were men. Even in acting women are a minority. Interviews with producers and actors (Peters and Cantor, 1978) revealed that many of those working in Hollywood express values different from those found in the content of their shows. Most of the time these expressed values are more liberal than those expressed in the drama. Actors, in particular, who make commercial advertisements for money or have to accept any part offered to them in order to work in the medium often appear in productions that express values opposite from theirs. A method of social control operates in the Hollywood studios that approximates that of the newsroom but is also different. Warren Breed's (1955) classic study showed that journalists were controlled by the reward system and the political preferences of their employers. There is a difference in working in a Hollywood studio, because the creative people in drama must bend to the pressures from the production company and from the networks as well. However, what is produced is determined by those above, leaving the creative people little control. There are individual differences in the way people adapt to these controls, and some people clearly agree with those in control. However, those who do not conform do not stay in the business. These findings are so consistent that it seems unlikely that women creators would act differently from men. It is possible that a greater number of women could make some difference in the way women interact within an organization, but content depends on the political orientation of those who control the marketplace and not on the sex of the communicator or the political values professed. Thus it is possible that a feminist woman could create an all-male show while a nonfeminist could create a show portraying women in a favorable way.

This does not mean that all women who work in bureaucratic structures act the same as men employees. There are both real and apparent differences between sexes. Rosabeth Kanter (1975, 1976) has persuasive evidence that organizational structure and not personal preference account for the apparent differences between men and women at work. The structural forces that support stereotyping women—blocked opportunities, powerlessness, and tokenism—affect not only women but members of any minority group who try to break into the majority system. Affirmative action programs are designed to change organizational conditions rather than people's attitudes or motivations; the assumption is that attitudes follow opportunity. One could also argue that because women are only tokens, they operate in a way that perpetuates the system rather than as agents of change because they accept organizational control and norms. The presence of a few tokens does not necessarily pave the way for others (Kanter, 1975). It is beyond the scope of this chapter to discuss the meaning of the word "token," but hiring one woman changes neither employment opportunities for women nor the orientation of the organization. If their numbers increase, women in organizations may benefit by becoming more powerful in controlling their own careers. The information on individual women in high positions shows that they are not likely to consider the issues raised by the Commission in the format of their programs. Joan Cooney, producer of "Sesame Street," is a case in point. A major criticism coming from women's group is that the program is sex-biased. All the moppets, for instance, are male (Lesser, 1974; Cantor, 1977).[5]

The evidence that females are a minority of characters in all fixtion has been well documented. The portrayal of females on television is not much different from the portrayal of women and girls in the movies, in television ads, and in children's books (Franzwa, 1978). Most screenwriters and advertisers who work in organizational settings are men. Few women become screenwriters or advertisers. However, the writers of children's books, in contrast, are likely to be women. A tabulation of female authors indicated that 41 percent of Caldecott and 58 percent of Newberry Medal winners analyzed by Lenore Weitzman et al. (1972) were written by women. Although they found that women authors were more likely than men authors to have female central characters in these story books, both sexes picture girls primarily in relation to the boys and men in their lives.

A study of journalism students found that women and men hold the same stereotypic views of the reading preferences of the audience (Orwant and Cantor, 1977). Both men and women in the sample believe that women prefer reading about fashions and cooking and that men prefer sports, business news, and public affairs. Also women students reported their own reading preferences as different from those they projected on the women audience. Thus it was concluded that both men and women stereotype their audience, regardless of their own views.

Conclusions and Discussion

Many researchers have theorized that the reason for the poor image of females in the mass media is that the media are male dominated. For instance, Judith Lemon (1978), who has been doing content analysis on how women are portrayed in prime-time television, implies that there is a direct relationship between the number and position of women in the industry and the way women are portrayed on the screen. Busby (1975), after reviewing the studies of sex roles in media research, states that since the same cultural forces have influenced writers, producers, directors, and publishers it is not surprising to find a similarity of sex-role imagery across media because all media have been dominated primarily by white American males.

Organizational and historical studies are necessary to understand the phenomenon criticized by both the FCC and the U.S. Commission on Civil Rights. Janus (1977) reviewed the same literature as Busby. Although Janus's language is explicitly Marxist, her conclusion is similar to Busby's: the problem is that almost everything we know comes from content analyses. The only other factual material is from the FCC Form 395, and those data have been criticized so frequently and by so many different groups that the forms are being changed. No one has done for commercial broadcasting at the station level what the Task Force on Women (Isber and Cantor, 1975) did for public broadcasting—a survey of individuals at the public stations. This survey gathered information on employment and career patterns for both men and women. Form 395 reports information only by job category, and even the job categories are not clearly differentiated.

Content analyses are important. They can show change in presentation over time if the same concepts and categories are used, but more studies on employment, control, industry practices, and the creative process are needed. All these studies should consider sex differences and similarities. To borrow from Paul Hirsch (1978), we can study the administration of the organizations producing television, social control of those employed by these organizations, organizational levels of authority, working conditions, organizational size, and occupational communities. However, to study these without including women, as in the past, would be unrealistic and would give none of the answers to the questions that arise over and over again. Until we know how content is created and selected, we cannot answer the question raised in this chapter. Although my research leads me to believe that the values of communicators are basically irrelevant in the selection of content and that the "industrial" setting and the economic and political orientations of employing organizations are more important, this question needs direct attention. Sexism in role portrayals should probably not be viewed as the oppression of females by males but within the framework of the historical development of business practices. While I am convinced that sex as a variable is always important, research on both the communicators and the content often ignores what is commonly discernible in the world around us: Not all women are the same. Therefore it is important to know when women are perceived as professionals and workers within an organization and when they are perceived as tokens or as stereotypes. To find this out it is necessary to study the production process and the control of industry practices, as the Commission suggests. What is needed is a combination of approaches to find out how each kind of production differs and in what ways they are similar.

The problem is not only that social scientists and policymakers have been limited in their perspectives; it is clearly related to advocacy and action. From control analyses and the few employment studies available, we know that women are neglected and sometimes even invisible on the air and that they are not employed in many important positions. The changes that have occurred in both employment and portrayal in the last ten years (U.S. Commission on Civil Rights, 1977) may not be directly attributable to citizen action but rather to larger societal changes: more women in the workforce and other social phenomena such as the high divorce rate and the increase of single women as heads of household. Because it was the citizens group who first brought these questions to the attention of policymakers and social scientists, citizen involvement in broadcasting is also part of larger social change. Further changes in media and particularly television will probably come only if the audience along with policymakers has an impact on decision makers and creators of both sexes at the highest level.

Notes

1. Direct intervention involves complex problems and issues, including the interpretation of the First Amendment, the role of the Federal Communications Commission (FCC) in the production process, the interpretation by the FCC and the courts of the Fairness Doctrine, and the regulatory laws and how they apply to the disseminators and broadcasters of television content.

2. The issue of the employment and portrayal of minorities is not being addressed in this chapter. Although women and minorities are often studied together in media research, the differences between the two statuses are greater than their similarities. *Window Dressing on the Sets* (U.S. Commission on Civil Rights, 1977) addresses each separately.

3. Although the Communications Act may be rewritten, it is unlikely that further controls will be enacted. In fact the opposite is true, and the Fairness Doctrine will probably be replaced with an equity principle requiring television stations to provide news, public affairs, and local programming and to treat controversial issues of public importance in an equitable manner. The networks would still be free from any control in the proposed legislation.

4. Even Hortense Powdermaker's (1955) famous study of Hollywood, which describes the occupationsl roles, structure, and culture of the movie industry in the late forties, had little that connected the people creating to the movies produced.

5. Paul Hirsch called to my attention that Faye Dunaway, in the role she played in *Network*, provides another example of a woman in a high posi-tion who behaved in accordance with the values and goals of the organization in which she worked. To her, ratings and audience were the most important criteria for selecting content. She was a "professional" first; her sex on the job was unimportant in relation to content selected.

References

Breed, Warren. 1955. Social control in the newsroom. *Social Forces* 33:326-335.

Busby, Linda J. 1975. Sex-role research on the mass media. *Journal of Communications* 25:107-31.

Cantor, Muriel G. 1971. *The Hollywood TV Producer: His Work and His Audience*. New York: Basic Books.

_____. 1972. The role of the producer in choosing children's television. In *Television and Social Behavior*, vol. 1: *Media Content and Control*, ed. George Comstock and Rubenstein Washington, D.C.: U.S. government Printing Office.

_____. 1977. Women and public broadcasting. *Journal of Communications* 27:14-19.

DeFleur, Melvin, and DeFleur, Lois B. 1967. The relative contribution of television as a learning source for children's occupational knowledge. *American Sociological Review* 32:777-789.

Franzwa, Helen. 1978. The image of women in television: An annotated bibliography. In *Hearth and Home: Images of Women in Mass Media*, ed. Tuchman, Daniels, and Benet. New York: Oxford University Press.

Head, Sidney W. 1954. Content analysis of television drama programs. *Quarterly of Film, Radio, and Television* 9:175-94.

Hirsch, Paul. 1978. Occupational, organizational, and institutional models in mass media research: Toward an integrated framework. In *Strategies for Mass Communication Research*, ed. P. Hirsch, P. Miller and F.G. Kline. Beverly Hills, Ca.: Sage Publishing Co.

Isber, Caroline, and Cantor, Muriel G. 1975. *A Report from the Task Force on Women in Public Broadcasting.* Washington, D.C.: Corporation for Public Broadcasting.

Janus, Norene. 1977. Research on sex-roles in the mass media: Toward a critical approach. *Insurgent Sociologist* 7:19-32.

Kanter, Rosabeth. 1976. Why bosses turn bitchy. *Psychology Today* 9:56-59.

_____. 1975. Women and the structure of organizations: Explorations in theory and behavior. In *Another Voice*, ed. Marcia Millman and Rosabeth M. Kanter. Garden City, N.Y.: Anchor/Doubleday.

Lemon, Judith. 1978. Dominant or dominated? Women on prime-time television. In *Hearth and Home: Images of Women in Mass Media*, ed. Tuchman, Daniels, and Benet. New York: Oxford University Press.

Lesser, Gerald S. 1974. *Children and Television: Lessons from "Sesame Street."* New York: Vintage Books/Random House.

McQuail, Denis. 1969. *Toward a Sociology of Mass Communications.* New York: Macmillan.

NOW Brief. 1972. Petition to Deny. In Re: Application of National Broadcasting Co., Inc. For: Renewal of License. Washington, D.C.

Orwant, Jack, and Cantor, Muriel G. 1977. How sex stereotyping affects perceptions of news preference. *Journalism Quarterly* 54:99-108.

Peters, Anne, and Cantor, Muriel G. 1978. Acting as work. Paper presented at the Southern Society Annual Meeting, New Orleans, La., April 1.

Screen Actors Guild. 1974. *The Relative Role of Men and Women in Television Commercials*. New York.

____. 1977. *Women: Committee Commercials Survey Update*. New York.

U.S. Commission on Civil Rights. 1977. *Window Dressing on the Sets: Women and Minorities in Television*. Washington, D.C.: U.S. Government Printing Office.

8 Influence of Community-Based Citizens Groups on Television Broadcasters in Five Eastern Cities: An Exploratory Study

William E. Hanks and
Terry A. Pickett

The typical mass communications research study has examined the effects of mass communications messages on audiences or on audience selection and use of media. Audience behavior, beliefs, and attitudes are usually the dependent variables, while exposure to media messages is usually the independent variable [40]. However, there is a paucity of research that takes audience influence on media as an independent variable (other than ratings and marketing studies) and the operation of mass communications systems as dependent variables. The citizens group movement in U.S. broadcasting provides an opportunity to study the influence of audience members on a mass medium.

In the last decade citizens groups have increasingly demanded special ascertainment of and programming to meet the interests and needs of minorities, women, and children and access to the airways through regularly scheduled citizen-initiated editorials. Broadcasters have frequently negotiated agreements with citizens groups to provide for increased employment of minorities and women and to provide special employment and programming in return for the groups' withdrawals of legal petitions against station licenses or license transfers. In some cities specific hiring policies and television programs have resulted from citizen-broadcaster agreements. Citizen access agreements have also been executed in several cities.

The Literature

Most of the literature on the citizens group movement is popular and anecdotal [20]. Few authors have generated research from theoretical models, although Honig [31] and Grundfest [26] have employed economic models of cost-effectiveness to predict broadcaster reactions to petitions to deny station licenses.

The most obvious deficiency in the literature is the lack of theoretically grounded research on long-range effectiveness that attempts to isolate the factors best predictive of change. There are no follow-up studies that track the activities of citizens groups over time, although Powell and Meek [60] and Levin [46] have described WLBT-TV under its interim licensee. Lewels [47] has, in part, followed Chicano agreements, and Shement et al. have described some effects of a license challenge in San Antonio, Texas [65].

Grundfest has called for measurement of satisfaction with agreements between citizens and broadcasters [26]. Certainly, such studies would be valuable. This chapter incorporates to some extent a follow-up on some of those agreements and tracks over a four-year period relationships between selected broadcasters and citizens groups.

This chapter reports on an exploratory study of the activities and of perceived influence of selected citizens groups on several television broadcasters in five major eastern cities.

Methods

To identify cities that could represent activity in the citizens group movement, four public interest communication attorneys and a legal assistant familiar with the movement were asked in 1974 to name the most active cities in the citizens group movement. The attorneys, who either worked or had worked at Citizens Communication Center, were Ellen Agress, Robert Stein, Frank Lloyd, and Ed Hayes; the legal assistant, also at Citizens Communication Center, was Susan Dillon. The ten cities named most frequently were divided into two sets, five east and five west of the Mississippi. The five eastern cities, which are the sample for this study, are: Rochester, New York; New York City; Boston, Massachusetts; Atlanta, Georgia; and Pittsburgh, Pennsylvania.

Citizens groups were identified for each city from a list supplied by the Citizens Communication Center, and each minority interest, women's interest, and access group was telephoned to set up face-to-face interviews. Many groups were consistently unavailable; they either did not answer the telephone or had telephones disconnected. Those who could be reached were scheduled for interviews in December 1974 and January 1975.

Television station managers or other knowledgeable station management representatives, for example, a public affairs director, were interviewed in the sample cities. Representatives of citizens groups and broadcasters were asked the following questions regarding the activities of citizens groups and broadcasters.

1. Who approached you? (Whom did you approach?)
2. What did they (you) want?
3. Were they (you) wanted?
4. How influential were they (you)?
5. What was the most effective thing they (you) did?
6. What was the least effective thing they (you) did?

Face-to-face interviews were conducted in December 1974 and in January 1975, and telephone interviews using these same questions were conducted in April 1978. Several stations, through their general managers or presidents, refused to be interviewed, some giving reasons such as WPIX-TV, New York: "We are currently in litigation and we are advised by our attorneys not to grant interviews with you." Other stations, such as WCBS-TV, New York, and WAGA-TV, Atlanta, simply refused without explanation. Except for cases in which station managers had been replaced, the same broadcast representatives were interviewed in 1974 and in 1978.

At least one citizens group representative in each city was interviewed face-to-face in 1974-1975, and the same group representative was interviewed by telephone in 1978 (except in Boston where a new director of the Atlanta Media Coalition replaced the original interviewees).

Employment data from the cities were collected from the FCC Office of Policy Analysis Branch, for 1972 through 1977. These data were to compare employment of women and minorities in the sample cities with employment of women and minorities in the nation as a whole.

Station files in the FCC Public Reference Room in Washington, D.C., were examined for 1972, 1975, and 1978, as were *FCC Reports* (1969-1976) and other available literature, to determine the amount and kind of citizens group activity in the sample cities. A summary of some of these activities is presented here to provide background for the interview findings.

Activities

Atlanta

In 1969 the Citizens Communication Center requested and was granted an extension of time to file petitions-to-deny on behalf of the Community Coalition on Broadcasting of Atlanta against Atlanta Broadcasters. Within thirty days twenty-two of the twenty-eight stations had signed agreements on hiring, ascertainment, and programming for the black community [22].

In 1970 the Coalition petitioned to deny WTCG-TV's license for failing to program to the needs of the local black community, for an alleged inadequate ascertainment survey, for presenting news from an exclusively white perspective, for employment discrimination against blacks and other minorities, and for conspiracy and antitrust violations. The FCC denied the merit of all charges and denied the petition in August 1971 [67].

A citizen's committee requested the FCC in 1968 to deny the license transfer and format change for WGKA-AM and -FM because the stations were the only ones devoted to the broadcast of serious classical music. The FCC approved the transfer, but in 1970 the citizens' committee won an appeal requiring the FCC to hold hearings on whether the license transfer and format charge were in the public interest [66].

Boston

In 1970 Anthony Martin-Trigona filed petitions to deny licenses against WBZ-TV (Westinghouse), WNAC-TV (RKO General), and six other stations in Connecticut and Rhode Island for "false and misleading spot political advertising." This petition was dismissed in 1972 because, according to an FCC majority, Martin-Trigona as a resident of Urbana, Illinois, lacked standing as a party in interest for Boston [56]. In March 1972 Evertt C. Parker, director of the United Church of Christ (UCC) Office of Communication requested that the FCC "initiate an inquiry into the employment practices of Massachusetts television stations and that pending the result of such inquiry all license renewals be deferred." The FCC at first denied the request, but acting on UCC's "Petition for Reconsideration" and its own interpretation of employment data for 1971-1973, it ordered WHTN-TV and WNLP-TV, Springfield, and WSBK-TV, Boston, "to conduct additional affirmative action efforts regarding the full-time female employment in the upper four job categories . . ." [13]. In 1973 the Boston Community Media Committee (BCMC) Minority Caucus negotiated an agreement with Heftel Broadcasting in the pending sale of WVOR-FM from RKO General, but the FCC held the agreement invalid, in part, for, improperly delegating license program responsibility to BCMC [62].

New York

In 1969 and in 1972 identical pleadings were filed by Anthony R. Martin-Trigona against the three network-owned and -operated television stations in New York (WNBC-TV, WABC-TV, and WCBS-TV), charging the stations with "antitrust, conglomerate and media concentration violations," but the petitions were denied because Martin-Trigona did not live in the service area of the station [55, 56]. In 1972 other unsuccessful petitions were filed by Friends of Earth and Citizens for Clean Air against WCBS-TV, WABC-TV, and WNEW-TV for failure to air anti-big engine ads on the air pollution issue [12] and by the National Organization for Women for failure to ascertain women, discuss as a controversial issue the role of women in society, and present programming responsive to women [6]. In 1972 the New Jersey Coalition for Fair Broadcasting filed petitions-to-deny against WNEW-TV, WCBS-TV, and WNET-TV for failure to ascertain and serve the programming needs of New Jersey. Petitions were withdrawn in return for extensive agreements that called for New Jersey news coverage and New Jersey ascertainment and for public affairs programming. The coalition negotiated similar agreements with WOR-TV, WNBC-TV, and WABC-TV [42]. Black Citizen's for Fair Media (BCFM) petitioned to deny the license of WCBS-TV in May 1972 for employment discrimination and failure to adequately ascertain and program to the needs and interests of blacks, but the group withdrew in lieu of an agreement. BFCM also negotiated agreements with WNBC-TV and WABC-TV in which the stations were committed to regular meetings with a black advisory board, hiring and training of blacks, and specific black-oriented programming [41].

WNET-TV (noncommercial) has been subject to several petitions. In May 1971 the Puerto Rican Media Action and Educational Council (PRMAEC) petitioned to revoke its license, to deny its application for renewal, to require the station to cease alleged discriminatory hiring practices, and to declare that WNET-TV was discriminating against Puerto Ricans. The FCC denied this petition [61]. PRMAEC had also filed an unsuccessful petition-to-deny against WNBC-TV and negotiated an agreement on hiring and programming in 1972 [41].

In May 1975 the New Jersey Coalition for Fair Broadcasting filed a "Petition for Declaratory Ruling and for Issuance of Notice of Apparent Liability," which claimed that WNET-TV had not maintained studio facilities in Newark despite FCC orders to do so and that the station failed to serve Newark with adequate programming. The coalition in 1974 had petitioned the FCC for inquiry into television service for New Jersey, and the FCC had determined that "additional locally oriented television service to the State of New Jersey is required" [17]. In May 1975 Governor Brendan Byrne and the state of New Jersey petitioned the FCC to defer action on license renewals for all New York City VHF stations and on WNET-TV pending "the results of the Commission's Proposed Rulemaking." This petition, too, was denied [58]. The FCC issued a memorandum in 1976, stating that it would act to provide WNET with VHF services through "drop in" of a channel, increased "presence" of New York City television stations in New Jersey, or the relocation of WABC-TV from New York City to Newark as a condition of licensing [17]. The coalition appealed the FCC action to the Third Circuit Court of Appeals, and in March of 1978 the court upheld the FCC [70]. One other unsuccessful petition was filed against WNET-TV in 1975, by the Congress of Afrikan People (CAP). CAP said: "The Commission cannot evaluate the New Jersey Coalition's Petition for Declaratory Ruling against WNET without concurrently taking into account the particular needs of Newark's Black and Latino residents." The petition was denied [17].

Pittsburgh

Create Coalition for Media Change (CCMC) was organized in 1972, and the group called on the Pittsburgh office of the Community Relations Service

(CRS) of the U.S. Justice Department. CRS brought an attorney from the Citizens Communication Center to Pittsburgh to set up a negotiating timetable with KDKA-TV, WTAE-TV, and WIIC-TV, but CCMC succeeded only in getting a general letter from the stations agreeing to continue efforts to serve the black community [43]. CCMC unsuccessfully petitioned in 1972 against relicensing WDUQ-FM, a noncommercial station licensed to Duquesne University, for employment discrimination, failure to program locally, and fraudulent promises to the FCC [72]. In 1972 the Media Awareness Committee of the Greater Pittsburgh chapter of NOW and the Committee for Open Media (COM), Pittsburgh, negotiated agreements with these same stations. Agreements with NOW for 1975-1978 called for women's advisory councils, a regular weekly news segment on women's issues, and several prime-time locally produced programs on women's issues [45]. The COM agreements required as a ninety day experiment the production, presentation, and promotion of citizen-access editorials, or free speech messages [44]. Finally, in 1972 extensive agreements were reached between Heftel and CCMC, COM, and NOW, in Heftel's purchase of WAJR (changed to WKTQ) and WSHH-FM from NBC. The groups agreed not to file against Heftel in retun for numerous concessions on employment of blacks and women and for access programming [4].

In early 1975 the NAACP Media Committee organized to assess broadcaster performance related to blacks and women. The group filed no petitions on racial issues but petitioned to deny the license of WQEX-TV (UHF) on grounds that Metropolitan Pittsburgh Public Broadcasting (MPPB), licenses of WQED-TV and WQEX-TV, operated as a public television monopoly and essentially duplicated 80 percent of WQED-TV's programming. In September 1976 MPPB submitted an amendment to WQEX's renewal application extending its programming to evenings, including separate programs from those on WQED-TV. In April 1976 the NAACP petition was denied for failure to demonstrate that program duplication and limited hours of operation deserved the public interest [16]. Also in 1975 the Association of Arab Americans filed an unsuccessful petition to revoke KDKA's AM radio license for alleged bias in presenting the Arab-Israeli issue on a talk show [8].

Rochester

Activity by citizens groups in Rochester has been mainly the work of Action for a Better Community (ACTION) and the Rochester Black Media Coalition (RBMC), its successor. The first legal action, however, was taken in 1969 by METRO ACT and by FIGHT, a coalition of Caucasion and minority groups that filed an informal objection against WHEC-TV's relicensing. METRO ACT charged WHEC-TV (Gannett Newspapers) with media concentration, news distortion, and inadequate ascertainment surveys. METRO ACT withdrew its petition in return for an agreement requiring the station to establish an advisory council to increase local public affairs programming, to expand its local news, and to maintain independence from the Gannett company which owned both Rochester newspapers) [74]. In 1972 ACTION petitioned against renewal of WHEC-TV's license, charging the station with inadequate ascertainment of the black community and employment discrimination. In May 1975 (a month before the expiration of WHEC-TV's license for the 1975-1978 term), after ACTION formally withdrew its complaint, the Commission noted that the 1975 renewal application, as amended, and the 1975 Annual Employment Report revealed "a substantial improvement in your minority profile and in your compliance with the Primer's requirements for relating needs to proposed programming" [75].

In 1972 METRO ACT and ACTION also filed petitions to deny license of stations WROC, WPXY-FM and WROC-TV. METRO ACT alleged that ascertainment of community needs had been inadequate and faulty and that the past public affairs programming of the stations were duplicative and lacked a minority perspective. ACTION based its complaint on an allegedly inadequate survey of the minority community. In 1975 these petitions were denied. In 1972, ACTION also filed, petitions-to-deny against WOKR-TV, WHAM, WHAM-FM, WMMF-Inc., WNYR, WEZO-FM, WSAY, and WSAY-FM [63]. WOKR-TV's license was renewed by the FCC in 1976, and ACTION and RMBC appealed the renewal to the District of Columbia Court of Appeals [1]. Meanwhile, in order to transfer control of its license in a pending sale to the Post Corporation, WOKR reached agreement to dismiss the appeal in return for an amendment of WOKR-TV's license promising "its best efforts to obtain a minority person as a management nominee to Post's Board of Directors." The agreement also called for minority employment plans, scholarships for minorities, programming, and support of minority business [2].

Findings

Probably the most important finding of this study is that the immediate goals of citizens groups significantly increased employment for minorities and women, increases in programming, coverage of events and issues relevant to these groups—were realized to varying degrees. Most interviewees felt, however, that little significant progress had been achieved toward long-range goals of social change through television, such as decreased sexism in television programming, improved image of minorities, and an increase in diversity of opinion and public debate on issues through citizen-initiated free speech messages. This opinion was most strongly expressed by broadcasters. Several generalizations were derived from these interviews. (Full texts of interviews are available on request).

In the period 1972-1975 minority groups, women's groups and access groups met periodically with broadcasters and demanded employment, special-interest programming, ascertainment of minorities, and free speech messages. For example, Al Rugless and Charles Hatcher from the Rochester Black Media Coalition (RBMC) were interviewed along with Jack Decker, general manager of WHEC-TV, G. Bennett Larsen, general manager of WOKR-TV, and William Pearce, president of WXXI-TV (noncommercial UHF) Rochester, New York. Decker and Bennett reported that they had been approached regularly since 1972 by the Rochester Black Media and irregularly by the National Organization for Women (NOW). Pearce said that the only pressure exerted on WXXI-TV was by "white politicians seeking TV exposure and threatening to withdraw support from the station." Decker and Bennett agreed that the main demands of RBMC were for increased employment of minorities, special prime-time black oriented programming, and programming that addressed problems noted in audience ascertainment surveys. RBMC agreed that these were the group's main goals. In New York City Kenneth MacQueen, vice-president and general manager of WABC-TV, and Lawrence Gershman, station manager at WNBC-TV, listed Black Citizens for Fair Media (BCFM), NOW, the New Jersey Coalition for Fair Broadcasting, and the Puerto Rican Media Action and Education Committee (PRMAEC) as the primary groups in New York that had regularly made demands for hiring, training, employment, programming, and ascertainment.

In the 1975-1978 period only black groups maintained regular contact with stations, although the groups representing black interests changed in two cities. For example, WTAE-TV and WIIC-TV, Pittsburgh, reported that the Women's Advisory Council, specified in 1972 agreements between NOW and broadcasters, had ceased to meet. Pittsburgh's Committee for Open Media (COM) and the Community Coalition for Media Change (CCMC) had dissolved, although Pittsburgh's NAACP became active in 1975. NAACP filed one petition to deny against WQEX-FM (noncommercial) and met irregularly with television stations. Atlanta's Community Coalition, an organization of black groups, had changed leadership since 1975, according to John Evans, executive director of the Atlanta NAACP. Evans said that the new name for the group was the Atlanta Media Coalition (AMC) which consisted of essentially the same organizations as the earlier Community Coalition. "Our plan now is to rate the stations on who does the best in hiring and public affairs programming," said Evans. He agreed with Jack Bonn, vice-president of personnel for Cox Broadcasting, who said that AMC had been irregular in meeting with the station. Mr. Bonn said, "John was out a couple of weeks ago going over an evaluation of our performance . . . that was the first time in a year."

In New York City Kenneth MacQueen, of WABC-TV, and Robert Williamson, general manager of WOR-TV, reported that the Puerto Rican group (PRMAEC) had apparently become inactive and that the New Jersey Coalition for Fair Broadcasting (unavailable by telephone in 1974 and 1978) had become less active; that is, they rarely contacted the station.

According to the 1974-1975 interviews, broadcasters and citizens groups generally agreed that citizens groups achieved at least some of their goals in the 1972-1975 period. In Atlanta, for example, Cox Broadcasting executives at WSB-TV noted that "the station employs 21 percent minorities presently." It had employed 9 percent in 1971. William Bonn, Cox's vice-president of personnel and public relations, said that WSB-TV had initiated paid internships for blacks. "We have taken the initiative to write into contracts with the unions provisions for training after Lonnie King, of the Atlanta NAACP, asked how blacks could get into the station. That was in 1968. In addition, we have the scholarship program with Clark College of Atlanta State University." Michael Kievman, Cox's president of programming, said that WSB-TV's improvement resulted frequently from station initiatives rather than from the black pressure group's (Community Coalition) demands.

Martin Cooke, in 1974 an assistant director of the Atlanta Urban League and a leader in Community Coalition, said that some changes were definitely attributable to the group's demands. Employment of blacks in Atlanta broadcasting had increased "but not significantly," he said. "We need more people in decision making in the stations." WAGA-TV had run special black documentaries, "Trouble in the Ghetto" and "A Case of Hunger," with Dick Gregory, "but things like this did not get initiated by station managements," he said.

In Rochester, New York, Jack Decker (WHEC-TV) said that documentary programs and the hiring of black news people resulted from the Rochester Black Media Coalition's (RBMC) pressure. He said that the group attained "some of what it wanted." RBMC leaders, Charles Hatcher and Al Rugless, agreed with this assessment.

Arthur Eskew, in 1974 the executive director of the Boston Community Media Committee (BCMC), listed some of the accomplishments of this joint organization of the media and citizens groups: the training of blacks at Boston stations during off-hours, "a dramatic increase in the numbers of blacks in the media," and a trainee from Boston, Steve Jones, "is now at NBC." James Coppersmith, then general manager of WNAC-TV, Boston, said that pressures from local Boston citizens groups had brought "good results."

According to Ann Lang, of NOW in Pittsburgh, and station managements at WIIC-TV, WTAE-TV, and KDKA-TV, agreements between the stations and NOW for programs about feminist issues were aired, a regular public affairs program on women's issues was initiated, and women were hired in specific job categories. NOW representatives, however, expressed dissatisfaction about the attainment of longer-range goals, such as quality feminist programming, although they admitted that stations had honored the letter of the agreements.

The Committee for Open Media, Pittsburgh, also expressed satisfaction that stations honored terms of a 1972 agreement to air free speech messages; but station managers, John Conomikes, WTAE-TV, Johnathon Hayes, KDKA-TV, and public affairs director for WIIC-TV, Marney Sarver, agreed with a codirector of the committee that long-range goals of greater public debate on issues were not achieved. Currently only KDKA-TV airs free speech messages, or "speak-out messages."

Interviews conducted in 1978 indicate that broadcasters and citizens groups generally agreed that citizens groups achieved only some of their goals in the 1975-1978 period, although some broadcasters and some citizens groups felt that the groups achieved most of their goals. Neil Kuvin, vice-president of programming for WXIA-TV in Atlanta, and John Evans, executive director of the Atlanta NAACP, agreed that a great deal of progress ahd been made in employment at the station, although Evans felt "there's a long way to go in programming." Charles Hatcher, of the Rochester Black Media Coalition (RBMC), agreed with Jack Decker, general manager of WHEC-TV, in 1978 that the station is "doing a good job." Hatcher, who had been party to a 1972 petition (still pending against the station when interviews were first conducted in 1974), said recently, "you wouldn't know it was the same station; nobody in town can touch WHEC-TV in local public affairs programming." RBMC has not visited WHEC-TV to press for concessions since 1975. Dennis Schatzman, city editor for the *Pittsburgh Courier* and 1977 Pittsburgh NAACP Media coordinator, said, "We've gotten pretty much what we've asked for. Of course, we haven't asked for much. We went to WTAE-TV and got a minidocumentary series on cable television." (The Pittsburgh city council will soon consider a CATV franchising ordinance, and NAACP is urging minority ownership). Emma Bowen of Balck Citizens for Fair Media (BCFM) added WNEW-TV, WPIX-TV, and WNET-TV to the list of stations that her group had contacted since 1975. "There's not much that you can ask of the others because there are not that many people there; I mean employees. But WNET-TV, they've got to hire some blacks." BCFM, which reached an agreement with WCBS-TV in return for drawing its 1975 petition-to-deny, has gotten "more or less what we wanted, but WCBS is low on blacks in managerial positions, especially black men."

According to 1974-1975 interviews broadcasters generally rated citizens groups as very influential, while citizens groups generally rated themselves as only moderately influential. The one group that rated itself as "very influential" was New York's Black Citizens For Fair Media (BCFM). Larry Gershman, of WNBC-TV, agreed that the group had greatly influenced him. Kenneth MacQueen, of WABC-TV, agreed that BCFM was most influential: "The main group is BCFM. They are the most reasonable and most active." In Pittsburgh all managers who were interviewed rated NOW as "very influential, "but representatives of NOW, Ann Lang and Kathy Bonk, felt that the organization was only "fairly influential" through meetings and through its 1972-1975 agreements with Pittsburgh's three UHF commercial stations.

In Atlanta the WSB-TV management said, "You can't deny the effectiveness of the groups . . . local pressures have influenced us." Martin Cooke, of Atlanta's Urban League, felt the Community Coalition had been "somewhat influential," but he also said, "Now a new strategy is needed . . . we can target the stations which do not measure up, go to their advertisers and tell them we'll advocate against their products."

Jack Decker (WHEC-TV) called the Rochester Black Media Coalition (RBMC) very influential. He said that the group "had generally done its homework." NOW, on the other hand, he said, "was less prepared, irregular, and far less influential." G. Bennett Larsen, of WOKR-TV, called the demands of RBMC and NOW "unreasonable" and said the groups had "no influence on me whatever." He noted, however, that a "sit-in" at the station had resulted in the addition of a black public affairs program, "Black Dimensions," and in the promotion of a black employee to the position of newsman. Charles Hatcher and Al Rugless, of RBMC, and James McCuller, then president of the National Black Media Coalition, summed up RBMC's influence: the group made "some progress" with the commercial stations and had high expectations for negotiations with stations prior to the 1975 relicensing.

According to the 1978 interviews, broadcasters and citizens groups generally regarded citizens groups as having been "somewhat" to "very" influential in the 1975-1978 period. Three representatives of citizens groups rated their groups as "very influential" in the 1975-1978 period. Charles Hatcher, of RBMC, and David Honig, a former advisor to RBMC and now a professor at Howard University, said they viewed the group as significantly more influential in 1975-1978 than they had in 1974, largely because of WHEC-TV's 1975 revised hiring and programming and because the group had secured an extensive agreement with WOKR-TV in 1977. The agreement, in part, called for nomination of a minority member to the Post Company's board of directors. (Post Company recently acquired WOKR-TV, and in April 1978 a black Rochester woman was elected to Post's board of directors.) Emma Bowen, of BCFM, regarded the group as "more influential" since 1975 because "Equal Employment Opportunity has advised us that we have grounds for a class action suit against NBC, Inc." (She suggested that NBC would probably consent to an extensive agreement on upgrading of minority salaries and affirmative action programs to prevent BCFM from filing a suit). WABC-TV's Kenneth MacQueen attributed the heavy influence of BCFM to its "strong, ongoing relationship" with the station. In Pittsburgh Dennis Schatzman felt that the NAACP had been strongly influential in securing coverage on issues, employment, and greater sensitivity to blacks. Schatzman said, "In March of 1977, we asked why they [WIIC-TV] didn't have more blacks, and they hired some blacks. On several occasions we talked with KDKA-TV about Evening Magazine when we heard that it was being developed, and we asked how NAACP could assist them in hiring some black people. I think we were responsible for getting some hired." Pittsburgh broadcasters, however, rated the NAACP as less influential than Schatzman did.

At the time of the 1974 interviews in Rochester, G. Bennett Larsen was general manager of WOKR-TV. Larsen, who had insisted that citizens groups had no influence, retired in 1976 and was replaced by Frank DiProsa. DiProsa, who felt RBMC had been very influential, said, "we look at things differently now."

In Atlanta William Bonn, of Cox Broadcasting, said that the Atlanta Media Coalition (AMC) had been very influential. "Two stations have black coanchors. We're moving in that direction. We have a female black on the noon news and we want more blacks in anchor positions as soon as they are ready." He attributed these changes to pressures by AMC, but he said, "With any pressure groups, you're not dealing with a satisfied position; it's their role to say that there's still more to be done." John Evans, AMC's director, seemed to play out the role Bonn envisioned. He said, "Since 1975, we haven't been as influential; we let it slip."

Interviews from 1974-1975 and from 1978 indicate that broadcasters generally regarded confrontation, petitions, lack of consistency, and "unreasonable demands" as the least effective tactics of citizens groups. Representatives of citizens groups generally identified lack of follow-up as the only ineffective thing done by the groups. Jack Decker, of WHEC-TV, said "confrontation as opposed to reasonable discussion" was counterproductive. Larry Gershman, of WNBC-TV, described "NOW's screaming session" as ineffective. Kenneth MacQueen, WABC-TV, agreed: "With NOW, well, they're just too unreasonable." Sy Yannoff, WBZ-TV, Boston, cited United Church of Christ's blanket petition against New England stations in 1972 as ineffective because United Church had "totally erroneous information on WBZ." A former WNAC-TV official said that picketing the station effectively retained the black news format, but it was "least effective in terms of long-term goals, we wanted to mainstream blacks in the 6 and 11 p.m. news." In Atlanta Jack Clifford, of WXIA-TV, complained of lack of contact with the Atlanta Community Coalition: "The Coalition will have to get off their asses and meet once a year." Among broadcasters, only William Pearce, manager of Rochester's public television station regarded legal petitions at all favorably. He said that RBMC's petitions in 1972 were having "a serious and positive effect."

Interviews from 1974-1975 and from 1978 indicate that groups that now regard themselves as very influential and who are thought by broadcasters to be very influential regard ongoing relationships with broadcasters as a most effective tactic. For example, Emma Bowen of BCFM, New York, said in 1975, "you have to be reasonable and you have to keep going back to them. They have to know you're not going away." In 1978 she said the same thing. Apparently Mrs. Bowen has been consistent, and according to some New York broadcasters, she proved that from 1972 to 1978 that she was not going to go away. Charles Hatcher, who has been active in Rochester's Black Media Coalition since at least 1972, summed up the main reason for the Coalition's success as "presence" and "doing your homework." In 1978 he felt that the Coalition had established an ongoing relationship with many broadcasters in Rochester and that as a result of extensive agreements and FCC hearings on relicensing of several radio stations in Rochester, the Coalition had become very influential.

Statistical Analyses of Employment Data

Employment statistics for these analyses were obtained from yearly averages for women and minority national averages from the Office of Communications, United Church of Christ; all other data are derived from 395 forms, Office of Research, Federal Communications Commission. These data (see tables 8-1 to 8-6) were analyzed from two perspectives. First, the five-city active sample and national industry were compared to determine whether the active sample diverged significantly from national trends. Second, comparisons were performed within each group (women and minorities) across time (1971 to 1977) to determine whether there were any unique trends associated with each group.

The analyses between the sample group and the national data utilized t-tests for comparison of a sample mean with a population mean. The analyses within each group utilized a one-way analysis of variance and the Tukey B-test for decomposition of significant F-ratios. The 0.05 level of significance was established for all tests. Results are summarized in tables 8-7, 8-8, and 8-9 and graphically represented in figures 8-1 and 8-2.

Total Employment

In any analysis of employment data and trends, especially those comparing two groups, it is of interest to know whether observed differences are due merely to economic factors (did group 1 hire more mean than group 2 only because group 1 hired more people overall than group 2). Table 8-7 summarizes a comparison of the five-city active sample and the industry as a whole on the basis of the percentage increase in total employment. The only significant difference was for the period 1974-1975 ($t = -2.193$, df = 29, $p < 0.025$), in which the national increase in total employment for commercial and noncommercial television stations (4.05 percent) was significantly higher than the increase for the five-city active sample (0.7 percent). At all other times, the sample and national rates of increase did not differ significantly statistically.

Table 8-1
Sources of Employment Data: Stations in Each City

Area	*Stations*
Atlanta	WAGA-TV WETV-TV WHAE-TV WSB-TV WTCG-TV WXIA-TV
Boston	WBZ-TV WCUB-TV WGBH-TV WGBX-TV WLUI-TV WNAC-TV WSBK-TV
New York City	WABC-TV WCBS-TV WNBC-TV WNET-TV WNEW-TV WNYC-TV WYNE-TV WOR-TV WPIX-TV
Pittsburgh	KDKA-TV WIIC-TV WPGH-TV WQED-TV WTAE-TV
Rochester	WHEC-TV WOKR-TV WROC-TV WXXI-TV

Women as a Percentage of the Workforce

Results of data analysis indicated that for every year between 1971 and 1977, the five-city active sample employed a significantly greater number of women as a proportion of total employees than did the industry as a whole (see table 8-8). In 1971 women accounted for 24.8 percent of the sample's workforce and 22 percent of the national workforce ($t = 2.095$, df = 27, $p < 0.025$); in 1972, 27.4 percent and 22 percent respectively ($t = 4.518$, df = 27, $p < 0.0005$); in 1973, 27.9 percent and 23 percent respectively ($t = 4.172$, df = 28, $p < 0.0005$); in 1974, 28.8 percent and 25 percent respectively ($t = 3.291$, df = 30,

Table 8-2
Employment Statistics for Atlanta Area

Period	*Women*	*Minorities*
Percentage of Workforce		
1971	26.4	11.0
1972	29.0	11.4
1973	29.5	16.9
1974	32.0	17.9
1975	31.7	17.5
1976	32.9	18.5
1977	33.4	19.3
Percentage Increases		
1971-1973	3.1	5.9
1973-1975	1.6	1.5
1975-1977	1.7	1.8

Table 8-3
Employment Statistics for Rochester Area

Period	*Women*	*Minorities*
Percentage of Workforce		
1971	24.1	2.2
1972	28.4	3.5
1973	29.5	6.6
1974	27.1	6.2
1975	27.8	9.7
1976	28.4	9.7
1977	29.1	9.3
Percentage Increases		
1971-1973	5.4	4.4
1973-1975	–1.7	3.1
1975-1977	1.3	–0.3

Table 8-4
Employment Statistics for Boston Area

Period	*Women*	*Minorities*
Percentage of Workforce		
1971	26.3	8.7
1972	26.6	11.3
1973	28.5	13.8
1974	30.0	14.2
1975	33.6	14.3
1976	35.6	15.0
1977	37.5	14.0
Percentage Increases		
1971-1973	2.2	5.2
1973-1975	5.2	0.5
1975-1977	3.9	–0.3

Table 8-5
Employment Statistics for Pittsburgh Area

Period	*Women*	*Minorities*
Percentage of Workforce		
1971	21.4	8.4
1972	22.5	9.3
1973	24.6	12.6
1974	26.4	13.3
1975	34.9	12.5
1976	30.7	14.0
1977	33.2	14.3
Percentage Increases		
1971-1973	4.4	4.4
1973-1975	5.2	1.2
1975-1977	–1.7	1.8

Table 8-6
Employment Statistics for New York City Area

Period	*Women*	*Minorities*
Percentage of Workforce		
1971	24.5	16.6
1972	28.6	16.6
1973	27.3	19.2
1974	27.6	19.8
1975	28.7	20.7
1976	28.6	20.2
1977	31.3	22.4
Percentage Increases		
1971-1973	1.2	2.7
1973-1975	1.4	1.5
1975-1977	2.5	1.7

Table 8-7
Increase in Total Employment, 1971-1977: National versus Active
(*in percent*)

	Active			
Years	*Number*	*Standard Deviation*	*Mean*	*National Mean*
1971-1972	25	8.8	1.8	–
1972-1973	27	13.3	5.6[a]	5.16[a]
1973-1974	29	8.9	4.6[a]	2.56[a]
1974-1975	30	15.2	0.7	4.05
1975-1976	31	15.8	0.8[a]	1.59[a]
1976-1977	31	28.8	8.4[a]	4.55[a]
1971-1973	27	18.43	9.63	–
1973-1975	29	18.92	5.89[a]	6.72[a]
1975-1977	31	19.92	6.15[a]	6.23[a]

[a]Means in the same row do not differ significantly at $p < 0.05$ using *t*-tests for a sample mean and a population mean.

Table 8-8
Women and Minorities as Percentages of Workforce, 1971-1977: National versus Active

	Active			
Year	*Number*	*Standard Deviation*	*Mean*	*National Mean*
Women				
1971	28	6.7	24.8[a]	22
1972	28	6.0	27.4[a,b]	22
1973	29	6.5	27.9[a,b]	23
1974	30	6.4	28.8[a,b]	25
1975	31	9.1	31.3	26
1976	31	8.8	31.3[b]	27
1977	31	8.8	33.1[b]	29
Minorities				
1971	28	5.6	10.4[d]	9[d]
1972	28	5.3	11.4[b,c,d]	10[d]
1973	29	5.3	14.9[a,b]	12
1974	30	5.4	15.4[a]	13
1975	31	5.4	15.9[a,d]	14
1976	31	6.5	16.4[a,d]	15[d]
1977	31	6.4	16.9[a,d]	16[a]

[a,b,c]Means for women and minorities do not differ significantly at $p <$ 0.05 using the Tukey *B*-test.

[d]Means in the same row do not differ significantly at $p < 0.05$ using *t*-tests for a sample mean and a population mean.

$p < 0.005$); in 1975, 31.3 percent and 26 percent respectively ($t = 3.299$, df = 30, $p < 0.005$); in 1976, 31.3 percent and 27 percent respectively ($t =$ 2.677, df = 30, $p < 0.01$); and in 1977, 33.1 percent and 29 percent respectively ($t = 2.552$, df = 30, $p < 0.01$).

The analysis of variance performed to determine whether there were any significant differences in the percentages of women employed by the five-city active sample between 1971 and 1977 was significant ($F = 4.213$, df = 6,201, $p < 0.001$). Decomposition of this F-ratio indicated the following differences. The percentages for 1975, 1976, and 1977 were significantly higher

Table 8-9
Increases in Employment of, Women and Minorities, 1971-1973, 1973-1975, 1975-1977: National versus Active
(in percent)

Period	*Active*	*National*
Women		
1971-1973	Mean 2.8%[a] Standard deviation 3.8% Number 27	Mean 1%
1973-1975	Mean 2.5%[a,b] Standard deviation 4.4% Number 29	Mean 3%[b]
1975-1977	Mean 1.8% Standard deviation 4.9% Number 31	Mean 3%[b]
Minorities		
1971-1973	Mean 4.4% Standard deviation 2.8% Number 27	Mean 3%
1973-1975	Mean 1.4% Standard deviation 4.0% Number 29	Mean 2%[b]
1975-1977	Mean 1.0% Standard deviation 2.7% Number 31	Mean 2%

[a]Means for women and minorities do not differ significantly at $p < 0.05$ using the Tukey *B*-test.

[b]Means do not differ significantly at $p < 0.05$ using a t-test for a sample mean and a population mean.

than for 1971. The percentage for 1971 was not significantly different from the percentages for 1971, 1972, and 1973. There were no significant differences among 1972, 1973, 1974, 1975, 1976, and 1977. In short, 1971's percentage accounted for the significance found in the initial analysis.

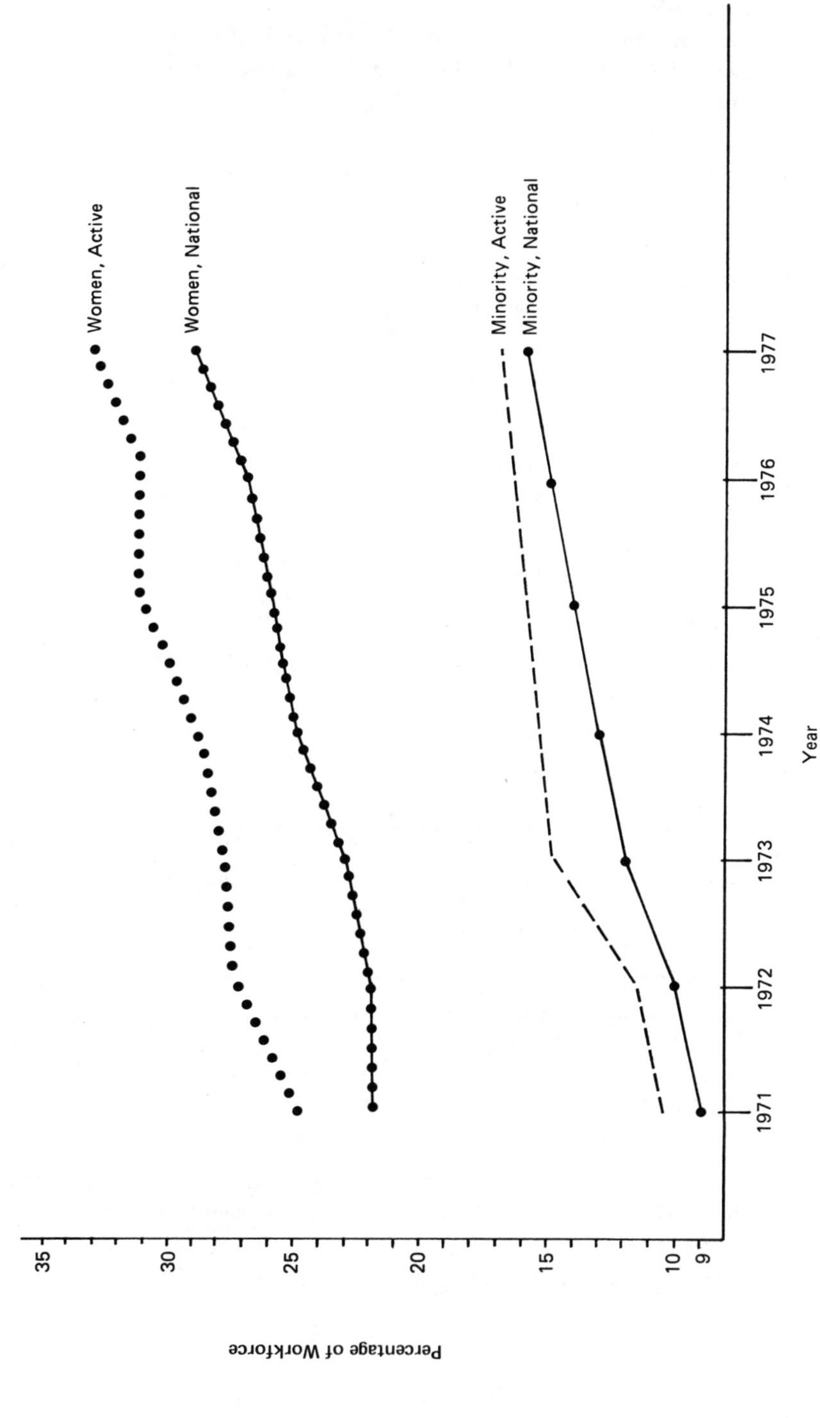

Figure 8-1. Women and Minorities as percentages of Workforce, 1971-1977: National versus Active

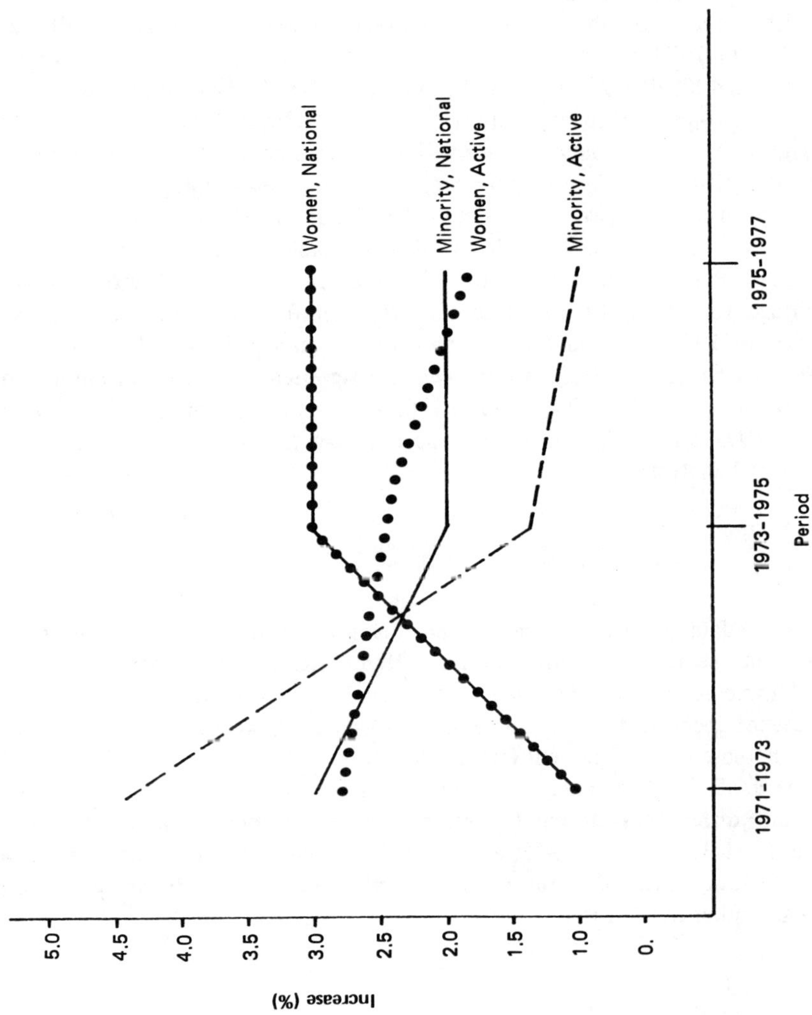

Figure 8-2. Increases in Employment of Women and Minorities, 1971-1973, 1973-1975, 1975-1977: National versus Active

Minorities as a Percentage of the Workforce

Results of the data analysis indicated that only in 1973, 1974, and 1975 did the five-city active sample employ a significantly greater number of minorities as a proportion of total employees than the industry as a whole (see table 8-8). In 1973 minorities made up 14.9 percent of the sample's workforce, compared with 12 percent of the industry's national workforce ($t = 2.751$, df = 28, $p < 0.01$); in 1974 the percentages were 15.4 percent and 13 percent respectively ($t = 2.400$, df = 29, $p < 0.025$); and in 1975 the percentages were 15.9 percent and 14 percent respectively ($t = 1.931$, df = 30, $p < 0.05$).

The analysis of variance performed to determine whether there were any significant differences in the percentage of minorities employed by the five-city active sample between 1971 and 1977 was significant ($F = 4.213$, df = (6,201), $p < 0.001$). Decomposition of this F-ratio indicated the following differences. While the percentage for 1971 did not differ significantly from the percentage for 1972, 1971's percentage was significantly lower than the percentages for 1973-1977, and 1972's percentage was significantly lower than the percentages for 1974-1977. There were no significant differences among the percentages for 1973, 1974, 1975, 1976, and 1977. In short, the low percentages in 1971 and 1972 appeared to account for the significant F-ratio found in the initial analysis.

Women: Percentage Increase

Results of data analysis indicated that the rate of increase in the hiring of women in the five-city active sample differed significantly from the national rate of increase only in the 1971-1973 period ($t = 2.465$, df = 26, $p < 0.025$). The rate of increase for the active sample was 2.8 percent while the national rate of increase was 1.0 percent (table 8-9).

The analysis of variance performed to determine whether there were any significant differences among the sample's rates of increase across the three periods 1971-1973, 1973-1975, and 1975-1977 was not significant ($F = 0.154$, df = 2, 84), indicating that the rates of increase were not significantly different from each other in the three periods.

Minorities: Percentage Increase

Results of data analysis indicated that the rate of increase in the hiring of minorities in the five-city active sample was significantly higher than the national rate of increase for the 1971-1973 period ($t = 2.605$, df = 26, $p < 0.01$). The sample rate of increase was 4.4 percent while the national rate of increase was

3 percent (table 8-9). However, for the period 1975-1977 the national rate of increase (2 percent) was significantly greater than the sample rate of increase (1.0 percent) ($t = -2.061$, df = 30, $p < 0.025$).

The analysis of variance performed to determine where there were any significant differences among the sample group's rates of increase across the three periods was significant ($F = 9.009$, df = 2, 84, $p < 0.001$). Decomposition of this result indicated the following differences. The rate of increase for the period 1971-1973 was significantly higher than the rates of increase for either the 1973-1975 or the 1975-1977 period. In short, the highest rate of increase in minority employment for these years was recorded in 1971-1973.

Discussion

This study has explored the extent to which activities of selected local citizens groups have affected programming and hiring by television stations in markets where the groups have been active. Apparently the activities of citizens groups—for example, meeting with broadcasters, petitioning, and obtaining agreements by broadcasters—have significantly affected the employment of women and minorities in selected television markets. The activities of citizens-groups have apparently affected programming for minorities and for women. Unfortunately, we have only anecdotal measures of programming, and no comparison was made between programming in the active markets and programming in other markets during the same period.

One particularly interesting fact emerged from this study. After the dust of petitions by citizens groups had settled and adversary relationships diminished (in Rochester and New York City, for example) a generally positive relationship seemed to have developed between some broadcasters and some citizens groups. Thus an important element in the perceived effectiveness of some citizens groups may simply be the duration and regularity of contact with broadcasters. This possibility does not, of course rule out the effectiveness of legal petitions, but it could help to explain why citizens groups and broadcasters in the active markets tended to see the groups as generally more influential at the very time when employment increases for minorities and for women did not differ significantly from national averages. (The overall lack of significance, however, does *not* mean that gains were not being made in higher-paid EEO categories through promotions of minorities and women.)

This study did not systematically assess the effect of a representative sample of any one kind of citizens group. Such studies are certainly in order. A major constraint to such a systematic study has been the reluctance of broadcasters and some citizens groups to respond to research questions on the topic of citizens groups, especially when many broadcasters and groups were in litigation with each other. A 1975 attempt at questionnaire surveys mailed to stations in thirty cities produced too few responses to be useful.

In addition to survey data, which should be collected to determine the origin, evolution, and progress of several kinds of broadcast citizens groups, case studies of groups (such as Shement et al., 1977) would also be valuable. For example, it would be instructive to determine why the Rochester group, RBMC, appears to have been so successful. RBMC seems to have made more progress in employment than any other citizens group in this study. Its success seems largely attributable to legal actions. It would be useful to analyze the legal briefs and petitions related to RBMC's activities and the decisions by the FCC to determine lines of argument and evidence that were effective before the Commission.

It would also be interesting to determine how much of the burden of writing petitions and replies was assumed by members of RBMC and what role its attorney, the Citizens Communication Center, played. One suspects that given the heavy case load on the center when RBMC was most active (1972-1975), a great deal of work had to be done by RBMC members themselves.

Employment data lend themselves more easily to hypothesis testing than does other information used in this study and are thus useful in developing a theoretical construct for predicting the future progress of citizens groups.

These data seem to suggest a stepwise process of influence within the broadcasting industry. From 1971 to 1974 the percentage of women and minorities employed in the active sample cities significantly exceeded national averages. It is clear that this was due, at least in part, to the activities of women and minority citizens groups. However, from 1974 to 1977 the percentages of women and minorities did not differ significantly from national averages. Hence while certain active markets may surpass the nation as a whole in certain indexes, the nation as a whole may be pulled to parity as certain behavioral norms and levels of awareness filter to the rest of the nation. The interview data indicate a decline after 1974 in aggressive activitist tactics (petitions-to-deny, sit-ins). In lieu of such tactics interviewees (representatives of citizens groups and broadcasters) indicated that relationships between the two groups were informal, ongoing, and relatively nonpressured. However, this should not be taken to mean that influence has necessarily declined. Rather, the change in relationship may imply a more subtle form of influence. Hence one might predict that while increases in the hiring of women and minorities have slowed compared with national trends, the number of women and minorities employed in higher-paying categories may have increased in 1974-1977. The effect of post-1973 influence may be subtle, as is the form of influence. Future research (now being initiated by the authors) should investigate the distribution of women and minorities across job categories to determine the effect of citizens group influence, as well as to further test the theoretical construct of citizens group influence in the broadcasting industry proposed here.

References

1. *Action, Inc., and RBMC* v. *FCC*, Case No. 76-1089, 1977, District of Columbia Court of Appeals.

2. *Action, Inc., and RBMC* v. *FCC*, "Agreement For Dismissal of Case with Prejudice," August 10, 1977, D.C. Court of Appeals.

3. Agreement between Cecil Heftel and Joyce Heftel as Assignees of Stations WJAS and WJAS-FM, Pittsburgh, Pennsylvania, and the National Organization for Women, Southwestern Pennsylvania Chapters, signed August 29, 1972, by Kathy Bonk and Phyllis Weatherby, NOW, and Cecil Heftel.

4. Agreement between Cecil Heftel and Joyce Heftel as Assignees of WJAS and WJAS-FM, Pittsburgh, Pennsylvania, and Create Incorporated Pittsburgh Community Coalition for Media Change, signed August 29, 1972, by Herman Stubblefield and Bouie Hayden for BCMC, Cecil Heftel, and William Hanks, August 30, 1972 for COM.

5. Amendment to Application to Golden Triangle Broadcasting, Inc., and Radio 1080 Corp. for approval of Assignment of License of Stations WEEP and WEEP-FM, Pittsburgh, Pennsylvania by Joseph Field, president, August (no date) 1977.

6. *American Broadcasting Co., Inc.*, 52 FCC 2d 98-125 (1975).

7. A Reply to WNET's Opposition to Petition to Deny, July 17, 1975, FCC File No. BRCT-329.

8. *Association of Arab Americans*, 56 FCC 2d 453 (1975).

9. Astor, Gerald, *Minorities and the Media*, Ford Foundation Report, November 1974, pp. 13-16.

10. Atkins, Jeanni. *Chicano Media Challenge: Basta Ya!* Freedom of Information Center Publication, No. 282 Columbia, Missouri, 1972.

11. Barron, Jerome. *Freedom of the Press for Whom?* Bloomington, Ind.: Indiana University Press, 1973.

12. *Columbia Broadcasting System, Inc.*, FCC 2d, 313-319.

13. Concerning 1972 Massachusetts Television Station License Renewals, *Memorandum Opinion and Order* 44 FCC 2d, 647-656 (1974).

14. Daly, W. "The FCC's Role in Providing Equal Opportunity." *Boston University Law Review.*

15. Daly, W. "Free Speech Messages: People's Access." *Access* 1 (1975): 9, 10.

16. FCC letter to William E. Hanks, from Richard Shiben, Chief Renewal and Transfer Division, Broadcast Bureau, April 9, 1976.

17. *FCC Report* No. 11743, FCC Acts on New Jersey Television Service (Docket 20350), March 19, 1976.

18. Friendly, Fred. *The Good Guys, the Bad Guys, and the First Amendment*. New York: Vintage Press, 1976.

19. Fry, Carlton F. "The Private Agreement and Citizen Participation in

Broadcast Regulation." M.A. thesis, Ohio State University, 1972.

20. Gandy, O.H. "Citizen Action and the FCC: The Decade of Dissent." Department of Communications, Stanford University, 1975.

21. Gandy, Oscar H., et al. "Citizen Action and Broadcasters Interests: The Record of the FCC." Paper presented at the Annual Convention of the International Communication Association, 1976.

22. Garland, Phyl. "Blacks Challange the Airwaves." *Ebony Magazine* November 1970, pp. 38-38+.

23. Gross, L. "Television under Pressure: Are a Handful of Activists Saving or Wrecking the Medium?, *TV Guide* , February 22, 1975, pp. 4-7.

24. Gross, L. "Television under Pressure: The Sweet Sell." *TV Guide,* March 1, 1975, pp. 28-30.

25. Gross, L. Television under Pressure: Citizens with Clout. *TV Guide,* March 8, 1975, pp. 31-34.

26. Grundfest, John. *Citizen Participation in Broadcast Licensing before the FCC.* Santa Monica, Ca.: Rand Corporation, 1976.

27. Guimary, Donald L. *Citizen's Groups and Broadcasting*. New York: Praeger, 1975.

28. Hanks, William E. "Review of the Literature: The Citizen-group Movement in Broadcasting, 1966-1977." Department of Speech, University of Pittsburgh, Pittsburgh, Pa. Available on request.

29. Hanks, W., and Longini, P. "Television Access: A Pittsburgh Experiment." *Journal of Broadcasting* 3, (1974): 289-298.

30. Harris, James. "Free Speech Messages: When the Public Get Access, What Does It Say?" *Access*, 34 May 17, 1976, pp. 20-21.

31. Honig, D. "Effects of the F.C.C.'s Petition-to-Deny Process on Minority Employment in the Broadcast Industry." Systems Analysis Program, Graduate School of Management, 1974. University of Pittsburgh.

32. Interview on women's programming in Pittsburgh TV since 1975, Frieda L. Lalli, April 5, 1977, pp. 5-7.

33. Jennings, R. *Television Station Employment Practices.* New York: Office of Communication, United Church of Christ, 1972.

34. Jennings, Ralph M. *Television Station Employment Practices: The Status of Minorities and Women, 1972*. New York: Office of Communications, United Church of Christ, 1973.

35. Jennings, Ralph M., and Jefferson, Veronica. *Television Station Employment Practices: The Status of Minorities and Women, 1974.* New York: Office of Communications, United Church of Christ, 1975.

36. Jennings, R., Kerr, M., and Parker, T. *Public Television Station Employment Practices and Composition of Boards of Directors. The Status of Minorities and Women.* New York: Office of Communication, United Church of Christ, 1973.

37. Jennings, Ralph M., and Tillyer, David A. *Television Station*

Employment Practices: The Status of Minorities and Women, 1973. New York: Office of Communication, United Church of Christ, 1974.

38. Jennings, Ralph M., and Walters, Allan T. *Television Station Employment Practices, 1976: The Status of Minorities and Women.* New York: United Church of Christ, Office of Communication, 1977.

39. Johnson, N. *How to Talk Back to Your Television Set.* New York: Little Brown Co., 1970.

40. Kline, F.G., and Tichenor, P.J. (eds.). *Current Perspectives in Mass Communication Research.* Beverly Hills, Ca.: Sage Publications, 1972.

41. Letters of agreement: (1) August 17, 1972, from Metromedia, Inc., by Thomas J. Dougherty, its attorney and signed for New Jersey Coalition for Fair Broadcasting by its attorney, Robert Jay Stein; (2) May 15, 1972, from WNET-TV, signed by James Day, president and trustee, Educational Broadcasting Corporation, and by Gordon A. MacInnes, Jr., New Jersey Coalition of Fair Broadcasting; (3) June 29, 1972, from WCBS-TV, signed by Robert L. Hosking, vice-president and general manager, and George MacInnes for the New Jersey Coalition for Fair Broadcasting; (4) letter May 15, 1972, Ben F. Waple, FCC, Washington, D.C., from WOR-TV Vice-President and General Manager Robert L. Glasser and signed by Gordon A. MacInnes, Jr.; (5) letter May 3, 1972, from WNBC-TV Lawrence Gershman, station manager, and signed by Gordon A. MacInnes, Jr., New Jersey Coalition for Fair Broadcasting. Letters to Committee for Open Media, Pittsburgh; from Leonard Swanson, vice-president and station manager, WTAE-TV, signed by Peter Longini COM, Pittsburgh; from James M. King, general manager, KDKA-TV, July 3, 1972.

42. Letters from American Broadcasting Co., Inc., to Ben Waple, FCC, Washington, D.C., *Re Amendment to WABC-TV License Renewal Application,* May 1972, from Lawrence Gershman, WNBC-TV, N.Y., to Ben Waple, FCC Washington, D.C., *Re Amendment to WNBC-TV License Renewal Application,* May 1972.

43. Letters to Create Coalition for Media Change; KDKA-TV, from James M. King, general manager; WTAE-TV, from John Conomikes, station manager; WIIC-TV, from Leonard Swanson, general manager, signed by Bouie Hayden and Herman Stubblefield (CMC), July 1972.

44. Letters to Committee for Open Media, Pittsburgh; from Leonard Swanson, vice-president, and general manager, June 16, 1972; John G. Conomikes, vice-president, and station manager, WTAE-TV, signed by Peter Longini (COM), Pittsburgh; from James M. King, general manager, KDKA-TV, July 3, 1972.

45. Letters to Pittsburgh chapters of the National Organization for Women, July 3, 1972; from John G. Conomikes, vice-president and station manager, WTAE-TV; July 12, 1972, from Leonard Swanson, vice-president and general manager; letter to Kathy Bonk, FCC Task Force.

46. Levin, E. "Under New Management." *TV Guide*, April 15, 1975, pp. 3-5.

47. Lewels, F.J.. *The Uses of the Media by the Chicano Movement: A Study in Minority Access.* New York: Praeger, 1974.

48. Lobsenz, N. "Everett Parker's Broadcasting Crusade." *Columbia Journalism Review*, Fall 1969, pp. 31-36.

49. Mayer, Martin. "The Challengers (part 1)." *TV Guide*, February 3, 1973, pp. 4-12.

50. Mayer, Martin. "The Challengers (part 2)." *TV Guide*, February 17, 1973, pp. 18-21.

51. Mayer, Martin. The Challengers (part 3)." *TV Guide*, February 17, 1973, pp. 18-21.

52. McLeod, J.M., and O'Keefe, G., Jr. "The Socialization Perspective and Communication Behavior." In *Current perspectives in mass communication research*, ed. F. Kline and P. Tichenor. Beverly Hills, Ca.: Sage Publications, 1972.

53. Methvin, E.H. "Minorities Gain in Broadcasting." *Business Week*, March 10, 1973, pp. 168-174.

54. Methvin, E.H. "What You Can Do about TV violence." *Reader's Digest*, July 1975, pp. 185-190.

55. *National Broadcasting Company, Inc., et al.*, 20 FCC 2d, 59-67 (1969).

56. *National Broadcasting Company, Inc., et al.*, 27 FCC 2d, 897 (1972).

57. Padden, P. "The Emerging Role of Citizen Groups in Broadcast Regulation." *FCC Bar Journal*, (1972): 82-110.

58. *Petition to Defer Action on License Renewals Pending Results of Proposed Rule Making* by Governor Brendan Byrne and the State of New Jersey, May 1, 1975.

59. *Petition for Declaratory Ruling and for Issuance of Notice of Apparent Liability* by New Jersey Coalition for Fair Broadcasting, May 1, 1975.

60. Powell and Meek. "Mississippi's WLBT, After the Challenge." *Columbia Journalism Review*, May/June 1973, pp. 50-55.

61. *Puerto Rican Media Action Council, Inc., et al.*, 51 FCC 2d, 1178-1200 (1975).

62. *RKO General, Inc., et al.*, 44 FCC 2d, 1004.

63. *Rust Craft Broadcasters of New York, Inc.*, 54 FCC 2d, 1222-1235.

64. Shayon, Robert L. *Parties in Interest*, New York: Office of Communication, United Church of Christ, 1974.

65. Shement, R.S., et al. "The Anatomy of a License Challenge." *Journal of Communication*, Winter 1977, pp. 89-94.

66. *Straus Broadcasting Company of Atlanta, et al.*, 32 FCC 2d, 644.

67. *WGAL Television, Inc., et al.*, 34 FCC 2d, 296-303.

68. *Turner Communications Corp.*, 35 FCC 2d, 433.

69. *United Church of Christ*, 44 FCC 2d, 647-657.

70. United States Court of Appeals, No. 76-2674, March 22, 1978.

71. Waters, Harry, et al. "TV: Do Minorities Rule?" *Newsweek*, June 2, 1975, pp. 78-79.

72. *WDUQ-FM*, 5 FCC 2d, 703.

73. *WGAL Television, Inc., et al.*, 34 FCC 2d, 297.

74. *WHEC-TV, Inc.* 52 FCC 2d, 1079.

75. *WHEC-TV, Inc.* 55 FCC 2d, 519.

76. Ziedenberg, L. "The Struggle over Broadcast Access, part 1." *Broadcasting*, September 20, 1971, pp. 32, 33, 36-41.

77. Zeidenberg, L. "The Struggle over Broadcast access, part 2." *Broadcasting*, September 27, 1971, pp. 24-29.

78. Zippay, A. "Update: Free Speech Messages Are Working." *Access* 4, 1975, 14, 15.

9 Mexico's Television Network in the United States: The Case of Spanish International Network

Félix Gutiérrez

Communications policymakers in the developed countries have become increasingly aware of the concerns that Third World leaders have about the impact of the free flow of information on their countries. The free flow, which many Third World spokespersons contend is actually a one-way flow from developed to developing nations, is painted as a clear threat to the communications integrity of the Third World nations.

The Third World nations point with justification to the domination of their airwaves by U.S. television programs and news from U.S. and European wire services. No doubt their alarm would increase if the U.S. broadcasters actually owned percentages of the television networks in their countries and the stations transmitting their signals. Their ire would most certainly increase if these same U.S. networks used the Third World stations for broadcasting reruns of programs made in the United States and began daily satellite programming from the United States without seeking approval of the national government.

Such a situation would surely raise the arguments against media imperialism to new heights and would be a cause for concern among both U.S. and Third World communications policymakers. Yet such a situation in international television communications already exists, and all signs are that it will continue to grow. The only difference is that the nation sending the television signals is in the Third World and the nation receiving them is the United States.

There is a commercial television network in the United States that is 75 percent owned by a foreign television network. It beams programs to stations from coast to coast, some of them 20 percent owned by the family that controls the foreign network. The network, which began by bicycling foreign-produced programs from station to station in the early 1960s, started a weekly twenty-five-hour live satellite feed of news, variety shows, and sports from Mexico City in 1976 without seeking the approval of the Federal Communications Commission. The network continues to bicycle some programs from station to station and sells advertising on the television stations. Its influence extends to four foreign-licensed stations of the network that include U.S. audiences through over-the-air and cable television.

The network is Spanish International Network (SIN), which is 75 percent owned by Mexico's Televisa television network. SIN links sixteen stations serving U.S. Spanish-speaking audiences with television programming produced at Televisa's giant Televicentro studios in Mexico City. A sister corporation, Spanish International Communications Corporation (SICC), holds the licenses to five of the SIN stations in the United States and is 20 percent owned by the family controlling Televisa.

The case of SIN is an intriguing chapter in the debate over the free flow of information because it represents a reverse in program flow between countries and because there is heavy foreign involvement in the ownership and management of the network and stations. The case raises obvious questions regarding the desirability of tighter regulation for television signals across international borders in the satellite age. It also represents an interesting case for the possible application to the SICC-licensed stations of the foreign ownership and control provisions of the Federal Communication Act's Section 310. Related issues are the potential regulation of networks by the Federal Communications Commission (FCC) and the presumed beneficial social effects of relaxed regulation of imported foreign-language programs; as expressed in the 1972 *Cable Television Report and Order.*

The economics of the daily satellite transmissions from Mexico's Televisa also represent a change in the relationship between the U.S. Spanish-language stations and their Third World program suppliers. SICC stations, which have long been charged for airing the Mexican programming bicycled to them by SIN, are now paid by Televisa through SIN for carrying the live satellite feed of programs and commercials from Mexico City. Between September 1976, when satellite programming began, and the end of 1977, Televisa had paid SIN $1,057,943 and SIN had paid $844,488 as part of the international telecommunications arrangement.

The programs, commercials, and payments indicated Televisa's continued interest in the development of U.S. Spanish-language television. It is an interest that began in the early 1960s when, at the same time that U.S. media barons looked to Third World countries for new program markets, the owner of Mexico's leading television network began expanding northward. The development of SIN and its projected growth parallel the rapid upswing of Latinos in the United States. Some people may find the dynamic and sustained growth of U.S. Latinos even more surprising than the tale of the television network addressing them.

Latinos in the United States

Latinos are the nation's fastest-growing racial population group. Their numbers are expected to grow at an even faster rate as the United States reaches the twenty-first century. According to the U.S. Bureau of the Census, which admits it has not accurately counted the group, there are about 12 million Latinos in the country. (References for each section are at the end of the chapter.) With an additional 3.1 million Puerto Rican citizens and an estimated 6 million to 8 million undocumented immigrants (the government's most conservative figures), the total easily tops 20 million. Between 1970 and 1975 the nation's Latino population grew by 20 percent, five times the national average, and steady growth is expected to continue for several reasons.

One reason is that Latinos have younger median age than the national average (20.9 years for Latinos and 28.9 years for the United States overall) and will thus be entering their most productive childbearing years when other sectors are passing out of it. Another factor is the larger anticipated family sizes of Latinos. Continued immigration from Latin America is a third factor. To these are added the political factors of possible statehood for Puerto Rico and some form of amnesty for undocumented residents. Even without the political factors Latinos at some time in the future will surpass blacks for the dubious honor of being the nation's largest minority group. Demographers predict this will happen sometime early in the next century, but some government officials predict that it could occur as early as 1990.

Long stereotyped as a regional or rural group found in large numbers only in the Southwest and a few urban areas such as Miami, Latinos are actually a nationally dispersed group with large population centers in the Midwest and the Northeast. The U.S. city with the largest Latino population is New York. New York, New Jersey, and Illinois each have more Latinos than Arizona, Colorado, or New Mexico. New York City, Chicago, Philadelphia, and Washington, D.C., rank among the nations top twenty-five centers of Latino population. About 85 percent of all Latinos live in urban areas, and Latinos have a lower percentage of their workforce employed as farm labor than the U.S. labor force overall.

Large numbers and urban representation have not resulted in a better life for U.S. Latinos. Most still face a stiff battle to improve their lives and communities. Median family income for Latinos is 25 percent below the U.S. average, and nearly one-fourth of all Latinos live in poverty. Other social indicators such as education, housing, health, employment, and political representation consistently show Latinos far below national norms.

Latinos and Broadcasting

Individual poverty can be measured as collective wealth by advertisers, marketing strategists, and media managers eager to penetrate this growing segment of the U.S. population. Since audience and marketing studies show that Latinos, like other low-income populations, tend to be heavy users of broadcast media and prefer Spanish-language broadcasting where it is available, the rise in the population has been accompanied by a steep growth in Spanish-language broadcasting in the past fifteen years.

The *1978 Broadcasting Yearbook* lists over five-hundred radio stations broadcasting Spanish programs, nearly one-hundred of them with Spanish as their primary format. In the 1973 edition less than three-hundred Spanish radio stations were listed. The 1978 *Yearbook* also lists twenty-one television stations with Spanish as their primary format, a listing its 1973 edition did not even carry. Five of the twenty-one, are Mexican border stations that beam to Spanish-speaking audiences in the United States.

Advertisers, attracted by an estimated $31 billion in annual earnings by Latinos, have learned to ride the airwaves into Latinos homes and have stepped up their efforts to reach that audience. In 1972 a media marketing magazine estimated that advertisers spent $40 million annually on Spanish-language media. By 1977 *Television/Radio Age* put the figure between $60 million and $70 million and said that "some estimates are much higher." Those estimates run to $125 million, with $90 million of that spent on Spanish-language broadcasting.

Spanish International Network (SIN)

Spanish International Network (SIN), which claims billings with twenty-three of the nation's top twenty-five advertisers, has reaped a good portion of the advertising coming into Spanish-language broadcasting. The network has gained this advertising through a hard-sell approach to advertisers and advertising agencies. The appeal stresses Latino dependence on television, Latino consumer patterns, SIN programming, and the cost-effectiveness of Spanish-language television.

"Spanish families are younger and bigger . . . they buy more of almost everything," says a brochure for SIN's New York station. An advertisement for the Miami station reads, "More than half of the City of Miami, 33 percent of Dade County and 23 percent of the entire ADI [Area of Dominant Influence] is Spanish and growing in size and affluence every day." At times SIN's appeals to advertisers have bordered on exploitation of their largely Spanish-speaking low-income audience. A SIN brochure issued in the early 1970s quoted a study noting that "the insecurity of newcomers" leads them to buy brand name products and added in all capitals "Latins are brand buyers because, for many, advertised brands represent a status symbol!"

The hard sell to advertisers and the hard work in stringing together a network of Spanish-language television stations have spurred the growth of SIN. The network now numbers sixteen stations (see figure 9-1). Five of the stations (Miami, New York, San Antonio, Los Angeles, and Fresno/Hanford) are owned by SIN's sister, Spanish International Communications Corporation (SICC). Two others (San Francisco and Phoenix) are owned by principals in the SIN/SICC ownership and management. Four (San Diego, El Paso, El Centro/Yuma, and Laredo) are actually Mexican stations owned by Televisa (which owns

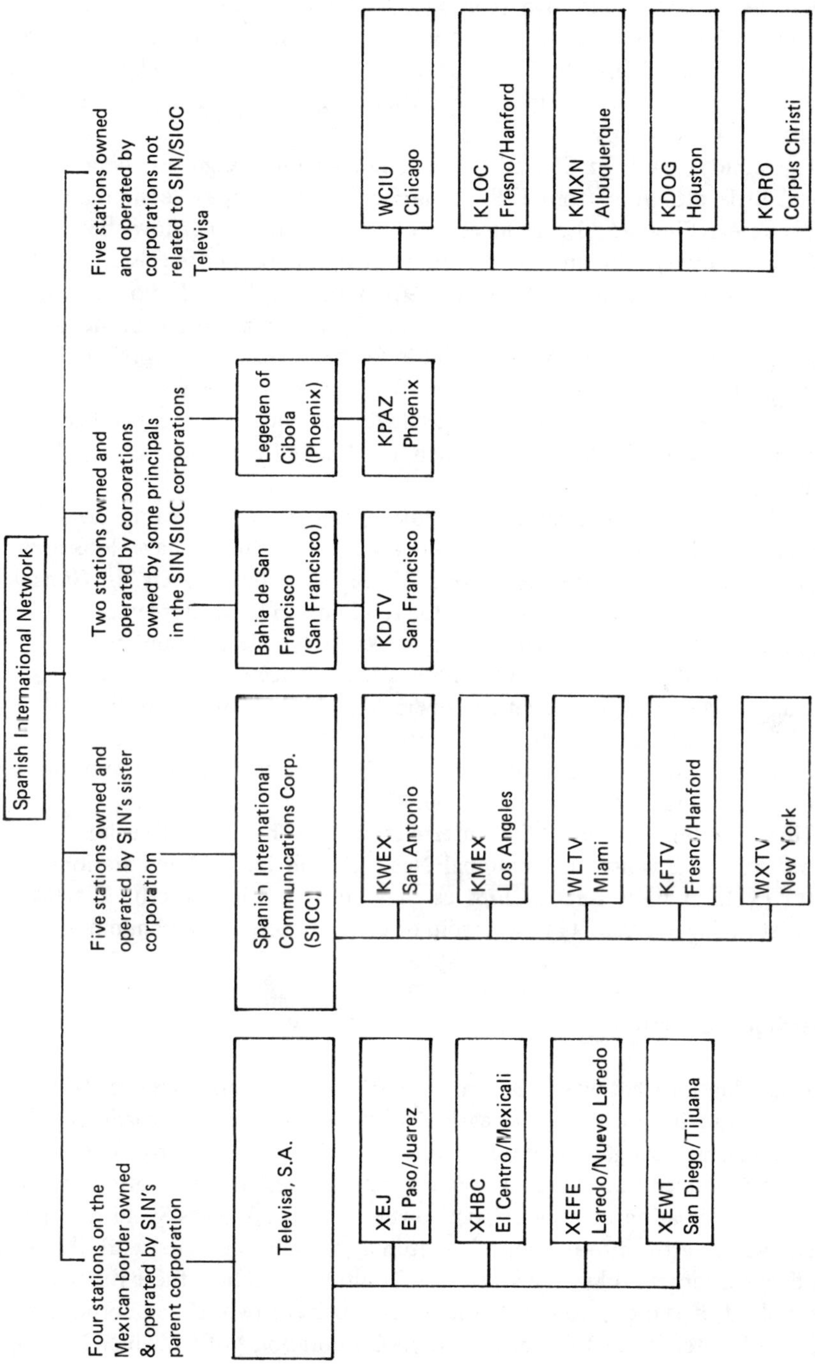

Figure 9-1. The Stations of Spanish International Network (SIN)

75 percent of SIN), and five (Chicago, Sacramento/Modesto, Albuquerque, Houston, and Corpus Christi) are affiliates. In addition to the bicycled and satellite programs from Mexico's Televisa, the network also plans to beam fifty hours a week from a network operations center in San Antonio, most of it also in Mexican programs.

The rapid growth of SIN and its dependence on foreign programming has received little attention from U.S. researchers. But European communications scholars researching worldwide investment and programming flows have noted the flow in ownership and programming from Mexican television to U.S. Spanish-language television stations. In 1974 Finnish scholar Tapio Varis wrote in his study of global television traffic that "progress produced in Mexico are widely distributed in Latin America and in Spanish-speaking areas of the United States." In 1977 Armand Mattelart, of the University of Paris, wrote on the large holdings of the owners of Televisa and added that they "even control Spanish-speaking TV chains in the States, which are directed to 'Chicanos' and Puerto Ricans."

Aside from numerous audience studies (which are used mainly to sell the audience to advertisers) few U.S. communication scholars have looked seriously at Spanish-language television. Those who have, such as Gonzales (1978) and Cox (1969) have outlined the relationship between Mexico and U.S. Spanish-language television. The other works in the field, found primarily in broadcasting trade journals, tend to emphasize audience demographics, programming, and advertising more than the operation of the network and its stations. Many of these articles note the links between Mexico and SIN stations but stop short of examining the extent to which U.S. Spanish-language stations are actually extensions of Mexican television.

This chapter examines the interrelationships between Televisa and its owners and the Spanish International Network and its affiliated stations. The chapter explores these relationships as they are found in two critical areas for broadcast policy analysts: (1) ownership or control and (2) programming.

Ownership and Control

The beginnings of what was to become the SIN/SICC Spanish-language television operations can be traced to the early 1960s, when Mexican broadcaster Don Emilio Azcárraga sought an export market for the television programs that his company produced and aired in Mexico. This interest in exportation, coupled with the participation of Frank Fouce, owner of a Los Angeles Spanish-language theater, was to grow into the complex international corporate network that now links SIN stations with Mexico's Televisa. To understand how these relationships have evolved, it is useful to review the development of two related corporations: the Spanish International Communications Corporation (SICC), which holds the licenses to five UHF Spanish-language stations, and Spanish International Network (SIN), which supplies programming to and is national sales representative for the five SICC stations and eleven other Spanish-language television stations.

Spanish International Communications Corporation

The beginnings of what has become SICC are tied to individuals or corporations with interests in Mexican media, a point made clear in trade press reports of the 1961 purchase of the group's first station in San Antonio. In identifying the buyers of what was to become KWEX-TV, *Broadcasting* listed Fouce, "Emilio Azcárraga, Mexico City broadcaster and theater owner; Edward J. Noble, Mexico City advertising agency executive; Julian Kauffman, general manager, XETV (TV) Tijuana Mexico." The group represented a combination of capital and program supply that was to set the pattern for the development of the San Antonio and future stations.

"I own 600 old Mexican movies outright—enough to keep a Spanish-language station on the air three years without repeating." Fouce said in a 1961 interview. "We also have rights to many of the Mexican TV series and the Spanish-language versions of American series."

"We" was the Spanish International Broadcasting Corporation, the company put together by Azcárraga, Fouce, and the others, for the purpose of owning the station. Because Federal Communication Act Section 310 prohibits the issuing of broadcast licenses to aliens, to the representatives of aliens, or to corporations in which aliens control more than one-fifth of the stock, the corporation was structured so that Azcárraga, a Mexican citizen, would own only 20 percent of the company that would hold the licenses. However, while the other stockholders were U.S. citizens, all had active business interests in Mexican media and two were employees of Azcárraga's television operation in Mexico.

The other SICC shareholders included Reynold V. Anselmo, a Boston-born U.S. citizen of Italian heritage, who was head of Azcárraga's Teleprogramas de México, S.A., an export company for Mexican-produced television programs. Another was Julian Kaufman, a U.S. citizen who was vice-president and general manager of the Azcárraga-controlled station XETV-TV in Tijuana. The last shareholder for the San Antonio station was Edward J. Noble, a U.S. citizen working as an advertising executive in Mexico City. When the group secured the license for Los Angeles television station KMEX in 1962, a sixth shareholder was added, Joseph S. Rank, a former account executive for station representative John Blair (see table 9-1 for a listing of owners).

All five original KWEX owners were involved in Mexican movies, broadcasting, or advertising; only Fouce had his primary base of operations in the United States. Eighty percent of the stock was held by Mexican citizens or U.S. citizens with primary business interests in Mexico and 60 percent of the stock

Table 9-1
Original Owners of KWEX (San Antonio) and KMEX (Los Angeles)

Name	*1961 KWEX Owner-ship (%)*	*1962 KMEX Owner-ship (%)*	*Citizenship*	*Occupation/Location*
Emilio Azcárraga	20	20	Mexico	Head of Telesistema Mexicano, San Antonio, Mexico City
Reynold V. Anselmo[a]	35	15	United States	General manager of Teleprogramas de Mexico, San Antonio, Mexico City
Edward J. Noble	20	15	United States	Advertising executive, Mexico City
Frank Fouce/Fouce Amusement Enterprises	20	35	United States	Owner of Spanish-language theater, Los Angeles
Julian Kaufman[a]	5	10	United States	Vice-president and general manager, XETV–TV, Tijuana
Joseph S. Rank	–	5	United States	Former John Blair account executive

[a]Employee of Azcárraga-controlled corporation.

was held by Azcârraga or employees of his corporations. Of the six original KMEX owners, only Fouce Amusement and Rank were based primarily in the United States. Sixty percent of the stock was held by Mexican citizens or U.S. citizens with primary business interests in Mexico, and 45 percent was owned by Azcârraga or employees of his corporations. Thus from the beginning of the group's first television stations there were clear ownership ties to Mexican media interests.

Fouce died before KMEX went on the air, and his son, Frank Fouce Jr., was later named president. But according to a 1967 *Television Magazine* article the position was largely a titular one. The U.S. stations were really run, the magazine said, by Azcârraga employees Anselmo and Kaufman. Of the two, Anselmo was to play the dominant public role in the growth and development of the Spanish-language television stations.

From its inception in 1954 until 1965 Anselmo was general manager of Teleprogramas de Mêxico, San Antonio, a company involved in exporting the programs of Azcârraga's Mexican television network, then called Telesistema Mexicano, San Antonio. During the latter part of 1961 and all of 1962, while still general manager of Teleprogramas de Mêxico, Anselmo worked full-time at launching the San Antonio and Los Angeles stations. In the latter part of 1963 Anselmo took a leave from Teleprogramas de Mêxico, took over the related

Spanish International Network operations in New York, and continued to be involved in the operation of the two stations. In 1965 he resigned from Teleprogramas de México to devote full time to the SICC stations and the SIN network.

During the first fifteen years of development (1961-1976) Anselmo says that he spent 80 percent of his time working for SICC or its predecessor companies and received no salary. His salary was paid by SIN, which was wholly owned by Azcárraga until 1971 and is still 75 percent owned by Televisa. Until 1973 he continued to receive monthly payments from Productora de Teleprogramas, which he says represented a continuation of the payments for his work for Teleprogramas de México, its predecessor. Thus for the first fifteen years of station ownership and growth, his income was largely derived from corporations owned or controlled by the Mexican television network. During these years Anselmo was either executive vice-president or president of SICC.

It was under Anselmo's leadership, backed by Azcárraga's program supply and financial guarantees, that SIN's Spanish-language television network in the United States grew and developed. The fledgling network and its stations did not have an easy time trying to sell the concept of a UHF foreign-language television network to the industry and to audiences. SIN and other Spanish-language broadcasters have fought a long series of battles with the audience-rating services for accurate sampling of the Latino community. They also faced an uphill struggle with advertisers who had negative or stereotyped perceptions of Latinos as consumers. Many in their potential audience could not get UHF signals in the early 1960s and had to buy set converters to receive the Spanish-language programming. Later, importation of live Spanish-language programming from Mexico by cable systems posed a threat to the growth of domestically licensed Spanish-language stations.

Despite all these problems and others, the number of stations continued to grow. The licensee corporation went through a series of expansions, mergers, and reorganizations in the 1960s and 1970s until the current SICC holdings were put together in a 1972 merger. These holdings include the SICC-owned and -operated stations in Miami, San Antonio, Los Angeles, Fresno/Hanford, and New York. Anselmo and other SICC principals also own SIN affiliate stations in San Francisco and Phoenix, which Anselmo says they plan to merge with SICC at a later date.

Through these growths new shareholders have been added to SICC; in 1976 shareholders numbered thirty-two. But the largest blocks of stock are still held by the original owners or their heirs. Laura M. de Azcárraga, widow of Don Emilio Azcárraga, retains his 20 percent. Fouce Amusement Enterprises, now controlled by Frank Fouce, Jr., owns 23 percent. Anselmo owns 22 percent. Mrs. Azcárraga is a Mexican citizen, and her SICC interests are managed by her son Emilio Azcárraga-Milmo, president and majority stockholder of Mexico's Televisa. The Azcárraga, Fouce, and Anselmo holdings make up over

65 percent of the current outstanding SICC stock. Another 25 percent is controlled by four more stockholders: Daniel Villanueva, Emilio Nicolás, James Jacobson, and Metropolitan Theaters. Villanueva and Nicolás run the SICC stations in Los Angeles and San Antonio respectively. The other 10 percent of the stock is divided among twenty-five shareholders.

The SICC corporation has had its financial ups and downs, but the value of the stations has steadily increased. A July 1976 appraisal of SICC by the broadcast brokerage firm of Hogan-Feldman placed the value at $22 million. Nearly $4 million of debts was deducted from this amount, for an equity value of $18 million. The same month the SICC board of directors declined to consider a telegram offer to buy SICC for "a price in excess of fifteen million dollars".

Spanish International Network

Although there is a limitation on the amount of stock a foreign national can hold in a broadcast licensee, there apparently is no such restriction on U.S. television networks. In the early 1960s Don Emilio Azcárraga established the Spanish International Network (SIN) to syndicate Mexican television programs in the United States and to represent to national advertisers the television stations in which he had an interest. Thus while he could not own the licensees himself, he could and did own the corporation that supplied the programming and sold the national advertising.

One researcher has written that Azcárraga started SIN "to handle a surplus market for the productions of Televicentro of Mexico City" (Telesistema's production center) and that the programming was initially sold to the San Antonio station at $40 per hour and to the Los Angeles station at $70 per hour. That rate remained unchanged for the first years of operation, but by 1976 it had increased to $150 per hour for the Los Angeles station and $75 per hour for San Antonio. The sales portion of SIN represented the U.S. stations and Azcárraga's Mexican border stations in securing national advertising.

The Azcárragas or a corporation they control have always had a controlling interest in SIN. According to Anselmo, Don Emilio Azcárraga directly owned 100 percent of SIN until 1971, when a 25 percent equity interest was transferred to Anselmo. In 1975 the 75 percent Azcárraga interest was sold to Televisa, of which Emilio Azcárraga-Milmo is the president and controlling stockholder. In a 1976 affidavit Anselmo, who is president of SIN, stated: "Televisa owns 75 percent of the stock in SIN and for all practical purposes controls the corporation." In May 1977 SIN was merged with Magnaverde, an Anselmo-owned corporation that promotes and sells television rights to sporting events, and the new corporation retained the SIN name and function. Televisa owns 75 percent of the stock in the new corporation, and Anselmo or his family own the remaining 25 percent.

SIN continues to supply Spanish-language television programming and provide national sales representation for the sixteen television stations affiliated with it. Most of that programming comes from Mexico, since Mexico's leading television network is also the controlling stockholder of SIN.

SIN-SICC-Televisa Interrelationships

Although the senior Azcárraga and Fouce died (their corporate roles were assumed by sons) and corporations on both sides of the border have expanded and reorganized, the links between SICC, SIN, and Televisa are similar to those that existed when the original corporations were put together in the early 1960s. SICC is the corporation holding broadcast licenses and is one-fifth owned by the Azcárragas. SIN supplies programming and sales representation to SICC and affiliated television stations and is 75 percent owned by Televisa. Televisa, through a subsidiary called Productora de Teleprogramas or Protele, provides television programs to SIN for broadcast on SICC and affiliated stations and is controlled by Emilio Azcárraga-Milmo.

Anselmo described the relationship between the various corporations in terms of programming in a 1976 affidavit (see figure 9-2). He said:

> To the best of my knowledge and belief, Televisa is a Mexican corporation which owns and operates a broadcasting system of approximately 95 television stations in Mexico. . . . Televisa exports Spanish-language programming to other parts of the world through its subsidiary, Protele. One of the purchasers of such Spanish-language programming from Protele is SIN, which in turn licenses such programming to SICC and other affiliated stations in the United States.

Over the years the relationships between the three companies have become strained. Anselmo, who is both president of SICC and president and treasurer of SIN, has been accused of not getting the best program prices when negotiating on behalf of SICC for programming from SIN and of committing other branches of management in a lawsuit filed by Fouce in November 1976. Azcárraga, whose companies are owed nearly $2 million by SICC for programming supplied but not paid for over the years, has pushed for a settlement of the debt. Anselmo claims that Fouce filed the suit in order to force the other stockholders to buy his holdings at an inflated price, while Fouce contends that the programming debts are invalid because the prices agreed on were not set through arms-length negotiations. The lawsuit is currently before the U.S. District Court for the Central District of California in Los Angeles. The materials filed raise the issue of foreign domination of SICC as a major contention.

Fouce, who was a member of the SICC board until September 1976 but was not involved with SIN, has charged that Azcárraga and Televisa, with

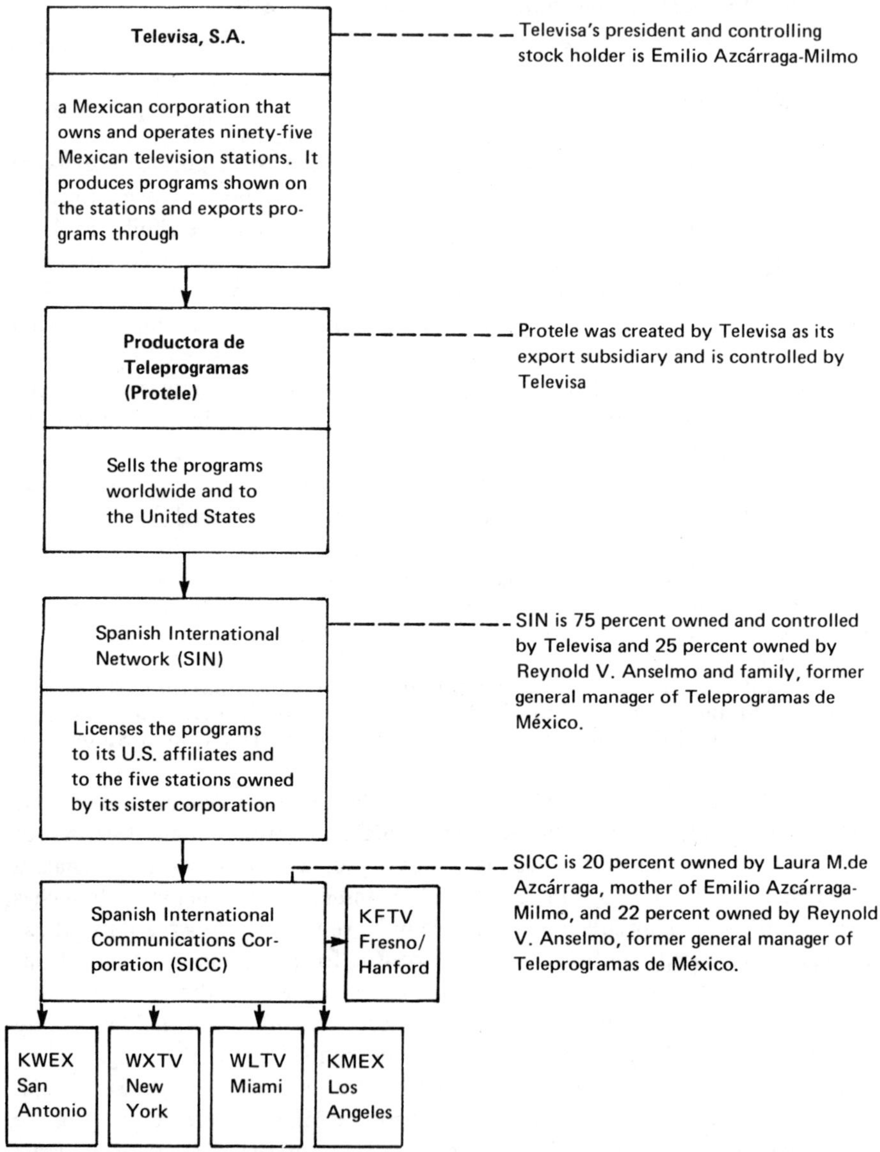

Figure 9-2. Export of Mexican Television Programs to the United States

Anselmo, "direct or cause the direction of the management and business of defendent SICC, directly or indirectly." This control, he argues, is in violation of Federal Communications Act Section 310 and has been exerted through several links.

Fouce's amended complaint contends that these links exert control in several ways. First, the complaint alleges that Televisa controls Anselmo "by virtue of various employment, business and personal relationships." These include a $300,000 debt that Anselmo owes Televisa and a $1,930,000 debt that Anselmo's Magnaverde owed Azcárraga's V.T. Latin. Second, the complaint notes that the Azcárragas own 20 percent of SICC and, under the terms of a convertible note issued September 1976, have the right to acquire an additional 20 percent of SICC stock. Third, Televisa owns 75 percent of SIN which "supplies defendent SICC with substantially all of SICC's television programming and acts as exclusive national sales representative for the SICC television stations." Fourth, SICC is purportedly indebted to Televisa for $1,935,000 in programming debts. Fifth, in 1976 the Azcárragas, Televisa, and Anselmo caused SICC to pay or agree to pay $328,000 to SIN for 1973 and 1974 programming that was not properly billed. Sixth, Televisa uses the SICC stations "to feed live news, advertising and other television programming into the United States." And seventh, in December 1975 Televisa, the Azcárragas and Anselmo caused SICC to issue 3 million shares of undervalued stock.

Anselmo, Azcárraga, and the defendant corporations deny charges of foreign domination of SICC in their responses to the charges, affidavits, and answers to interrogatories. They also indicate that Fouce was an officer and/or board member at the time that SICC entered into most of the arrangements to which he now objects. Azcárraga contends that neither he nor Televisa dominates or controls SICC, although he does not dispute that Televisa controls SIN.

In his own affidavit Anselmo acknowledges that a relationship has existed between the Azcárragas and SICC but paints a picture of Mexican beneficence toward the U.S. stations, not domination and control. Anselmo argues that the availability of low-cost Mexican programming and deferred payments was critical to the development of the SICC licensees. He states that payment for programming was made between Azcárraga-owned distribution corporations even when SICC had not paid for the programming and that between 1962 and 1967 neither the Los Angeles nor San Antonio station made any payments to Teleprogramas de México. It was only after Don Emilio Azcárraga died and his son assumed control that the relations became strained. According to Anselmo, Azcárraga-Milmo saw SICC "as a poor investment and continuing source of problems and headaches" and wanted to sell his family's SICC stock. Anselmo also points to long-standing program debts and a net loss of $200,000 in SIN's national sales representation of SICC stations between 1965 and 1975 as further evidence that SICC has benefited from the relationships.

The Convertible Note

A major portion of Fouce's allegations of foreign control of SICC involves a $1,935,000 convertible subordinated promissory note issued by SICC to Televisa in August 1976. The note was issued to cancel SICC's programming debts to SIN and V.T. Latin, an Azcárraga-controlled Panamanian television program distributor that formerly supplied programming. Fouce charges that the note, which permits Televisa to collect 8 percent annual interest on the debt until 1986 and then convert the note to 20 percent of SICC's outstanding stock, violates Section 310 of the Federal Communications Act. It was to avoid such conflict that the SICC board, after Fouce was removed in September 1976, approved the creation of a wholly owned subsidiary of SICC, to which all assets including broadcast licenses would be transferred. The plan, which was proposed by SICC's Washington attorney James McKenna of McKenna, Wilkinson, and Kittner, allows Televisa to own 20 percent of the holding corporation and the Azcárragas to own 20 percent of the operating corporation. This, according to McKenna, was permissible under the FCC's 1975 Datran ruling. Fouce and his Washington attorney, Stanley B. Cohn of Cohn and Marks, argue that the issuance of the note itself and its conversion is a de facto transfer of control to aliens in violation of Section 310.

In an affidavit filed by SICC Norman Leventhal, partner in McKenna, Wilkinson, and Kittner, argues that the FCC has the authority to permit more than 25 percent of a controlling corporation to be held by an alien or foreign corporation if it is in the public interest and that in its 1975 Datran ruling the FCC allowed a foreign national to hold 9.5 percent of the stock in a licensee corporation and 22 percent of the stock in a controlling corporation. Furthermore, Leventhal states, his informal consultations with the chiefs of the Broadcast Facilities Division and Transfer Branch of the Renewal and Transfer Branch of the FCC's Broadcast Bureau confirmed his interpretation of the Datran ruling. In its responses to Fouce's interrogatories, Televisa states that "even when Televisa exercises its right to convert the convertible stock to SICC, it will not dominate or control SICC."

Cohn contends that the conversion of the stock would result in an amount of foreign ownership in violation of Section 310 and that the complete chain of ownership must be traced and foreign ownership aggregated. Through aggregation there would be alien control of 36 percent of the corporations, violating the Federal Communications Act. Because of this, Fouce has asked that the federal court place SICC into receivership so that the licenses are not jeopardized by an action based on Section 310.

The allegations and responses regarding foreign control of SICC are currently before the Federal District Court and cannot be adjudicated here. However, the undisputed facts in the development of SICC indicate strong interest and participation by Mexican nationals, the Mexican television network, and their employees and former employees in the founding and growth of the SICC stations and SIN. Such relationships can result in the continuing dependency of a domestically licensed string of television stations to a foreign-owned network and its suppliers. Such indebtedness and the attempts to resolve it by the licensee management have resulted in a lawsuit by the largest single stockholder, alleging that the operation of the corporation holding the licenses is controlled by a foreign network in violation of Section 310 of the Federal Communications Act.

Programming

Whatever charges have been made against the current SIN and SICC managements, the two should never be accused of concealing their dependence on Mexican programming for their television stations. Their links to the Azcárraga television empire have been a matter of public record since the beginning of programming, and their heavy dependence on Mexican sources of programs has been amply documented by SICC lawyers in pleadings before the FCC.

Programming History

The development of SIN and the U.S. television stations was part of an international marketing strategy for Azcárraga's Telesistema Mexicano, predecessor of Televisa. The export company, which Anselmo headed from its inception in 1954 until he joined SIN permanently, distributed programs to other countries in Latin America, Europe, and the Far East, as well as to the United States. A 1964 *Sponsor* magazine article on SIN and the Los Angeles and San Antonio stations noted, "There is a family tie between this group and Mexico's highly successful Telesistema Mexicano, with a number of Telesistema's programs (soap operas, bullfights) carried on the U.S. stations."

In 1967 *Television Magazine* reported KMEX and KWEX had been conceived "to establish a market for export product in the U.S." and that Telesistema Mexicano decided to "create its own export market" in order to "get more mileage out of its some 80 taped shows produced each week." The magazine then asked itself what the stations showed, and it answered in the following words:

> Just about everything, but sardines, that comes out of a can, Telesistema Mexicano, the huge Mexico City-based tape production center is supplier. From this ever-cranking source comes a torrent of product to fill most of the stations' 60-odd hours of weekly air-time: telenovelas, or soap operas Latin-style, packed with emotion; musical

> varieties, heavy on heavy-limbed girls; comedy, pie-in-the-face and sight gags to be sure; folklore, sombreros and horseback serenades a must; bullfights, kills and all from corridas held in Mexico City's Plaza de Toros and from Tijuana; wrestling, starring some of Latin America's hamiest ring performers; feature films, pretty much of a recent addition to the schedule; soccer, including World Cup matches; boxes, from ringside (a week or more delayed) in Mexico, and frequent special events of a public service nature, such as exclusive live coverage of Mexican Independence Day ceremonies from that country's capital.

The article said that in 1967 KMEX was receiving between seventy thousand and eighty thousand feet of videotape and film from Mexico each week. Most of the programs were first aired in Los Angeles and then shipped to KWEX in San Antonio, the beginning of the SIN station-to-station bicycle network. The network still bicycles a good deal of its programming, but it is also connected by Westar and Intelsat satellites. The four California stations began interconnected transmission from Los Angeles in the early 1970s, and the national satellite coverage began later. Through these developments the SIN stations have continued to depend on Mexican television programming.

Televisa, along with Brazil's Television Globo, is the top exporter of television programs in Latin America. The corporation's export distribution arm, Protele, sells an estimated seventeen thousand hours of Televisa programming annually, and its biggest customer is SIN. Mexican-produced programming dominates SIN's feeds to the SICC and affiliated stations. A 1976 content analysis of network programming on KMEX found that about 96 percent of the network hours for a composite week of that year originated in Mexico. Along with the programming has come indebtedness to the Mexican program suppliers and the deferral of payments that led to the 1976 shareholders' lawsuit.

But the Mexican-produced programs and the ability to defer program debts also served the needs of the new stations. Except for a few hours of locally produced news and public affairs, required by the FCC, and perhaps a locally sponsored live program, the stations could practically eliminate studio production costs by airing videotapes of programs or films first produced and aired in Mexico. The management, facing great difficulties in selling advertising on a foreign-language UHF station, could thereby operate the station at a far lower cost than if it produced its own programs. In 1967 KMEX ran the station with thirty-six full-time employees at a monthly budget under $100,000, and KWEX's budget was less than $40,000. KMEX reported its first profit in 1965, three years after beginning operation, and KWEX posted its first profitable quarter in 1966.

Although both stations and SICC have experienced financial ups and downs since then, Anselmo stated in his affidavit that the "agreement and principle of providing the other SICC stations with [Mexican] programming at a low cost until they could get on their feet has been honored to this day." KMEX continued paying the $70 per hour price from 1962 through 1968 and KWEX paid $40 per hour from 1962 through 1972. In 1976 the hourly charges were Los Angeles, $150, San Antonio, $75; New York, $150; Miami, $100; and Fresno, $25; thus the total SICC charge was $500 per hour. The prices charged for Mexican programs and the accumulation of program debts are two of the main reasons for the 1976 stockholder's lawsuit.

FCC Acknowledgment of Dependence on Mexican Programming

Just as the trade press made no secret of SICC's dependence on Mexican-produced television programs, SICC made the relationship clear in repeated actions before the FCC in the early 1970s. The actions grew out of the issuance of the FCC's *Cable Television Report and Order* in 1972. The order indicated that the importation of foreign-language stations from Mexico and Canada would serve foreign-language groups in the United States and ruled that it would "permit cable systems to carry non-English-language programming without limitation" unless a local foreign station "demonstrates that such importation will adversely affect its ability to serve the public." SICC, which had failed in an earlier attempt to modify the ruling, petitioned for reconsideration and asked that imported signals not be allowed "where U.S. Spanish-language programming is available to the cable operator." In its petition SICC demonstrated its dependence on Mexican programs:

> For example, Spanish International relies to a great extent on the same programming carried by many, if not all, of the Mexican border stations. Many of the best programs carried on the Mexican network stations do not become available to Spanish International for a year or more after the date of first transmission in Mexico.

The FCC rejected this and other appeals by SICC, restated its original policy, and said that it could afford relief "in unusual situations where a domestic Spanish-language station makes a compelling demonstration for relief with respect to a particular [cable] application." SICC then filed a series of challenges against cable systems as far away from Los Angeles as Colorado, New Mexico, and Arizona. In these challenges SICC argued against the proposed carriage of the signals of Mexican border stations by the cable operators, noted the availability of KMEX programming, and restated its dependence on Mexican programs. For example, in a petition against a California cable operator who proposed carrying a Tijuana station, SICC stated:

> KMEX-TV obtains significant quantities of its Spanish language programming from Spanish International Network Sales, which is the

> United States representative of V.T. Latin Programs, whose associated company also supplies programming to XEWT-TV, Tijuana. Thus, both stations receive major amounts of programming from the same source and thus carry similar program schedules.

SICC then compared programs on the prime-time schedules of both stations and concluded that "over 10½ hours of KMEX-TV's weekly *prime-time* programming—which is obviously crucial to its continued viability—would be duplicated by the signal of XEWT-TV." Having demonstrated its own dependence on imported television programs, SICC concluded its petition against cable importation by arguing that Latinos "desire and need U.S. -originated television service" and saying that it hoped the FCC would not "unjustly 'cripple' domestic Spanish language stations by permitting unfair competition from Mexico."

In its 1975 ruling on the case the FCC denied the SICC's petition and noted that "KMEX-TV obtains a significant amount of Spanish-language programming from the same affiliated sources as XEWT-TV and, consequently, has similar television program schedules as the Mexican station." But it ruled that such a show of duplication was not, in itself, a compelling reason to prohibit carriage of Mexican signals. This reasoning, first expressed in the 1973 *In Re Mickelson Media* ruling, acknowledged the dependence of U.S. Spanish-language broadcasters on Mexican programming. It was used by the FCC in turning down SICC petitions against cable operators in New Mexico, Arizona, Colorado, and California.

Thus the dependence of SICC's stations on Mexican television for their programming has been amply demonstrated in SICC's own petitions and arguments to the FCC and noted in the rulings on those petitions. It is a dependence that has developed from the expansion of Mexican television into export program sales to the United States in the 1960s and continues as SICC moves toward the 1980s.

Alleged Televisa Control of SICC Programming

The amount of control exerted by the Mexican interests on SICC's operation is one factor in a complicated lawsuit by stockholders. Fouce has alleged that since Televisa owns 75 percent of SIN and SIN supplies SICC stations "with substantially all of SICC's television programming" that Televisa, the Azcárragas, and Anselmo "dictate the television programming and national advertising that appears on SICC stations." Anselmo and Azcárraga deny the charge.

But it is undisputed that almost all of the SIN/SICC program schedule, particularly the prime-time programming, originates witn Mexico's Televisa. The proposed schedule for the Corpus Christi station that opened in 1977 indicated local or live programming only in the early afternoon and the evening news. Almost all the other programs would come from Mexico by either satellite or videotape. A 1977 content analysis of the programming on the Los Angeles station revealed less than three hours per day of local programming, with most of the rest coming from Mexico. The only KMEX prime-time SIN program that originated in the United States was wrestling from Los Angeles' Olympic Auditorium. All but two of the SIN programs that dominated the station's schedule came from Televisa. The analysis of programs by type showed that novelas (soap opera-type dramas) accounted for the greatest percentage of programming. Also high were musical and variety shows, movies, and religious programming.

Televisa influence on U.S. Spanish-language television programming is also destined to increase with the direct satellite broadcasts from Televisa's Mexico City studios each day. Some stations carry Mexico City's "24 Horas" as part of their weekday news programming and the remainder of the satellite feed. Although the network cuts out commercials deemed unsuitable for the United States (such as cigarettes), regular SIN satellite viewers have become accustomed to seeing commercials for products widely available only in Mexico and for Mexico City business establishments. In a 1976 interview Televisa group vice-president Alejandro Sada said, "An advertiser who buys spots on Mexico's Channel 2 automatically also appears throughout the U.S."

Other Spanish-Language Programming

Along with the entertainment programming from Mexico, SICC stations and SIN affiliates feature regular local newscasts and public affairs programs. As in most commercial stations the public affairs programming is often seen on weekends and, in some stations, in the early or mid afternoon. Newscasts, supplemented by Mexico's "24 Horas," are aired in the early evening and sometimes later at night. Although largely dependent on wires for national and international coverage, some SICC stations have won a reputation for outspoken reporting on issues of collective concern to the Latino community at a local level. KMEX won a Peabody Award for its coverage of rioting in East Los Angeles in 1970 in which its news director, Rubén Salazar, was killed by a sheriff's deputy. The network has also fed election night coverage in 1976 and live reports of space flights, national political conventions, and other newsworthy events. In 1977, two days after going on the air, the Corpus Christi station doubled the nightly half hour it had allocated for news when it decided that more time was needed. The station says one-fifth of its weekly air time is devoted to news and public affairs.

In 1977 SIN specials included salutes to Latin American countries produced on location by Radio Televisión Española, New Jersey's Puerto Rican Day parade, a Fourth of July Special from Miami and New York, and live coverage of the OTI Latin song festival. The network led off 1978 with live coverage of Pasadena's Tournament of Roses Parade. Sports coverage, including weekly boxing matches, is also strong on SIN stations. Soccer is especially popular with Latino sports fans, and SIN plans to carry the World Cup competition.

The initial and continued dependence of SIN-affiliated television stations on imported Spanish-language television programs has been shown through references to articles in trade journals and SICC's own petitions to the FCC. This dependence, which has resulted in an accumulation of debt that partially triggered the current stockholders' lawsuit, has developed from station-to-station bicycling of videotaped programs to direct satellite transmission of programs and commercials from Mexico City. As a result, SIN-affiliated stations will probably continue to rely on Mexican programmers for a large portion of their entertainment and news programming, although local newscasts and public affairs programs are also broadcast. Thus the initial lines of development of the network to provide an export market of Mexican-produced television programs have continued into the satellite era.

Discussion

This review of the growth and development of the SICC/SIN stations is likely to provoke a number of questions among communications policy researchers and analysts. Since I can mention only a few of these questions in this chapter, it might be useful to state some of the major issues raised and the arguments for both sides of each issue. The following presentation of these positions should not be construed as the position that I accept or advocate.

The Application of Section 310

A major question raised by the case of Spanish International Network is the application or desirability of application of the provisions of Section 310. Some might favor a strong application in this case, arguing that the intent and effect of the provision should be to retain control and use of national broadcast resources for citizens. They would argue that the elaborate corporate structures of SIN and SICC indicate conformity with the letter but not the spirit of the section. And they might also indicate that the programming dependency and indebtedness that has forced SICC to issue the convertible note is an illustration of the inherent dangers in loose interpretation or application of the law.

Others might argue that Section 310, which was originally designed to avert takeover of broadcast facilities by aliens during time of war, has no legal application in this case since station ownership conforms with the letter of the law.

They would argue that licensees should be able to do business with whatever program suppliers they see fit and that looking for "flow-through" influences and aggregating alien stock ownership in operating and holding corporations does not necessarily indicate that influence is being exerted by aliens or, if it is, that it has a negative effect. They might also argue that Section 310 is outdated in this era of transnational corporations and that a rigid application could have negative effects for U.S. communications corporations with interests in other nations. The advent and use of communications satellites can be seen as making the 1934 statute even more archaic and difficult to enforce. Finally, they might argue that in the interests of hemispheric unity the United States should recognize that the importation of television programming can have a beneficial effect on language minorities in this country.

Network Regulation

The ownership of the Spanish International Network and the indebtedness to it by licensees is another issue that could be raised. Some might use the case of SIN to indicate the abuses of law that can take place when the FCC regulates licensees but not the networks supplying their programs and selling their advertising time. Since the federal government now regulates the importation of color television sets, one could argue that it should also turn its attention to the importation of programming for those sets by television networks. Such a regulation could limit the amount of foreign programming a network could offer, or it could require networks doing business in the United States to supply domestically produced programming along with imported programming.

Those arguing against network regulation in this and other cases would probably contend that federal regulation of networks should be through the licensees not the networks themselves. The regulation of networks would add another layer of bureaucratic responsibility to a regulatory commission already overworked in the regulation of stations. There is no evidence in this case, they would offer, that indicates that any federal laws have been broken by the operation of these networks. Therefore no additional regulation is warranted.

Beneficial Effects

A third position might be that whatever the alleged transgressions, the growth and development of SIN has had overriding beneficial effects for broadcasting and the public interest. They would point out that the development of SIN has proved the viability of UHF stations in certain markets and the viability of a UHF network. They might argue that the network has been able to extend television service to a non-English-speaking segment of the population that has been virtually ignored by commercial and public television stations. They could contend that financing by the Mexican networks and the availability of Mexican programs has facilitated the expansion and growth of the network at a far lower cost than the cost of conventional financing and programming. They could argue that there is no inherent threat to the public interest in having television stations depend on foreign sources for programming and that such stations might even ease the transition of immigrants to this country since they offer a blend of foreign-based programming and local news and public affairs. Finally, they might argue that the payments by Televisa and SIN to licensees carrying satellite broadcasts even makes a small contribution to easing this nation's foreign debt.

Those in opposition to this position could acknowledge the short-term beneficial effects of SIN in developing Spanish-language television service and the UHF spectrum while arguing that the short-term gains have been at the expense of longer-term goals. They might see the long-term dependence on foreign program sources as cultural imperialism that retards the development of domestic Spanish-language production capability and siphons profits out of the country. The dependence on foreign program sources might also reinforce the alienation of Latinos in this country, since it focuses their attention on events and personalities elsewhere. This dependence also limits access to the broadcast licensees by members of the audience and to Latino professionals desiring positions in broadcasting. Thus the short-term gains must be weighed against the long-term effects of allowing Spanish-language television to develop without a strong domestic production base to produce programming that reflects the needs and realities of Latinos in the United States rather than in Latin America.

Conclusion

I hope that this chapter will stimulate discussion of Spanish-language television by policymakers and members of the public. Both the current and projected growth of Latinos in the United States and the existence of the Spanish International Network are realities. Given the failure of commercial and public broadcasters to adequately address the opportunities presented by the growing Latino audience, other networks have developed and grown. It is through continued examination and reexamination of the growth and operation of such networks that their role for the future can best be defined.

Research Notes and Primary References

Except where otherwise noted, all legal documents cited are from the following case: *Fouce Amusement Enterprises, Inc., et al.* v. *Spanish International Communications Corporation et al.*, United States District Court, Central District of California, Los Angeles, California; Case No. CV 76 3451 IH, filed November 4, 1976.

Primary Sources

Les Brown, "Seven U.S. Stations to Broadcast TV Shows Directly from Mexico," *New York Times*, September 9, 1976, reprint; "Spanish-Language TV Net Starts with Seven Outlets," *Advertising Age*, September 13, 1976, reprint; "Answers and Objections of Defendant Televisa, S.A. to First Set of Interrogatories Propounded by Plaintiffs," *Fouce et al.* v. *SICC et al.*, p. 46, and the KORO, Corpus Christi, Texas, proposed programming log.

Latinos in the United States

For current demographic and media use statistics on Latinos see "Spanish-Language Market Study," *Television/Radio Age*, November 7, 1977, pp. S-1-S-24; Joseph Aguayo, "Latinos: Los Que Importan Son Ustedes," *Sales and Marketing Management*, July 11, 1977, pp. 23-29; and "So They Speak Spanish," *Media Decisions*, May 1977, pp. 68-71, 116.

Latinos and Broadcasting

For information on broadcast stations see the television and radio listings of *Broadcasting Yearbook* under Spanish-language programming. For audience studies on Latinos see Nicholas Valenzuela, *Media Habits and Attitudes of Mexican Americans*, and Nicholas Valenzuela, Frederick Williams, and Pamela Knight, *Prediction of Mexican-American Media Habits and Attitudes*, University of Texas Center for Communication Research, June 1973; and Ronald López and Darryl Enos, "The Role and Functions of Spanish-Language-Only Television in Los Angeles" (Claremont Colleges Center for Urban and Regional Studies and Chicano Studies Center, 1973). For advertising data see articles listed under Latinos in the United States.

Spanish International Network

For background information see "Spanish International Network," *Cablines*, July-August 1975, p. 17; "What's the Future of Spanish-language Television?" *Media Decisions*, May 1977, p. 71; "The New York Spanish Market," WXTV-TV, undated sales promotional brochure; "The U.S. Spanish Market,"

Spanish International Network, undated sales promotional brochure; and references to broadcast marketing publications previously cited. For references to SIN by European communication researchers see Tapio Varis, "Global Traffic in Television," *Journal of Communication*, Winter 1974, p. 103 and Armand Mattelart, "Cultural Imperialism in the Multinationals' Age," *Current Views on the World Information Order*, 1977, p. 68. For research by U.S. scholars see López and Enos, "Role and Functions," Dorrit Sue Cox, "Spanish-Language Television in the United States: Its Audience and Its Potential" (master's thesis, Univeristy of Illinois, 1969), and Arturo Gonzales, "A Case Study of KMEX," (master's thesis, California State University, Northridge, 1978).

Ownership and/or Control

Primary sources cited were "Approved," *Broadcasting*, December 11, 1961, p. 69; "Mexican TV Mogul to Invade U.S.," *Sponsor*, June 4, 1962, p. 8; Lincoln Haynes, "Energetic Frank Fouce Just Too Busy to Quit," *Los Angeles Mirror*, September 28, 1961; Sect. 310, "An Act to Provide for the Regulation of Interstate and Foreign Communications by Wire on Radio and for Other Purposes," *The Status at Large of the United States of America from March 1933 to June 1934* (Washington, D.C.: U.S. Government Printing Office, 1934), p. 1086; "Yanqui," *Television Magazine*, May 1967, pp. 30-31, 57-61; Cox, "Spanish-Language Television"; "Complaint," "Affidavit of Reynold V. Anselmo in Opposition to Motion for Preliminary Injunction," "Answers of Defendant Reynold V. Anselmo to First Set of Interrogatories," "First Amended and Supplemental Complaint," "Answers and Objections of Defendant Televisa, S.A. to First Set of Interrogatories Propounded by Plaintiffs," "Affidavit of Norman Leventhal in Opposition to Motion for Preliminary Injunction," "Reply of Plaintiffs Re: FCC Issues," and related exhibits and documents, *Fouce et al* v. *SICC et al.*

Programming

Primary sources cited include "Want Spanish-Language Availabilities? Here Are the Top Reps in this Field," *Sponsor*, October 19, 1964, p. 46; "Spanish-Voiced TV: New Boost for UHF," *Sponsor*, October 18, 1965, p. 57; "Yanqui," *Television Magazine*, May 1967, pp. 30-31; "Spanish International Network," previously cited; "Rivals Challenge Mexican TV Czar," *Business Week*, March 15, 1969, p. 52; "Spanish U's Move into Black," *Variety*, July 20, 1966, p. 35; "Cable Television Report and Order," *FCC Reports*, 36 FCC 2d, p. 180; "Petition for Reconsideration," filed by Spanish International Communications Corporation," FCC Docket Number 18397, March 13, 1972; "Memorandum

Opinion and Order on Reconsideration of Cable Television Report and Roder," *FCC Reports*, 36 FCC 2d, pp. 334-335; "Petition for Special Relief, in the Matter of United Cablevision, Inc., Loma Linda, California, and United Cablevision, Inc., Riverside, California," filed by Spanish International Communications Corporation; CAC-445, CA442 and CAC-739, CA468, October 27, 1972; "In Re Applications of Riverside Cable TV, Inc., United Cablevision, Inc., Loma Linda and Riverside," *FCC Reports*, 51 FCC 2d, p. 551; "In Re Mickelson Media, Inc., Las Vegas, New Mexico," *FCC Reports*, 39 FCC 2d, p. 603; "First Amended Complaint," Affidavit of Reynold V. Anselmo," "Answers of Defendant Reynold V. Anselmo," "Complaint," "Answers and Objections of Defendant Televisa, S.A.," *Fouce et al* v. *SICC et al*; "Spanish-Language TV Net Starts with Seven Outlets," SIN advertising and programming promotional materials, undated and untitled; and "Espejo Latino," *Nuestro*, March 1978, pp. 55-56.

10 Research and Children's Television Advertising Policy

Richard P. Adler

In a paper prepared for the 1977 Airlie House conference, George Comstock observed that the "emergence of the possibly detrimental effects of television advertising on children as an issue" represents "one of the major developments in policymaking for television in the 1970s." With the recent launching of a rule-making inquiry into children's television advertising by the Federal Trade Commission, it is certain that the issue will be receiving even greater attention in the immediate future.

Concern about the effects of television advertising on children is not new. It dates back at least to 1961, when the National Association of Broadcasters (NAB) adopted guidelines governing the advertising of toys to children. Since 1961 the extent of industry self-regulation of children's television advertising has grown considerably, as has the amount of such advertising. In 1973, for example, the NAB guidelines were extended to include all advertising directed to children. In 1974 the Children's Advertising Review Unit of the Council of Better Business Bureaus' National Advertising Division (NAD) issued its own set of Children's Advertising Guidelines that are similar but not identical to those of the NAB.

Both NAB and NAD codes prohibit certain categories of products (drugs) from being advertised directly to children, prohibit the use of certain presentational techniques (host selling, exaggerated portrayal of product performance), and impose several positive requirements (disclaimers in certain toy commercials, statement about "balanced breakfasts" in cereal commercials). In addition, the NAB code specifies time limits for advertising during children's programming. At present, the limit is 9½ minutes per hour, down from a permissible maximum of 12 minutes per hour in 1974.

Over the years various government agencies have also expressed concern about children's television advertising. Although both the Federal Communications Commission (FCC) and the Federal Trade Commission (FTC) have undertaken inquiries, neither has enacted any significant policies or regulations. (The FTC has acted against a few specific advertising campaigns.) Rather, both commissions have traditionally demonstrated a preference for industry self-regulation to taking major steps against advertisers or broadcasters. Whether the same pattern will be repeated in the current FTC proceeding remains to be seen. Sentiments are already being expressed among broadcasters in favor of a voluntary reduction in the amount of advertising time permitted in children's programming from the current 9½ minutes per hour to 6 minutes per hour.

One relatively new element in the continuing debate over this issue is the availability of a growing body of empirical findings documenting the effects of television advertising on children. Almost all this research has been conducted since 1970, and only in the past few years has it begun to play a visible role in policymaking. Thus research results were taken into account in both the FCC's consideration of television advertising of over-the-counter drugs and the FTC's inquiry into the use of premiums in children's advertising. In both cases the commission decided against taking action, in part because the available evidence failed to identify a clearly harmful effect. On the other hand, the results of a study commissioned by the FTC played a positive role in helping the FTC to decide against a specific advertising claim (*FTC* v. *ITT Continental Baking Company*, 1973).

The New FTC Inquiry

The most extensive use of empirical research to date is clearly the current FTC rule-making inquiry. The FTC Staff Report (February 1977), which provided the starting point for the proceedings, argues that regulatory action is called for in two areas: (1) advertising directed to young children and (2) advertising for sugared food products. In making its case the report devotes some fifty pages to citing research findings documenting the nature of children's exposure to television advertising, the content of food advertising directed to children, and the impact of this advertising on children.

No matter how this proceeding is ultimately resolved, the staff report represents an important case study in the use of empirical research in policymaking. Even at this relatively early stage, it is apparent that research findings—and their interpretation—will continue to play a prominent part in the debate as the rule-making process unfolds.

What can research contribute to policymaking on this issue? I will summarize the results of the existing research relevant to the two areas identified in the FTC report, then offer some observations on the contribution of research and some persistent limitations on its role in policymaking.

The Significance of Age in Children's Responses to Television Advertising

The FTC staff report argues that "televised advertising directed to children too young to understand the selling purpose of, or otherwise comprehend or evaluate, commercials is inherently unfair and deceptive" and ought therefore to be banned.

The variable that emerges most strongly from the existing empirical studies as a determinent of children's perceptions of and responses to television advertising is the child's age. In fact, fully half of the twenty-one "key studies" identified by a National Science Foundation-supported review of research on the effects of television advertising on children dealt at least in part with age-related differences. Research has shown that compared with older children, younger children

pay more visual attention to commercials (Atkin, 1975)

Are less able to recall specific features and information from commercials they have seen (Rubin, 1972)

Are more likely to believe claims they hear in a commercial (Ward and Wackman, 1973)

Are more likely to be confused by or fail to understand the meaning of disclaimers contained in commercials (Liebert et al., 1976)

Most significant from a policy standpoint are the findings regarding children's ability to discriminate between programs and commercials and to understand the persuasive intent of commercials. The key study in this regard is "Information Processing of Television Advertising" by Ward and Wackman (1973). The study found that a majority of five-to-eight-year-olds exhibited a "low level differentiation" between commercials and programs, while 75 percent of nine-to-twelve-year-olds exhibited a "high level differentiation" between the two. Younger children tended to identify commercials on the basis of affect ("commercials are more funny") or perceptual characteristics ("commercials are short and programs are long"); older children were more likely to distinguish them on the more abstract basis of intent ("programs are supposed to entertain you . . . commercials try to sell you things")

Robertson and Rossiter (1974) examined children's ability to understand commercials in terms of six variables: (1) discrimination between programs and commercials; (2) recognition of an "external message source or sponsor; (4) identification of informative intent in commercials; (5) identification of persuasive intent; and (6) understanding of the symbolic nature of the representation in commercials. The authors found total cognitive understanding of commercials to be "highly correlated" with age and, they suggested, with total cumulative exposure to television advertising. Thus 99 percent of the study's fifth graders exhibited an understanding of the selling intent of a commercial, compared with 53 percent of the first graders.

That children become more sophisticated in their understanding of television advertising as they grow older is hardly surprising. In fact, it would be surprising if it were otherwise, in the light of the extensive body of evidence from developmental psychology documenting the importance of cognitive and other age-related changes. These findings indicate, at the least, that to treat all children from age two to twelve as a single homogeneous group masks important, perhaps crucial, differences. Yet both industry and government have largely failed to date to make any distinctions between children of different ages in formulating policies governing advertising directed to them. One of the few places that a distinction is made among children two to eleven is in the Nielsen demographic ratings, which breaks down the child viewing audience into children two to five and children six to twelve.

The Effects of Children's Food Advertising

The FTC staff report claims that the effect of television food advertising directed to children is to "include [them] to take health risks which they are not equipped to assess." Specifically, the report argues that children's food advertising consists almost entirely of commercials for "sugared foods" (candy and presweetened cereals) and that exposure to advertising for these products promotes their increased consumption which, in turn, produces increased tooth decay and other health problems. The staff recommends that the FTC consider banning advertising to children of the products (candy) that pose "the most serious health risks," and requiring that advertising for other sugared foods (cereals) "be balanced by nutritional and/or health disclosures funded by advertisers."

A large element in the debate over this issue involves the actual health implications of consuming sugared foods. This is a complex question in itself, and is a matter for nutritionists, dentists, and other health specialists. However, behavioral science research can document the role of television advertising in influencing children's nutritional knowledge, attitudes, and food preferences.

Comparatively less research has been conducted to date on the impact of television food advertising on children than on the issue of age-related differences in children's responses to advertising. The existing evidence can be summarized as follows.

Incidence. There is little question that children are exposed to a considerable amount of advertising for sugared foods. A 1975 analysis of advertising on weekend morning children's programming (Barcus) found that 68 percent of the commercials were for food products or eating places. Twenty-five percent were for ready-to-eat cereals; 25 percent for candies and sweets; 10 percent for eating places; 4 percent for other "snack foods"; and 4 percent for "miscellaneous" (including milk and dairy products, fruits and juices, and bread).

Children's learning from commercials. There is evidence from several experimental studies that children do learn brand names and other information presented in commercials. There is also evidence that children are capable of learning "hard" nutritional information when included in commercials (Atkin, 1975; Feshbach et al., 1976). Experimental studies have demonstrated shifts in children's beliefs about advertised foods following their exposure to specific commercials. One study found that younger children are more likely than older children to be persuaded by claims made in a commercial (Haefner, Leckenby, and Goldman, 1975).

Impact on attitudes and behavior. Evidence from several studies indicates that children play an active role in the selection of foods purchased by their mothers. Thus one national survey found that 75 percent of mothers of six- to fourteen-year-old children interviewed indicated that they were influenced by their children in their choice of cereals, cookies, fruit drinks, peanut butter, gum, and candy—all heavily advertised products (Gene Reilly, 1973). In a study based on observations made in a supermarket Atkin (1976) found that in more than two-thirds of the 516 families observed, children initiated the selection of a cereal. In a majority of these instances, parents acquiesced to their children's requests. In 25 percent of the cases "conflict" between the parent and child was observed following parental denial of the child's cereal request.

To what extent is television advertising responsible for stimulating these requests? One study found the frequency of preschool children's requests for specific brands in a supermarket to be positively related to the amount of television that children were reportedly exposed to at home (Galst and White, 1976). A survey by Atkin (1975) found a positive relationship between reported exposure to television advertising for cereal and candy and the consumption of these products among a group of fourth to seventh graders.

A relatively recent experimental study (Goldberg, Gorn, and Gibson, 1977) has tested the effect of both food commercials and pronutritional messages in influencing children's food preferences. The study found that first graders exposed to commercials for sugared snack foods and cereals chose significantly more of these foods (compared with more healthful, less sugared alternatives) than children exposed to pronutrition PSA's (public service announcements). However, most effective in promoting the selection of non-sugared foods was exposure to a twenty-four-minute animated program ("Fat Albert") which dealt with the negative consequences of eating junk foods. The authors concluded that the results suggest "that the type of snack foods and cereals children select can be affected by the type of food messages to which they are exposed."

Conclusions

The current FTC inquiry demonstrates the growing importance of research in policymaking on children's television advertising. Empirical research seems to have played a general role in helping identify the areas of concern (vulnerability of young viewers), as well as a specific role in supporting a case for regulatory action. It seems likely that this research will continue to be cited and discussed as the rule-making inquiry proceeds.

It is unlikely, however, that research will play a dominant role in determining what specific regulatory actions (if any) ultimately emerge from the FTC. First, although research can and does document various effects of television advertising on children, it cannot tell us whether these effects are "good," "bad," or neutral. Even when there is no dispute over the accuracy of a particular finding, there can be considerable disagreement over its meaning. For example, Pillemer and Ward (1977) cite the case of a previous FTC hearing at which the same finding from a study was used to support two opposite sides of an issue:

> The finding demonstrated that older children [over eight years of age] are more skeptical of advertising claims than younger children (Blatt, Spencer, and Ward). A consumer group cited the finding to support their position that advertising "breeds cynical attitudes" among children; an industry trade association lawyer cited the same finding to argue that advertising to children was legitimate since these attitudes among older children suggest they can cognitively filter advertising claims.

Similarly, whether a particular effect documented by empirical studies warrants regulatory action is to a large degree a legal and political question. Thus much of the FTC staff report is devoted to constructing a case for action based on legal precedents and interpretation of its authority to act on these issues.

The policymakers' definition of "evidence" is considerably different from that of the behavioral scientist. To the latter evidence is the empirical result of a scientifically conducted study. To the policymaker evidence includes the opinions and conclusions of expert witnesses. Given the almost inevitable ambiguity of empirical research, the testimony of expert witnesses can be extremely important in decision making.

A final limitation on the research currently available is that it was all conducted before the current rule-making proceeding. Therefore none of the existing studies was designed to address the specific policy issues being considered by the FTC. In the best of all possible worlds sufficient time would be allotted in the rule-making process for research to be designed and conducted to help answer the kinds of questions raised in the FTC's notice of inquiry.

References

Adler, Richard, Friedlander, B.Z., Lesser, G.S., Meringoff, L., Robertson, T.S., Rossiter, J.R., and Ward, S. *Research on the Effects of Television Advertising on Children.* Washington, D.C., National Science Foundation, 1977.

Atkin, Charles. "Effects of Television Advertising on Children." Michigan State University, June 1975.

Barcus, F. Earle. *Weekend Commercial Children's Television.* Newton, Mass.: Action for Children's Television, October 1975.

Comstock, George. "New Roles for Social and Behavioral Science in Television Policymaking." Paper submitted to the Telecommunications Policy Research Conference, Airlee, Va., April 1977.

Federal Trade Commission. *Staff Report on Television Advertising to Children.* Washington, D.C., February 1977.

Feshbach, Norman, Jordan, C.S., Dillman, A.S., Choate, R., Feshbach, S. and Zolotow, M. "The Design of a Graphic to Convey Nutritional Information to Children: Pilot Studies." Council on Children, Media, and Merchandising, Washington, D.C., October 1976.

Galst, Joann P., and White, Mary Alice. "The Unhealthy Persuader: The Reinforcing Value of Television and Children's Purchasing Influence at the Supermarket." *Child Development* 47 (1976): 1089-1096.

Gene, Reilly Group. "Meals and Snacking: The Child and What He Eats." *Child* 2 (1973).

Goldberg, Marvin E., Gorn, G.J., and Gibson, W. "TV Messages for Snack and Breakfast Foods: Do They Influence Children's Preferences?" McGill University, November 1977.

Haefner, James E., Leckenby, J.B., and Goldman, S.L. "The Measurement of Advertising Impact on Children." Paper presented at the American Psychological Association, Chicago, August 1975.

Liebert, Robert M., Liebert, D.E., Sprafkin, J.N., and Rubinstein, E.A. "Effects of Television Commercial Disclaimers on the Product Expectations of Children." Brookdale International Institute for Applied Studies in the Mental Health Sciences, Stony Brook, N.Y., 1976.

Pillemer, Daivd, and Ward, S. "Investigating the Effects of Television Advertising on Children: An Evaluation of the Empirical Studies." Paper presented at the American Psychological Association, San Francisco, August 1977.

Robertson, Thomas S., and Rossiter, J.R. "Children and Commercial Persuasion: An Attribution Theory Analysis." *Journal of Consumer Research* 1 (1974): 13-20.

Rubin, Ronald S., "An Exploratory Investigation of Children's Responses to Commercial Content of Television Advertising in Relation to Their Stages of Cognitive Development." Doctoral dissertation, University of Massachusetts, 1972.

Ward, Scott, and Wackmann, D.B. "Children's Information Processing of Television Advertising." In *Sage Annual Review of Communication Research,* Beverly Hills, Calif.: Sage Publications, 1973, ed., R. Gerald Kline and P. Clark, pp. 119-146.

11 Issues in Advertising Policy: Children and Nutrition Information

Emilie Griffin

I have spent most of my adult life in the business of designing advertising messages, and most of my advertising life has been lived at Compton Advertising. But there has been one important interlude: one in which I had three children of my own, spent some years at home when they were small, and then helped to establish a special program at the Council of Better Business Bureaus for nongovernmental regulation of advertising directed to children.

During my three years as director of that program I became interested in nutrition issues, both as they relate to children ad to adults. And during those years I came to believe that in designing policy and in specifying policy-relevant research, some valuable lessons can be learned from the accumulated experience of advertising and marketing people and their collective wisdom.

I am not only suggesting that advertising techniques might be put to work to communicate nutrition messages–in much the same way that "Sesame Street" borrowed the style of commercials to teach letters and numbers to children who were once thought too young to learn these cognitive skills. I am convinced that advertising techniques can be used in this way; indeed, to some extent they are now being used in some public service announcements. I am also confident that advertising know-how might be adapted to make some program-length nutrition segments particularly interesting and effective, again taking a lead from "Sesame Street."

I am also suggesting that there are other lessons to be learned from advertising and marketing, lessons that may be of particular value in the development of what might be called a nutrition communications policy for children and their parents, indeed for Americans of all ages. In offering these thoughts and observations I am of course speaking in my own behalf; these are my thoughts rather than a policy statement of the Compton agency.

From advertising and marketing experience I think we could learn valuable lessons.

Investigate first the realities of consumer behavior. "Hunches" and "suppositions" about how consumers are behaving are often proved wrong by the facts. Advertising can teach us to get the facts about consumer behavior before we develop strategy or policy. Even factually based advertising campaigns fail; basing campaigns on "hunch" alone is an even higher risk.

Understand first what consumers already know and believe about a given topic or product and how they are likely to think about it. Interviews with children about what they now know, for example, about the role of different foods in the diet, would indicate which lessons have already been learned and which remain to be taught.

Understand how food habits are developed and influenced by home, school, peers, society, and the media. Only a thorough knowledge of this process will prevent us from wasting our efforts. It's conceivable, for example, that parents need information and education more than children do. If the goal is to influence adult behaviors over a lifetime of eating, it may be that messages should be directed not to younger children but to teen-aged children. Ethnicity might need to be taken into account. Some nutritional counseling work has established the need to reinforce the good food habits of a particular ethnic group. Only by understanding the process of food selection and preparation can we influence it favorably.

Understand the differing information needs and comprehension levels of different age groups. Undoubtedly some good messages could be designed for preschool children. These might differ from messages directed to older children, teens, and parents. Both the content of messages and their scheduling should be guided by age-related viewing patterns.

Understand the limitations of advertising, what it can't do as well as what it can. Some of the popular assumptions made by nonadvertising people need to be challenged if effective campaigns are to be designed. It's wrong to overestimate what can be conveyed in disclosures, what burden can be carried within product advertising, how much information can be packed into short messages, the importance of heavy media weight.

Identify specific goals that are realistic and within reach. A campaign might be designed to encourage consumers to eat less meat and more of other forms of protein. Parents and children might be encouraged to consume different kinds of snacks or to snack less. Perhaps the goal could be as general as making nutritional literacy the "in" thing. But in any event the goal should be clear and the strategy specified before creative efforts begin.

Differentiate advertising goals from educational goals. It's important to remember that while advertising can "whet the appetite" for more nutrition information, it can never do the whole job of nutrition education. Educators can and should provide a deeper understanding of this complex subject than advertisements can ever do; but efforts can be made to coordinate educational efforts with advertising efforts and make them enhance one another. Advertising might suggest that nutritional awareness and self-discipline are desirable, healthy, and attractive. This heightened awareness may contribute to more successful learning.

Demonstrate the desired behavior and its benefits. Nutritionists often shrink from the "oversimplification" demanded by advertising communications. Most good ads make one simple point and make it very effectively. It's better to show consumers what to do than to lecture them about what not to do.

The rewards of good behavior should be demonstrated. "What's in it for me?" is the most important question any advertisement has to anwer. Nutritional advertising is no exception. And emphasizing the "do's" rather than the "dont's" is far more likely to shift nutritional behavior as well as nutritional knowledge.

The policy issues surrounding the question of children and nutrition information are difficult, and there is no consensus either on the nutritional status of American children or on their need for nutrition information and education, let alone the specific needs that ought to be addressed first. As a result, those who are hopeful of using the media for education about nutrition sometimes become discouraged. But I continue to take heart from the success of small-scale efforts by individual groups. As an example of one successful piece of nutrition communications to a child, I'd like to offer the following excerpt from a conversation I had recently with my own 8½-year-old son:

Mother: I am interviewing Henry Francis Griffin. And how old are you, Henry?

Henry: Eight and a half.

Mother: Eight and a half. Okay. I'm going to ask you a couple of questions about nutrition.

Henry: Okay.

Mother: What is nutrition?

Henry: Well, nutrition is like, all about, like, good foods that you should eat to keep you healthy.

Mother: Okay. Now what is good nutrition?

Henry: Well, eating things like fruits and vegetables, meat and milk and cheese, like those kind of dairy products, and let's see now. Oh, no, I said that. Those are all I can think of now, but I think there are a couple of others.

Mother: Okay. Now what is bad nutrition?

Henry: Bad nutrition is eating candy bars and soda pop and all that stuff. No, it's okay to eat them once in a while, but you can't live on them.

Mother: I see. Now where did you get these ideas about good nutrition and bad nutrition?

Henry: Well, I got them on this commercial, like.

Mother: You got them from a commercial?

Henry: I got the part about you can eat some, not all, on a commercial about, like, another commercial with these puppets.

Mother: Mmm-hmmm. Puppets.

Henry: And another one—

Mother: Well, now wait a minute. Let's stop on that commercial with the puppets. Tell me about the commercial with the puppets.

Henry: Well, two animals are going down a hot lunch line. And they just finished, and the ostrich eats all good food. And the abominable snowman eats nothing but bad nutrition.

Mother: Oh, So then what.

Henry: And the snowman says, "Boy, this is gonna be good, ice cream, soda pop, candied apples and stuff like that." Then the ostrich says, "You shouldn't eat all that, Abominable. It's okay to eat sweets once in a while, but you can't live on them." He says, "Okay, I'll go back and get another. "And then . . . . Well, then the ostrich says, "I'm having, well, good foods like orange juice and this lovely apple." And then the abominable snowman comes back with his lunch on his head. And the ostrich says, "What are you doing that for, Abominable?" And Abominable says, "I'm balancing my meal!"

What Henry didn't say, and what I think is very important to this discussion, is that the "commercial" that he described is a public service announcement produced by the Council of Better Business Bureaus and aired on all three networks over the last several years. My son's recall and playback of that message encourages me to believe that public service announcements, even those that are modestly produced and aired in times already available, can do quite a bit to further children's nutritional literacy. But I should hasten to add that one of the lessons I've learned from advertising is not to generalize to a national population from a sample of one child, who happens to be my own. (I do want to remind you, however, that even the great Piaget based much of his work on observations of his own children!)

But the one thought I can quite confidently take away from informal conversation with my own children is that some valuable nutrition messages are already being communicated, by such pilot efforts ad the CBBB's "Junior Consumer Tips" and other such messages. This suggests that even more can be done. It also suggests that we should keep our lines of communication with children wide open.

With some lessons from "Sesame Street" (like collaboration between educators and creators of good television) and some from Madison Avenue,

I believe that children could learn a good deal about good foods to eat. Starting small, but enthusiastically, may be the very best way to go about it.

I think it only right to add that most of the lessons I've learned from advertising I've learned at Compton, an agency whose wisdom is daily put to work for a good many major consumer products, and in behalf of a number of worthwhile causes, among them the campaign to inform consumers about the American Economic System, the Eye Bank for Sight Restoration, the Business Council on the Arts, and Religion in American Life. It's my hope that the talents of my industry, and perhaps even of my agency, may be useful in conveying sound nutritional knowledge to children and to the public in general.

12 What Are the Relative Contributions of Family and Television to a Child's Food Preference?

Norge W. Jerome and
Donna J. Frese

Although there is incontrovertible evidence that children's food choices can be influenced by television advertising, we lack information on whether or how this development has altered the family's traditional role as transmitter of dietary practices and nutrition ideology. It is generally recognized that children's food habits are shaped by the nutritional ideals and dietary styles of the immediate family. The family functions as a micro culture by filtering from the larger culture the nutritional ideals and diet patterns that it deems appropriate for the nutritional health and gustatory satisfaction of its members. Within this context the young child cultivates the food tastes and diet patterns of the family and learns strategies of food selection, procurement, preparation, distribution, and exchange. The child also learns social rules governing food etiquette and eating schedules, meal eating and snacking, helping and serving sizes, and the full range of rules and roles associated with food and deemed important by the family.

This basic concept of the family's role in influencing children's food habits was confirmed in a study conducted by Emmons and Hayes [1]. The authors reported that dietary customs and nutritional knowledge of mothers and their children were more often in accord in areas pertaining to habit, customs, likes, and dislikes than for "nutritional reasons." Litman, Cooney, and Stief [2] also confirmed the role of the family in influencing children's food behavior. Their study substantiated the salience of the family in the sanctioning of food habits and behavior.

By observing the food restrictions of parents and their children it is often possible to identify the origin of specific dietary patterns and preferences within families. In general, foods disliked by parents are also disliked by their children [3-6]. Fathers, especially, have been found to influence their children's dietary preferences through "dietary omission": food items disliked by fathers were usually omitted from their diets. Children often followed their example [7,8].

The transmission of food ways and diet patterns are far from unidirectional, however. It follows multiple pathways from the larger external macro culture to the family as a micro culture, from the older to the younger generation, and from young members of the family to adults. Developments in the macro culture are first filtered through the individuals who make up the corporate structure of the family before they become established as a family diet pattern.

One element of the macro culture, commercial television, is said to have assumed a "new" role in formulating the dietary practices of viewers. Jerome states that the child's developing sense of what the culture deems fit to eat is also influenced by the foods and beverages he or she sees in television commercials [9]. Gussow sees a direct connection between the types of food presented on the television screen and those that young children perceive as being desirable and appropriate to consume [10, 11]. Although these observations are true, they do not take into account the family's contribution in either mitigating or reinforcing those externally derived values and concepts. Recently, however, some investigators have taken this factor into account by examining interactions between children's responses to television advertising and the influence of social variables such as parent-child interactions and by viewing behaviors of children and care givers [12-15]. However, these studies still do not take into full account the family's contribution in reinforcing or mitigating the effects of television food advertisements on the child.

Since the diet patterns, nutritional ideals, and food habits of the contemporary American child are shaped by multiple forces transmitted by the macro culture, including television food advertising, it is important to know how these forces converge with the micro culture of the family to influence the diet patterns of individual family members. More specifically, the investigator should determine whether certain dietary practices of young children and their families represent the melding of forces from the macro and micro cultures. A study was designed to address this issue.

The Study

As part of a larger study on the influence of television food advertising on the fruit and cereal selection and consumption patterns of preschoolers, we determined family-child preference profiles for these products. Between February and April 1977, 778 families who had a three-, four-, or five-year-old child enrolled in one of twenty-two participating preschools or day-care centers in metropolitan Kansas City participated in this study. These children, boys and girls, were from a variety of racial and socioeconomic backgrounds. Parents were asked to respond to the following questions on the fresh fruit and ready-to-eat (RTE) cereal consumption practices of the family and the child:

> Which *ready-to-eat* cereals does the family [all members of the household, including parents and siblings] eat most often? (List all cereals.)
>
> Which *ready-to-eat* cereals does your child [the preschooler participating in the study] eat most often? (List all cereals.)

Which are his/her *favorite ready-to-eat* cereals? (List all cereals.)

Which *fresh* fruits does the family eat most often? (List all *fresh* fruits.)

Which *fresh* fruits does your child eat most often? (List all *fresh* fruits.)

Which are his/her *favorite fresh* fruits? (List all *fresh* fruits.)

These questions sought to determine similarities and differences between the fruit and cereal consumption patterns of the children and their families. This objective was not disclosed to the parents. The secondary goal of asking these questions was to ensure that the fruits and cereals used in the larger study were popular with the children.

Fresh fruits and ready-to-eat cereals were selected for two reasons: (1) they are liked by both children and adults; that is, they exemplify core foods of the society, and (2) one, RTE cereals, is highly advertised to children by television while the other, fresh fruits, is rarely advertised [16, 17]. We hypothesized that selection patterns showing strong similarities between children and their families would indicate the strength of family influences, while differences in selection patterns would indicate either the influence of an external force such as television advertising or the weakness of family influences.

Results

Cereal Consumption Patterns

The study reports data given by 778 parents. To ensure that answers to these questions applied to parents as well as siblings, the questionnaires of families with an only child (the child participating in the study) were closely examined. This examination indicated that these parents do eat much the same cereals as their preschool child. Eighty-nine different RTE cereal products were listed for the children and their families. Those eaten most often by at least 5 percent of either the children or their families are listed in table 12-1. Consumption patterns of the two groups are compared.

Sixteen products are shown in the child column and twenty-two in the family column, indicating some differences in child-family consumption patterns. With the exception of two products, the proportion of children selecting RTE cereals is less than that of families. Children also select from a more limited range of RTE cereal products than do families. In general, the six products shown in the family but not the child column are not presweetened items. Apparently these items are conceptualized and culturally defined as cereals for adults but not for children. Table 12-1 also shows that products 5-16 shown in the child column are positioned differently in the family column, illustrating another difference in child-family selection and consumption patterns.

Table 12-1
Consumption of RTE Cereals by Children and Their Families

	RTE Cereals Eaten Most Often			
	Child (n = 778)		*Family (n = 778)*	
RTE Cereal	*Number*	*Percent*	*Number*	*Percent*
1. Rice Krispies®	333	42.8	368	47.3
2. Cheerios®	286	36.8	346	44.5
3. Raisin Bran	198	25.4	270	34.7
4. Corn Flakes	130	16.7	189	24.3
5. Froot Loops®	130	16.7	132	17.0
6. Sugar Frosted Flakes®	128	16.5	154	19.8
7. Cap'n Crunch®	78	10.0	89	11.4
8. Honeycomb®	68	8.7	80	10.3
9. Alpha Bits®	66	8.5	71	9.1
10. Super Sugar Crisp®	66	8.5	87	11.2
11. Apple Jacks®	54	6.9	57	7.3
12. Trix®	54	6.9	54	6.9
13. Sugar Smacks®	41	5.3	35	–
14. Grape-nuts®	40	5.1	113	14.5
15. Life®	37	5.0	60	7.7
16. Sugar Corn Pops®	37	5.0	60	7.7
17. Wheaties®	34	–	58	7.5
18. Rice Chex®	31	–	44	5.7
19. Shredded Wheat®	29	–	84	10.8
20. Granola	23	–	40	5.1
21. Special K®	18	–	42	5.4
22. Wheat Chex®	16	–	44	5.7

Note: Dashes indicate less than 5.0 percent.

There is one major similarity in child-family consumption patterns. Four cereal products appear in the same order (first through fourth place) in both the child and the family columns. Presumably these products are rated similarly by the two groups. It appears that forces within the family setting (similar taste preferences, family schedules) engender these similar patterns of consumption. The four products are probably conceptualized as cereals for all family members.

Table 12-2 shows a list of the child's favorite cereals; the same criterion of a 5 percent response was applied to develop this table. The sequence of favorite RTE cereals differs in some respects from that shown in the child column in table 12-1. Comparison of tables 12-1 and 12-2 shows that children do not always eat their favorite cereals "most often." Some guidance is provided by the family. For example, three favorite presweetened cereals, items 4, 5, and 6 in table 12-2 are eaten as often or less than a cereal that is not presweetened (item 4 in table 12-1). However, since the child's favorite cereals

Table 12-2
Child's Favorite Cereals (*n = 778*)

RTE Cereal	*Number*	*Percent*
1. Rice Krispies®	247	31.7
2. Cheerios®	196	25.2
3. Raisin Bran	140	18.0
4. Froot Loops®	121	15.6
5. Sugar Frosted Flakes	92	11.8
6. Cap'n Crunch®	73	9.4
7. Corn Flakes	63	8.1
8. Honeycomb®	62	8.0
9. Alpha Bits®	59	7.6
10. Super Sugar Crisp®	53	6.8
11. Apple Jacks®	44	5.7
12. Trix®	42	5.4

also appear in the "most often eaten" product list, the preferences that they demonstrate obviously influence the purchasing decisions of their parents.

Table 12-3 shows the sugar content of the RTE cereals eaten by more than 5 percent of the children or their families. Cereals 1-4, which are conceptualized as cereals for all family members, have approximately the same sugar content as those conceptualized as cereals for adults but not for children. Apparently sugar level is not associated with preference for RTE cereals by either children or their families. On the other hand, comparison of tables 12-2 and 12-3 shows that the sugar content of the children's favorite cereals is much higher than that of cereals for all family members and the cereals for adults but not for children. As viewed by the parent-respondent, then, the child's favorite RTE cereals are relatively higher in sugar content than those eaten most often or those regarded as cereals for the family, thus providing further evidence that family forces continue to shape the diet of the young child in contemporary American society.

In general, the similarities and differences shown in the RTE cereal consumption patterns of children and their families indicate that influencing forces are present both within and outside the family setting. Preschoolers participating in this study are allowed to eat their favorite RTE cereals; however, a majority of their favorite cereals are also shared by the family as a whole. Apparently it is at this point that the two sets of forces from the macro culture of society and the micro culture of the family converge to influence the consumption patterns of family members. Perhaps, too, this juncture depicts the multidirec-

Table 12-3
Sugar Content of RTE Cereals Eaten by Children and Their Families

RTE Cereal	*Total Sugars in Grams per Ounce of Cereal Serving*
1. Rice Krispies®	3
2. Cheerios®	1
3. Raisin Bran	6-9, depending on brand
4. Corn Flakes	2
5. Froot Loops®	14
6. Sugar Frosted Flakes	11
7. Cap'n Crunch®	12
8. Honeycomb®	11
9. Alpha Bits®	11
10. Super Sugar Crisp®	14
11. Apple Jacks®	16
12. Trix®	10
13. Sugar Smacks®	16
14. Grape-nuts®	3
15. Life®	5
16. Sugar Corn Pops®	13
17. Wheaties®	3
18. Rice Chex®	2
19. Shredded Wheat®	1
20. Granola	6-8, depending on brand
21. Special K®	2
22. Wheat Chex®	2

tional and multigenerational transmission of food likes, preferences, and dislikes within the family. Preferences for particular RTE cereals are apparently established and reinforced in the family setting, although values, information, and experiences on which such preferences are based may originate outside that setting.

Patterns of Consuming Fresh Fruits

Table 12-4 shows the eight fresh fruits eaten most often by the children and the ten eaten most often by their families. A total of twenty-eight fruits were recorded for the children and their families but the 5 percent criterion for inclusion reduced that number to the ten shown in the table.

Table 12-4
Consumption of Fresh Fruits by Children and Their Families

	Fresh Fruits Eaten Most Often			
	Child (n = 778)		*Family (n = 778)*	
Fresh Fruit	*Number*	*Percent*	*Number*	*Percent*
1. Apples	655	84.2	695	89.3
2. Bananas	559	71.9	635	81.6
3. Oranges	550	70.7	669	86.0
4. Grapes	199	25.6	259	33.3
5. Grapefruit	96	12.3	193	24.8
6. Pears	91	11.7	144	18.5
7. Strawberries	77	9.9	143	18.4
8. Peaches	75	9.6	169	21.7
9. Melons (cantaloupe)	22	–	41	5.3
10. Plums	16	–	44	5.7

Note: Dashes indicate less than 5.0 percent.

All the fresh fruits shown in table 12-4 were selected by a greater proportion of families than children. Children select from a slightly more limited range of fruits than do families. Melons and plums apparently fall in the "adult fruit" category. Despite these differences the consumption patterns of fresh fruits by children and families are quite similar.

Of the eight fruits eaten most often by both children and families, apples, bananas, and oranges are highly favored by both groups, but peaches are eaten more often by families than by children. A relatively higher percentage of families than children eat oranges and grapefruit, the only two citrus fruits listed. This finding is consistent with that of Lamb and Ling [18], whose study indicated that preschool children liked noncitrus fruits somewhat better than citrus fruits.

Children's favorite fresh fruits are shown in table 12-5. A comparison of tables 12-4 and 12-5 shows that children's favorite fruits are those eaten most often in sequential order. Strawberries, the only item out of position in table 12-4, have probably been reordered to accommodate seasonal availability of the product.

In general, children and their families show similar patterns in their consumption of fresh fruits. Preference for fresh fruits is established and reinforced in the family setting although some products appear to have age-status connotations.

Discussion

The question posed in this study was, Could forces from the macro culture (societal forces, including television advertising) and the mini culture of the

Table 12-5
Child's Favorite Fruits (*n = 778*)

Fresh Fruit	*Number*	*Percent*
1. Apples	498	64.0
2. Bananas	438	56.3
3. Oranges	402	51.7
4. Grapes	186	23.9
5. Strawberries	92	11.8
6. Grapefruit	64	8.2
7. Pears	57	7.3
8. Peaches	56	7.2

family manifest themselves in the dietary practices of young children and their families? Comparing child-family preference profiles for two popular food categories that receive differential treatment and attention in television advertising makes it possible to assess whether consumption patterns of products in target food categories are associated with the family or with some external, societal force.

Unquestionably, as the data for RTE cereals indicate, children and their families select substantially more of those food items that are advertised on television. An external or societal force therefore exerts a strong influence by increasing an individual or group's awareness of new and available products. However, the family's influence mediates between the societal forces and those under its direct control by selecting and frequently consuming those items preferred by its members.

In general, for both fresh fruits and RTE cereals, family preferences parallel child preferences except for items that apparently are either not liked by young children or have adult-status connotations. This finding upholds the traditional role of the family in transmitting its food habits and dietary practices to its young members. It appears that while many sources of dietary information are new and in some cases pervasive, families continue their role of defining the parameters of appropriateness for their young members. The combined influencing forces operate at the dietary level and can be observed by comparing child-family dietary practices.

The Federal Trade Commission *Staff Report on Television Advertising to Children* [19] uses extensive empirical research as a basis of its rule-making inquiry. The report makes its case on the impact of television advertising on children by citing a number of studies that deal with the impact of television advertising on many aspects of children's attitudes and behaviors. Reference is made to the negative impact of television advertising on parent-child relationships, particularly as they are affected by children's purchase demands. The *Staff Report* cites results of studies that indicate that children not only are induced to demand products advertised on television but also play an active role in the selection of foods purchased by their mothers. This sometimes leads to conflict between children and their parents [20, 21]. None of the studies deals with methods utilized by parents in resolving these conflicts, nor do they discuss what strategies parents develop and utilize in order to prevent parent-child conflicts from children's purchase demands. These issues must be addressed.

Our study provides evidence that while parents are likely to be influenced by their preschoolers' stated preferences for products advertised on television, they exercise control by selecting and serving products that they consider appropriate. Our study therefore does not support the charge of parental neglect of child-rearing responsibilities. This study and the FTC *Staff Report* point out the need to learn more about the methods, skills, and strategies used by parents in rearing their children and in communicating sound dietary and nutritional practices in contemporary American society.

Summary and Conclusions

Questionnaires completed by 778 parents were used to compare the consumption patterns of fresh fruits and RTE cereals of young children and their families. The results indicate that consumption patterns of these foods by the two groups are similar in many respects. Young children and their families tend to select and consume similar products and to position many products in the same order. This suggests that the contemporary American family exerts a strong influence on the food habits of young children.

The data also show that there are some differences in the consumption patterns of young children and their families. The latter group tends to select from a wider variety of products and to select most products in greater proportion to children. Variations in consumption patterns may be attributed to genuine differences in taste preferences and to differential concepts and images of particular foods.

RTE cereal selections included three times as many products as fruit selections. This difference could be attributed to factors in the external environment, for example, marketing strategies such as television food advertising, which can increase an individual's awareness of the availability of certain products.

The study indicates the need to determine how and where forces in the larger society converge with those of the family to effect change in dietary patterns. The study also points out the need to identify and quantify the relative contributions of societal and familial factors that influence the establishment, persistence, and alterations of the food habits and dietary patterns of young children.

References

1. Emmons, L., and Hayes, M. Nutrition knowledge of mothers and children. *J. Nutr. Ed.* 5 (1973): 134.

2. Litman, T.J., Cooney, J.P., and Stief, R. The views of Minnesota school children on food. *J. Am. Dietet. Assn.* 45 (1964): 433.

3. Brechenridge, M.E. Food attitudes of 5- to 12-year-old children. *J. Am. Dietet. Assn.* 35 (1959): 704.

4. Metheny, N.Y., Hunt, F.E., Patton, M.B., and Heye, H. The diets of preschool children. II: Factors in food acceptance. *J. Home Econ.* 54 (1962): 303.

5. McCarthy, D. Children's feeding problems in relation to food aversions in the family. *Child Develop.* 6 (1935): 277.

6. Hellersberg, E.F. Food habits of adolescents in relation to family, training, and present adjustment. *Am. J. Orthopsychiatry* 16 (1946): 34.

7. Bryan, M.S., and Lowenberg, M.E. The father's influence on young children's food preferences. *J. Am. Dietet. Assn.* 34 (1958): 30.

8. Eppright, E.S., et al. Eating behavior of preschool children. *J. Nutr. Ed.* 1 (1969): 16.

9. Jerome, N.W. Children's television advertising: A vehicle for pro-social nutritional messages. In *1975 Combined Proceedings of Marketing in Turbulent Times and Marketing: The Challenges and the Opportunities,* ed. E.M. Mazze. New York: American Marketing Association, 1975, pp. 511-513.

10. Gussow, J. Counternutritional messages of TV ads aimed at children. *J. Nutr. Ed.* 4 (1972): 48.

11. Gussow, J. Using and abusing the mass media. *J. Canadian Dietet. Assn.* 38 (1977).

12. Clancy-Hepburn, K., Hickey, A.A., and Nevill, G. Children's behavior responses to TV food advertisements. *J. Nutr. Ed.* 6 (1974): 93.

13. Ward, S., and Wackman, D.B. Television advertising and intrafamily influence: Children's purchase influence attempts and parental yielding. In *Television and Social Behavior,* vol. 4. Washington, D.C.: U.S. Dept. of Health, Education and Welfare, 1972.

14. Prasad, V.K., Rao, T.R., and Sheikh, A.A. Can people affect television?/Mother vs. commercial. *J. Communication* 28 (1978): 91.

15. Liefeld, J.P., and Norsworthy, F. Preliminary research report: The influence of television advertising on children compared to peer and parental influence. Dept. of Consumer Studies, University of Guelph, Ontario, Canada, n.d.

16. *Advertising Age*, March 8, 1976.

17. Choate, R.B. Testimony before the Federal Trade Commission, Fall 1976.

18. Lamb, M.W., and Ling, B.C. An analysis of food consumption and preferences of nursery school children. *Child Develop.* 17 (1946): 187.

19. Federal Trade Commission. *Staff Report on Television Advertising to Children,* Washington, D.C., February 1978.

20. Reilly Group. The gene, meals and snacking: The child and what he eats. *Child* 2 (1973).

21. Atkin, C. Effects of television advertising on children. Michigan State University, June 1975.

13 To Share the Air

Richard K. Manoff

It was a hot August day in 1922 and the builder of the new Hawthorne Court cooperative apartments in Jackson Heights, Queens, hadn't seen a sales prospect all day. Despite boldly lettered signs around the property extolling the delights of a life at Hawthorne, the builder was undergoing hallucinations of invisibility. The high-pitched monotone of a human voice from the crystal radio was the only sound in the office.

While it rarely happens, the quiet of ordinary moments can produce a historic convergence of circumstance. The heat of that day, the beseeching apartments of Hawthorne, the want of people, the baffled builder, the rattle of the radio all came together in an explosive awakening.

The next day he was at the offices of WEAF in New York. Would they sell him the time for an advertisement for his apartments? Torn between an offer of hard cash for a cold commercial announcement and the fear of breaking with precedent, WEAF turned him down. But the lure of the market was too strong. An ingenious euphemism saved the day. They would sell him a ten-minute segment for $50 to deliver a lecture in memory of Nathaniel Hawthorne and to talk about the joys of owning an apartment at Hawthorne Court, named in memory of the famous author.

Fred Friendly refers to this incident as "the birth of the commercial" and its happening as "almost an accident"[1]. The man, the time, and the place may have been an accident, but the strategic event was inevitable, the logical conclusion to a laissez faire free-market-system policy in the operation of the public media.

The advent of the electronic marketplace was no accident. The market system needed it to expand. The real estate operator might have been a purveyor of canned peas or of used cars. Regardless of who he might have been, there was no doubt what he would be up to. He had to be a merchant because, given the nature of the broadcasting system, the most insistent demand for its facilities had to come from the commercial advertiser who needed larger audiences and could afford to pay for the privilege. No other suitor would prove as attractive to the licensee. The rest, as they say, is history—the evolution of the network mechanism to supply the major portion of media content and the proliferation of television advertising at an annual rate of $7 billion.

No one could have anticipated this commercial potential and the benefits it would produce for the marketplace. Nor could more than a few have foreseen the problems. With demand rising on all sides that we reclaim the "vast wasteland" of television, we have only recently begun to deal with them. So far we have harvested a cornucopia of enchanting and varied regulatory solutions, but for the most part they may provide about as much nourishment as a bowl of plastic fruit.

The difficulty is not so much with the solutions as with our failure to ask the right question. The real estate dealer from Jackson Heights was in fact raising the issue for all to consider. But nobody was listening, not to him, nor to WEAF as it oscillated between cash and commitment.

The foot-in-the-door development, which led eventually to explicit advertising arrangements, was a decision at the crossroads of social policy. We went the way of the market without setting limits.

The fundamental question is one of social policy: to what extent should ownership and operation of the airwaves be left to a free-market system as it is today, and to what extent is that system failing to realize our hopes and intentions for it?

All our coping since then, all our solutions, have been reactive regulatory palliatives for the symptoms. But the symptom is not the trouble; it is the sign of trouble. What television does not need is another dose of analgesia to make it appear that our troubles have vanished. Only change, not regulation alone, can resolve the conflict between the intent of policy and the structure and system of its execution. Without such systemic change, the issues of advertising, children's advertising, the presence of sound nutritional information either inside or outside television advertising, violence in programs, indeed, the whole question of program, will continue to bedevil us no matter how many regulatory palliatives are prescribed.

We must therefore discuss the question of nutritional information in the light of regulatory actions and changes necessary to articulate policy. Our policy thrust has been clear from the beginning: "in the public interest." Only the definition and standards for implementation remain vague.

We have to consider the need for change in addition to regulation. Regulation presumes the workability of a system. It seeks to keep that system effective until the need to replace or modify it becomes evident. It also provides a "holding action" until the change is in place.

But the need for new initiatives in social policy is usually indicated long before we care to acknowledge it. Instead, we continue piling rule on regulation and regulation on interpretation in a dizzying maze in which neither the regulated nor the regulator can clearly find the way. At its best, regulation is a deterrent, a containment. Yet it frequently fails, often with disastrous side effects. Hardly a regulation exists that has not opened new loopholes while closing old ones. Thus regulation can create problems. Regulatory response to a pressing social need depends on existing conditions. But while social conditions change, human institutions (with their rules and regulations) have a tendency to become rigid and to grow unresponsive to emerging needs. These baroque bureaucracies with their rubrics of rules and regulations, originally dedicated to the solution of a problem, can in time become the problem itself. The regulator becomes the recreant; the ruler, the unruly. It can happen.

I know of a food marketer who, responsive to the rising clamor against sugar, was able successfully to reformulate his grape jelly with one-third less sugar and more fruit. It took years and a sizable investment. He cannot call it grape jelly. The regulation, a standard of identity for grape jelly, requires a minimum amount of sugar. The new product was below this minimum. So while several government agencies with their regulations are demanding less sugar in foods, Food and Drug Administration regulation is insisting that it may not be done unless called something else. But a new unfamiliar product designation confuses the consumer, demeans the product, and saddles the marketer with a Herculean sales task.

Nor can regulation make up for ineptness in social policy. Example: contrast the surgeon general's ruling against cigarette smoking, his identification of it as one of the major health-risk factors, with the Department of Agriculture's continuing subsidies to tobacco farmers. No amount of antismoking regulation can rectify bad policy.

In other cases, regulation becomes a mask for ambiguity in the implementation of social policy to give the impression that a policy is in effect, when, in fact, the system itself makes a policy virtually impossible. Our food stamp programs have been prime examples of this regulatory perversion of policy.

The Federal Communications Commission (FCC) is another example. One consensus description of FCC's legislative mandate is that it "was deliberately left ambiguous." FCC commissioners "were at once critics and accomplices of the new broadcast industry; they were regulators, but they could not be censors; the public tended to charge them with responsibility for the quality of a universal public service, and yet the Commission was denied the rate-making and standard-setting authority applied to public utilities by other regulatory agencies."[2]

Regulators have also been known to develop a proprietary interest in failed systems that tends to prolong a problem long after its antipathy to social policy is well known and a new initiative is required to reunite program and policy.

These inherent shortcomings of the regulatory process over time make continuing review of our systems essential. No more appropriate situation for review exists than the system under which the public media operate. We need to identify in the current system the sources of policy abuse, either to remove these abuses from the ramshackle structure we permitted lawlessly to rise and then hastened to patch and prop up or to replace it with one that conforms with policy.

Someday, even these will need to be discarded. The process will have to be repeated. Call it, if you will, the molting of human institutions, the casting off of a rigid outer skin to permit internal structural change and the regeneration of a new one.

This periodic molting of the media has been long delayed. Few areas of our national life are ruled by institutions as antediluvian as the Communications Act of 1934 with its FCC and layers of ineffectual rules and regulations. No one should be fooled by this. The emperor may think he is bedecked in a new haberdashery, but isn't there someone among us still innocent enough to notice his nakedness—or, at least, modestly, to raise the question?

This is no diatribe against regulation. On the contrary, it is a reaffirmation of regulation's real function and an acknowledgment of its essential attachment to status quo systems and its ironic potential, therefore, for frustrating the aims of policy. A good regulation can be a bad system's most ardent defense and social policy's worst enemy.

In its famous "Blue Book" entitled "Public Service Responsibility of Broadcast Licensees" (March 7, 1947), the FCC asserted its new policy of carefully reviewing stations' on-air performance when licenses come up for renewal. This had become necessary, the FCC argued, because stations were not fulfilling programming promises made with original license applications. The FCC has failed to fulfill its own promise; as one television observer has written:

> In some ways the whole history of the FCC since the publication of the Blue Book can be seen as a gradual withdrawal from interventionism on so comprehensive a scale. A 1946 position was rapidly seen to be unworkable; the sheer profusion of stations, the demands of advertisers, the diversity of control, made it impossible for the FCC to conduct an investigatory operation on a scale which could have made stations responsive. Despite sudden eruptions of interventionism during the FCC Chairmanship of Newton Minow, and earlier during that of Clifford Durr, the FCC has necessarily sunk back into a more purely regulatory role, renewing licenses in geographical grouping and issuing rules governing the conduct of business at network level or between networks and stations. The FCC has legislated on the question of fairness, on the rights of reply, and on ethical questions in advertising. It has tried to force more local programs to be made. It has pushed stations and networks into providing more coverage of news, and of better quality. But whenever it summons up the will to take far-reaching action, it is defeated either by its adversaries or by the gradual weakening of its will.[3]

Public criticism—Newton Minor's 1961 "wasteland" speech, for example—and exhortation have proved to be feeble instruments of change. Yet as Anthony Smith says:

> The system whose heart is the FCC can approach its audience only along such lines. To attack the broadcasters for exercising their full energies within the system provided is merely to perplex them with inexplicable enmity. It is extremely difficult to try to change the nature of a competitive system by simply demanding self-denial or forbearance.[4]

The malaise of the system is not in its working but in its works, which have been designed entirely by free-market forces. Government's main responsibility under the law is to police the spectrum and regulate the traffic to assure the freedom of the market above all else. This free-market system has become the de facto communications policy of the United States. No amount of regulatory intervention will change that. Given the system, we have no right to expect greater social accountability of it than its design provides. What we might have expected has come to pass: extraordinary expansion of the consumer marketplace through television advertising and the fortune of Croesus for its network and station licensees.

The 1934 Communications Act was a missed opportunity to frame a policy that would have given it more control to structure and run the system "in the public interest" and to moderate the force of the market. By leaving such matters vague and undecided, the act was an open invitation to the market to set the policy by its practice.

No other country let the free market set policy. Maybe we have allowed this to happen because we were the first and lacked a precedent, a basis to anticipate that eventually the power of the media would rival the power of government. Whatever the reason, a policy of primary accountability to the market and to profits is the seed that has produced the bitter crop of issues in advertising, programs, nutrition, and the like.

The law, from concerns of free speech and censorship, makes sure that the FCC has no positive powers, no authority for stipulating standards of licensee conduct, no right to impose structure and form or to decide, for example, whether to license all or only part of the broadcast day. Yet, these are not free speech concerns.

To all other Western countries, if not to all countries of the world, it was obvious that the public media were "more than instruments of *expression* [emphasis added], that they would become concentrations of social power, which rendered the process of regulation (which had to be undertaken by the state) inseparable from the duty of imposing forms and structures of social accountability."[5] The problem for them was how to accomplish this within the framework of a democratic system. But in the United States we never got that far. It was not until recently that we began to raise the question whether we should even try.

Television's "market" accountability is to its customers who are not to be confused with the accountability to its *audience*. The system depends on the advertiser, not the listener. The real *measure* of accountability, therefore, is the rating, which answers the question, How many? but not, What do they want? When program ratings fall too low and advertisers lose interest, the program is cancelled. Only the most naive could call such a system responsive to audience desires. Nicholas Johnson, as FCC commissioner, cautioned us not to "confuse the public interest with what can be made to interest the public.

Les Brown, television editor of the *New York Times* and of *Variety* before that, has written:

> This emphasis on the popularity of shows has made television appear to be democratic in its principles of program selection. In truth, programs of great popularity go off the air without regard for viewers' bereavement or because the people it reaches are not attractive to advertisers.[6]

None of this should amaze us because it is consistent behavior for a free-market system. The licensees, the networks, the advertisers cannot be expected to ignore the dynamics of the system out of sheer social philanthropy. Market systems do not work that way.

Given the way they do work, it was inevitable that television would become almost exclusively an entertainment vehicle. Even its public interest obligation is discharged through news programs that are more entertainment than enlightenment. This led David Brinkley once to remark that "if most of the people in this country are getting most of their news from television, they are getting damned little news."

Now, what is true of program content is also true of the advertising. Market system dynamics predetermined the advent of advertising and has gone on to decide who those advertisers will be, what they will advertise, and how they will be executed.

The astronomic cost of television favors big advertisers. Only broadly distributed and widely consumed products with generous margins between cost and selling price can afford the ongoing television investment. The ultimate justification is the building of a strong proprietary brand name. Many vital categories of food products (fresh fruits and vegetables, fresh meat and poultry) are immediately eliminated because their low profit margin is incapable of generating the funds. By the same token, the opportunity for television exploitation has accelerated the development of products eligible to satisfy television's economic demands. Thus, the frivolity of television program fare is matched by the frivolity of the foods we can afford to advertise there. If the medium is market, then message must make money. It goes for programs. It goes for ads.

Even the style and content of advertising has been shaped by the environment of the medium. While advertising has grown more responsible over the years, it has also become more humorous, less specific, more involved with image (and less with reality) more with mood and manner than with value—more, in short, like the entertainment it seeks to interrupt with minimum pain and maximum pleasure.

What more could we expect? Who is to blame? The licensee, the advertiser, the programmer? The market system is a kind of anonymous dictatorship. We are all its vassals. Though no one is in charge, there are entrenched interests eager to preserve the status quo. But they are not productive targets for attack.

The system, not its operators, is the cause of our complaint. We need to change it so that it reflects more accurately what we once intended for it. Now, effecting change in a system is a fearsome enterprise. Keeping the old system and trying to make it work under a new regulation is the easier route. But in television, we would be deluding ourselves and our constituencies.

The public interest and such public conerns as nutrition education demand a mandatory share of television time and control over its content. There is no other answer. Licensee voluntarism has not worked well and gives no promise of working better in the future. Social responsibility in television has become a philanthropy of the licensee. The result is that nutrition education and other vital concerns are badly neglected. This, as one social critic has noted, has had a deleterious effect on education in general.

> Since television viewing occupies more time of Americans than any other single activity, and since it has taken the place of many of the learning experiences of the young people of an earlier period, its potential for constructive education should be developed and exploited. At present, most commercial television appears to distract young people from educational activities or to result in their learning things that conflict with accepted educational goals.[7]

Television as an extension of the classroom is an eminently preferable to the classroom as an extension of television. Television's entertainment vogue is finding sardonic mimicry in pedagogical style. Teachers confess a need to be more entertaining to engage student interest, to borrow fun and entertainment devices from television that have no integral connection with subject matter. The degrading effect on educational quality and learning is acknowledged, yet we seem helpless to oppose the trend.

Even in its demand on students' time, television preempts the school. The average child between ten and fourteen spends about fifteen hundred hours a year with television compared with about eleven hundred hours in school.[8] Turning off the set is one solution. A proper social policy solution is to control what's on the set. If the child is transferring from school to set, let's move the school with the child to the set.

This means a mandatory share of the air which can only be effected by returning to public ownership through government authority some stipulated portion of the schedule. The 1969 White House Conference on Food, Nutrition, and Health made such a recommendation. It urged that 10 percent of all radio and television time be set aside from commercial broadcasting. The operation of this time bank has been previously described:

> In 1969, at the White House Conference on Food, Nutrition and Health, a recommendation was made for our country that 10% of all radio and television time, in all time periods, be retained for public service.

On November 16, 1971, the *New York Times* reported that 'The National Broadcasting Company has expressed fear that the stability of radio and television is threatened by challenges from citizen groups and has proposed reluctantly that the Federal Government adopt a quantitative standard in reviewing licenses to use the air. (Any licensee of a television station who devotes 10 to 12 percent of his program schedule to non-entertainment material (sic) should be presumed to have reached a performance level entitling him to retain his frequency, NBC said.)'

NBC's recommendation could also apply to the amount of time stations and networks should be obliged to set aside for public service announcements. Were this to be a flat 10% of all the time allocated for commercials, then we can begin to see the tremendous changes that could take place in television's content and its impact on our country.

The special reserve could be a Time Bank for a public service campaign on nutrition education–and others. It could be administered by a special public corporation established by Congress and independent of any political control.

Let us examine the specific example of nutrition education to explore how the Time Bank would work.

Under most enabling legislation there usually is a mandatory provision for information and education activities. This calls for a budgetary appropriation and, budget appropriations being what they are, there is never enough to do the job right. There is precious little for mass media.

But now it could be different. There is time on deposit in the Time Bank. Lots of it and not dependent on some station's largesse or imperiled by some station executive's whimsy.

Congress could insist that the Executive Branch indicate in its budget an expert estimate of how much radio and TV time is required to do a proper job of informing the American public about the nature and availability of food and health service and vital nutrition information and where to go to obtain them.

There could be no justification for pennypinching because no money is actually required. The time is there in reserve. The only question to be answered is how much of it is required to do the job.

Once approved, this Time Bank draft is at the disposal of the responsible agency to use for education.

Think of all the many programs that could now be brought to the people in all their living color. Food stamps, nutrition education, health programs, aid to the aged and so forth.

Certain safeguards would be necessary. Stations would have to be secure against unreasonable losses from such a forced reallocation of such time to public service.

> The Time Bank would have to be free of any political control. Local communities should be assured of some share of the Time Bank's reserve for local public and private campaigns.
>
> The Time Bank would bring to nutrition education the status and the prominence it deserves on the airwaves of our country.
>
> The Time Bank would enable us to bring national wisdom to the American people via the public media without which the task is really impossible.[9]

Last year, the dollar value of television advertising amounted to $7.5 billion or about 20 percent of all national advertising. If 10 percent of this time were available to mandatory public use, it would be worth $750 million. And if nutrition education were to receive 10 percent of this, or $75 million, it would rank among the major food advertisers.

The White House Conference recommendation did not distinguish between program and advertising time but meant to embrace both. They represent radically different management problems. Managing 10 percent of program time is by far the more formidable task because producing programs is obviously a gargantuan undertaking compared with advertising. But it is a worthwhile long-term aim. It would make possible the realization of a distinctive program source in the form of a fourth network within the schedules of the existing three.

This Inner Fourth Network would air on every station and through each network, providing educational programs in health, nutrition, family welfare, and formal education within fixed positions, in all time periods of the schedule. It would not be another Public Broadcasting System which, as an addition to the network system, must unfairly compete for audience. Such head-on competition makes it yet another entertainment contest. Educational programs must either give way or be resigned to minuscule audiences composed of the already motivated for whom such programs are not primarily intended. The inner fourth network would be a part of each commercial network and each station's schedule at identical times. It would occupy noncompetitive blockeds of time.

Public television is not an alternative, for as Herbert Schiller has pointed out "the illusion of informational choice is more persuasive in the United States than anywhere else in the world. The illusion is sustained by a willingness . . . to mistake *abundance of media for diversity of content* [emphasis added]."[10] PBS is an addition to the system, not a change in it. Alternatives must be designed to alter the system and to become a fixed, permanent part of it.

Both the time bank and the inner fourth network are consistent with the intent of our social policy and with restoring that policy to practice. They leave the free-market system inviolate but under well-defined restrictions to protect the public interest. Our hindsight reveals that the original legislation could have accomplished this by partial licensing of the air for commercial use. This would have been consistent with public policy in all other sectors of our national life, such as land use, national seashores, waterways, personal and business incomes, and even manpower in times of threats to national security.

The principle of public claim to a share of our property, our income, our time, our service is deeply imbedded in our national purpose and integral to the policy of free enterprise and its system of a free market. The system could not survive without it. Restricted access to space and time is an integral support of the system and not, as short-sighted exponents of expediency insist, the instrument of its demolition.

Our concept of national security embraces the health and education of our people. We must now learn to accept the indispensability to health and education of the public media. Then we shall have no difficulty in accepting the need for mandatory public reserves of our air as we do of our land and our natural and human resources.

The notion of fixing a time commitment has even occurred to the industry. The figure of 10 percent was actually suggested by NBC in 1971 during one of the recurrent periods of intense public criticism in an apparent attempt to mollify it.[11] Even station licensees talk of the need for more control, but in their case the control they want is of the networks. Donald H. McGannon, long-time head of Westinghouse Broadcasting, an owner and operator of five television station licenses, in petitioning the FCC to take steps designed eventually to restrict networks' domination of television time, declared:

> The networks dominate the television industry. They exercise effective control over most of the time on affiliated stations; they influence over-all advertising rates and practices; they absorb a disproportionate share of revenue and profits and they work a major impact on economic conditions in the industry. The total effect of these actions and practices is inconsistent with the spirit of the antitrust laws which is reflected in the Federal Radio Act of 1927 and the Communications Act of 1934. This petition is prompted neither by a desire to injure the networks nor by a conviction that their role in TV broadcasting should be drastically altered. Rather it is an effort to call attention to the imbalances and inequities existing today between the networks and their affiliated stations.[12]

This position was motivated by both economic and program content concerns. Station licensees would like more money for network time, the right to give up less of it and to have some review and decision over content.

In principle, this is the same struggle that the public interest has waged with licensees and networks. The public interest bewails the total expropriation of television time by network and licensee, deplores the ineffectual philanthropy of public service time, would like to have some greater control over content short of actual censorship, and demands a fair share of the air for its own use. The validity of the petition is not marred by a change in the roles of the petitioners.

The judicial history of the issue makes clear that absolute right either to all the air or to total control over content has not been reserved to the licensee. What is clear is that the law has not made it clear. What is not so clear is whether the law could make it so. What is clear, again, is that no legislation has ever attempted it.

There is solid judicial opinion to justify making the effort. Supreme Court Justice Felix Frankfurter, in the majority opinion that forced NBC to divest itself of one of its two radio networks in 1943, declared:

> The Act itself established that the Commission's powers are not limited to the engineering and technical aspects of regulation of radio communication. Yet, we are asked to regard the Commission as a kind of traffic officer, policing the wavelengths to prevent stations from interfering with each other. But the Act does not restrict the Commission merely to supervision of the traffic. It puts upon the Commission the burden of determining the composition of that traffic.[13]

In 1946 the FCC issued its famous Blue Book, officially entitled "Public Service Responsibility of Broadcast Licensees." Three years later it issued its "Report on Editorializing by Broadcast Licensees," which is generally considered to be the first articulation of the Fairness Doctrine. It directed licensees "to operate in the public interest" through reasonable coverage of important controversial public issues and to provide fair and reasonable opportunity for opposing viewpoints. While principle was reaffirmed, stipulated performance standards as to structure and air share were deliberately avoided. We must assume that political inhibition, not legal prohibition, prompted such evasiveness.

When the Fairness Doctrine was finally established in the famous *Red Lion* landmark case before the Supreme Court in 1969, some of the key language of that decision opened up even greater possibilities for the future. Justice Byron R. White, who wrote the majority opinion, said:

> When there are substantially more individuals who want to broadcast than there are frequencies to allocate, it is idle to posit an unabridgeable First Amendment right to broadcast comparable to the right of every individual to speak, write or publish. . . .
>
> By the same token, as far as the First Amendment is concerned, those who are licensed stand no better than those to whom licenses are refused. A license permits broadcasting, but the licensee has no constitutional right to be the one who holds the license or to monopolize a radio frequency to the exclusion of his fellow citizens. There is nothing in the First Amendment which prevents the government from requiring the licensee to share his frequency with others and to conduct

> himself as a proxy or fiduciary with obligations to present those views and voices which are representative of his community and which would otherwise by necessity be barred from the airwaves.[14]

The opinion also reaffirmed that "it is the right of the viewers and listeners, not the right of the broadcasters which is paramount."

When the tobacco industry invoked the First Amendment against the FCC ruling in the historic *Banzhaf* case before the U.S. Court of Appeals, Justice Bazelon's decision rejected its claim, declaring:

> Some utterances fall outside the pale of First Amendment concern . . . promoting the sale of a product is not ordinarily associated with any of the interests the First Amendment seeks to protect.[15]

Bazelon went on to say that more than First Amendment issues were at stake here. On a key question of "public interest versus public health," he was unequivocal:

> Whatever else it may mean, however, we think the public interest indisputably means the public health . . . the power to protect the public health lies at the heart of the state's political power . . . public health has in effect become a kind of basic law.[16]

The Supreme Court subsequently permitted the Bazelon decision to prevail by refusing to review it.

Regulatory agencies offer a contrast in approach. They propose remedial actions, such as "affirmative nutritional disclosure," and presume these to be the obligation of the advertiser, not the medium. Now regulations and remedies aimed at misleading advertising are one thing, but responsibility for affirmative nutritional disclosures would seem more appropriately assigned to the medium. A regulatory philosophy that holds the renter liable for an owner's responsibility obfuscates the issue and diverts attention from the problem. Advertisers must be held responsible for advertising trespass; the medium, for the medium's.

If the FTC has no authority to deal with the media problem as a media problem, then converting it into an advertising problem over which it has authority is a low form of political forgery. It would be more constructive to refer the problem to the appropriate governmental turf for action.

Certain aspects of the problem are clearly advertising problems. The content of the advertising and the relationship of health-risk factors in our foods to the audience focus of their advertising are proper FTC concerns. But the issue of *access* for nutritional education requires modifications such as the time bank and the inner fourth network to bring the system into conformity with policy. These are not advertiser issues. They are media issues and must be addressed as such.

Reserving parts of the broadcast schedule for public use obviously has profit implications for licensees and networks. They can be expected to enter vigorous objections on the grounds that reducing the amount of time for sale will reduce their income. There is obvious merit to the case. But examination of the extenuating revenue and profit circumstances of the industry leaves little room for sympathy.

Television station and network profits are extraordinary whether measured by return on investment or as a percentage of sales. Smith observes:

> By the middle of the 1950's, the highly profitable nature of the new television medium was already clear. The right to transmit, given away by the FCC, became an extremely valuable fixed asset of the fortunate bidder, who could sell his holdings on the market as if the public resource element of the station were a wholly-owned investment. Stations rocketed in value by ten times the actual amount of cash invested in them. A new television station was transformed artificially and overnight into an extremely large block of capital from a small one.[17]

Nicholas Johnson, a former FCC commissioner, once declared:

> A broadcasting license has, understandably, been characterized as a "license to print money." There is no question about the financial ability of this industry to provide the American people what the national association of broadcasters characterized in the early days as a "considerable proportion of programs devoted to . . . activity's concern with human betterment." Is television providing such a service? Is it serving "the public interests"? Virtually every independent critic has given the industry failing marks. Not only has the industry failed miserably in its great opportunity, and obligation, to contribute to "human betterment," it has actually done great harm.[18]

What are the facts as reported by the FCC in its 1977 annual financial review of the television industry?

1. Total broadcast pretax profits for stations and networks grew from \$492 million in 1966 to \$1.25 billion in 1976, an increase of 150 percent.
2. Pretax profits in 1976 represented approximately a 68 percent return on original cost of investment accumulated to that year but more than a 152 percent on the depreciated investment.
3. Over the five-year period 1971-1976 television networks alone increased their earnings by 192 percent compared with all corporation earnings, which increased only 89 percent.[19]

Few businesses aspire to, let alone achieve, such returns. A reduction in such extraordinary profits would be a hardship only in the light of historic earnings but not on the basis of reasonable profit expectations. No other enterprise licensed to operate a public trust can demonstrate such spectacular profit performance nor, as in the case of public utilities, would it be permitted to do so.

Vigorous objection to proposals such as the Time Bank and the Inner Fourth Network will also cite the limited supply of commercial time and the rising demand for more. This cry of havoc is inappropriate and unbecoming. Licensees have the legal obligation to provide for the public interest. If they are already doing so to a reasonable degree, then the proposals are merely a formality for the fact. If, on the other hand, performance is not forthcoming, as the evidence indicates, then what justification can there be for opposing the necessity for the arrangement? The time is owed, and the debt to the public is long overdue.

It is specious to insist that time slots are so scarce. How can networks and licensees explain the superabundance of station and network promotional announcements every night and in all time periods for their forthcoming programs? It has been estimated that as many as three-hundred of these are telecast by each of the networks each week.

In addition, the reservoir of time slots for commercial announcements has increased markedly over the years. Between 1967 and 1977 alone, the total number of commercial announcements on the networks increased from 1,921 to 3,743 in an average week. This is an increase of 51.3 percent in ten years. Between 1974 and 1976 alone, the total number of commercial announcements on the networks increased from 3,286 to 3,644 in an average week. This is an increase of 9.8 percent in two years alone.[20] This could have put the time bank into operation if the number of commercial announcements had been frozen in 1974. Hidden even behind these figures is the virtual doubling of time available for commercial announcements when the industry converted from sixty-second to thirty-second message lengths. Today, between 80 percent and 90 percent of all announcements are the shorter length, thus almost doubling their number.

The alternative to the time bank or the inner fourth network is for the government to pay for access. Others have suggested this possibility. The Federal Trade Commission, in proposing that nutritional messages be "presented in other contexts and/or under other auspices than those of the advertisers," obviously had this alternative in mind. The Department of Agriculture and the House Subcommittee on Nutrition have made similar proposals. We have no problem with this alternative because it puts nutrition education on the air. But it does skirt the central issue of policy.

Wisdom would advise that no alternative be disregarded and that all three should be undertaken at the same time. The end is more important than the means, though some means are more to be desired than others. Sound social policy should be our number one criterion, for the underlying premise is, of course, that the advertising technique can be as effectively employed in behalf of proper food and nutrition habits as against them.

The premise has been proved beyond a doubt. Nutrition education projects employing the advertising technique over radio and television in developing countries have demonstrated beyond any doubt that remarkable educational progress can be made in relatively short time. In the Philippines, Ecuador, Nicaragua, and the Dominican Republic Manoff International has carried out such radio nutrition and health education projects with results carefully monitored by periodic tracking studies of target populations samples.[21] Others have conducted similar programs. The record books contain enough testimony that the technique is not only valid and useful for nutrition but also indispensable in a world in which the mass media exert such a dominant influence.

A free and open society has no choice. To breed freedom and strength in our people requires us to fortify our policies of openness, of which the most important may be the right to the power of knowledge, the right to put that knowledge constructively to work for oneself (for one's role in society), and the right to the tools to make it happen.

Nutrition education is knowledge, and television is an essential medium for for its transmission. The time to accomplish the obvious is here. We are involved in an enterprise to strengthen the capacity of people to be the custodians of their own health, to be armed with the self-confidence and self-reliance that are, finally, the only true defenses against the wrong, the misleading, and the deceptive.

> A society based upon the sum of its peoples' capacity, is a democracy . . . our democratic prospect depends upon the ability to distinguish between good work and good works. A service society based upon the "humanistic" ideology of service is sewing the seeds of tyranny. It teaches our people that they will be better because someone else knows better. A nation of clients is fertile ground for totalitarianism.
>
> In a society of clients, our problems are understood as technical issues resolved by the good works of elite professional service. In a nation of citizens, our problems are understood as political issues dealt with by the good work of people who have the capacity and competence to make a human community.[22]

Notes

1. Fred Friendly, *The Good Guys, The Bad Guys and the First Amendment* (New York: Vintage Books, 1977), p. 16.
2. Nicholas Johnson, as cited by Harry S. Ashmore, *Fear in the Air.*
3. Anthony Smith, "American Television," *Daedalus,* Winter 1978, pp. 201-202.

4. Ibid.

5. Ibid, p. 199.

6. Les Brown, *Television*, pp. 59-60.

7. Ralph Tyler, "Learning in America," *The Center Magazine*, a publication of the center for the Study of Democratic Institutions, Santa Barbara, Calif. November-December 1977. Reprinted with permission.

8. Wilbur Schramm, and E.B. Parker, as cited in ibid.

9. Richard K. Manoff, "You May Be Teaching Nutrition—But Are They Learning?" Paper presented to the American Dietetic Association fifty-seventh Annual Meeting, October 10, 1974.

10. Herbert I. Schiller, *The Mind Managers*, p. 19.

11. J. Gould, *New York Times*, November 16, 1971.

12. Donald H. McGannon, *Petition before the Federal Communications Commission*, August-September 1976.

13. U.S. Supreme Court, *National Broadcasting Company* v. *U.S. (1943).*

14. U.S. Supreme Court, *Red Lion Broadcasting, Inc.* v. *Federal Communications Commission*, 395 U.S. 367, May 9, 1969.

15. Daivd L. Bazelon, U.S. Court of Appeals, *Banzhaf* v. *FCC*, 405 F. 2d 1082, 1099 (1968).

16. Ibid.

17. Smith, "American Television," p. 204.

18. Johnson, as cited by Ashmore, *Fear in the Air*, pp. 138-139.

19. *U.S. News and World Report,* "Television: A Gold Mine," February 7, 1977, quoting U.S. Department of Commerce reports.

20. TV Bureau of Advertising: *Research Department Report on Network TV Commercial Activity, 1976.*

21. Manoff International, Inc., *Radio Nutrition Education: A Test of the Advertising Technique: Philippines and Nicaragua Final Report,* December 1977.

Manoff International, Inc., *Mass Media and Nutrition Education: Ecuador, 1975.*

Manoff International, Inc., *Mass Media and Nutrition Education: Report of the Assessment Visit to Pakistan, December 9-21,* 1974.

Manoff International, Inc., *Final Report on Nutrition Education Strategy: Dominican Republic, August 1977.*

22. John L. McKnight, "Good Work, Good Works," *New York Times*, November 16, 1977.

14 Toward the Public Telecommunications Center

Herbert S. Dordick

The Public Broadcasting Act of 1967 states that diversity of services is both an explicit goal and a working concept of public broadcasting in this nation. The extension of such services—in defining, demonstrating, and developing new markets—remains a constant challenge implicit in the language of the act. To encourage "programming which will be responsive to the interests of the people and constitutes an expression of diversity and excellence" requires courage in identifying and tapping new markets. For public television stations, a potential market exists in the area of public communications, specifically in the interaction between government and citizens as well as among citizens in deciding public issues and setting long range goals.

A market also exists in expanding the educational services offered by the stations beyond what is now offered to the public schools. In particular, the provision of specialized educational services for industry, medical institutions, as well as government is a need that has been felt in many cities, a need to which the public television stations as a public telecommunications center will respond.

But the single over-the-air broadcast channel severely limits the scope of services the stations can provide. Fortunately, the last few years have seen the emergence of new broadcast and nonbroadcast technologies. Instructional television fixed services (ITFS), multipoint distribution services (MDS), cable television, and videotape-videodisc technologies can increase the range of services delivered to the community. There exists the distinct possibility that new stations, perhaps in the VHF band, can be dropped in, thus making available potential new over-the-air channels for delivery of public services to the community.

The Corporation for Public Broadcasting is funding several demonstrations to determine just what new services could be delivered by the public television stations, how these services can best be performed by the station in cooperation with its clients, what kind of partnerships might be developed so that these services can be delivered and funded, and finally how best to utilize the mix of broadcast and nonbroadcast technologies. Here is a brief summary of these demonstration projects, as an introduction to chapters 15-17.

Toledo, Ohio

In Toledo, Ohio, station WGTE in cooperation with the cable television system and local government has designed a public communications system that will deliver live council meetings and special information concerning these meetings and their outcomes to all cable television subscribers through a channel identified as the government channel. In addition, most of the high schools within the city will receive these programs for possible classroom use through an instructional television fixed services link. Council meetings will be taped in their entirety and rebroadcast during evening hours for those who are unable to view such programming during the day. In addition, information in the form of character-generated data will appear on the television screens, on the cable systems, and on the school systems so that meeting agendas are known in advance. In time, it is expected that other video programs, such as interviews with officials and public events, prepared by the city will be made available to the system through this communications link established by station WGTE operating as a public telecommunications center.

An intensive evaluation design has been prepared to enable city government and the public station to determine the effectiveness of these type of services for improving communications between the citizens and the government as well as among citizens and the impact of the programming on city politics. Several significant issues concerning the role of local government as a broadcaster have emerged and are now under study to determine the interaction between Federal Communications Commission (FCC) and locla government legal regulations and practices.

Spokane, Washington

Prior to awarding a franchise for cable television in Spokane, local government authorities with the assistance of a consortium known as CABLE, which included representatives of the public television station, schools, colleges, and medical institutions, prepared a prospectus for the utilization of up to six cable channels for the delivery of educational and other public services. Contrary to previous policy decisions and actions, the FCC ruled on the basis of this prospectus that the cable operator could provide up to six cable channels for the services proposed by CABLE. Cox Cable, the winning bidder, contributed access equipment so that services to the subscriber can be programmed from station KSPS. These additional channels have enabled school district 81 to increase its instructional programming for the secondary schools, provide more flexible programming schedules to the primary schools, offer a preview channel, assist the local city government in providing public communications services on the government channel. Public communications services include live council meetings as well as taped replays and character-generated information concerning agendas and meeting outcomes. Another channel is now being used to provide adult educational programs prepared by the community colleges. Finally, a channel is being prepared for providing formalized education and training for the medical community in Spokane.

KSPS-TV has acquired the licenses for the instructional television fixed services in the area, and it will use these to provide educational services in schools outside the city of Spokane that are not yet wired. These ITFS channels will also be made available with two-way interactive capability for programs as needed by the many small educational communities within the city of Spokane. Studies are now under way evaluating the impacts on the community of these public telecommunication activities.

Los Angeles, California

A four-way partnership between public television station KCET, the city of Los Angeles, Theta Cable of Southern California, and the Annenberg School of Communications at the University of Southern California is working to provide public communications services to the West Los Angeles area of the city. These services will offer information specifically related to the affairs of that large district in which approximately 40 percent subscribe to cable television. Public facilities, including libraries, councilman's offices, shopping centers, and other places available to the public will also provide viewing for those not having access to the cable.

The services to be provided are similar to those being offered in Toledo. The major difference is that station KCET is a major producer of public programming for the Public Broadcasting System and with facilities and talents not available in other stations which can be focused on the provision of public communications. Two-way communications will be delivered through the use of open telephone lines, and plans are under way to offer question-and-answer sessions on the cable.

The city of Los Angeles sees this demonstration as the first step in implementing the provisions of their recently adopted master plan for cable television. If the experiment is perceived as being successful in West Los Angeles, it will be expanded into other cabled areas of the city and will be required of other operators awarded franchises for servicing Los Angeles. Alternate means of providing public communications, including the use of MDS, ITFS, and the videotape-videodisc technologies in public places, will also be explored in order to expand these services in Los Angeles. Los Angeles is not an attractive cable television market now.

San Diego, California

San Diego is a telecommunications-conscious community, with the largest cable system in the country and several providers of ITFS services. The public television station, KPBS finds itself in both an attractive market and a competitive one. The demonstration grant to KPBS seeks to examine how a state university station can become a public telecommunications center.

The station now provides FM-SCA services for the blind in the form of talking book programming and the county supervisoral meetings are provided through standard FM broadcast. A full complement of school services is also provided.

Studies are under way defining the many specialized audiences in San Diego, including the aged, the Navy families, and the Mexican farm workers, all of whom need specialized broadcast and information services. The basket of telecommunications offerings is full: ITFS, MDS, CATV, and the videodisc-videotape technologies. The objective of this demonstration is to determine the best mix of delivery modes for serving the highly diverse audiences in the city and in a manner that does not in any way detract from useful competitive services now under way.

The CPB demonstration strategy is to fund projects that utilize the widest mix of technologies for the most diverse range of services capable of being offered by different station licensees. Toledo and Los Angeles are community licensees, purportedly having the greatest flexibility for action. Spokane is a school licensee with certain limitations imposed by its board but nevertheless exhibiting a wide range of service delivery successes. San Diego is a state college licensee with certain limitations imposed by its board. These experiments will provide models for other licensees as they move toward becoming public telecommunications centers.

Additional demonstration sites are now being selected. It is expected that within the next several months, experiments will be undertaken in Lincoln, Nebraska; Atlanta, Georgia; Bowling Green, Ohio, and Wichita, Kansas.

15 Cable Television: An Idea Whose Time Has Come? Gone? Not Yet Come?

Walter J. Schaar

During the past several months I have been involved in long-range planning sessions and seminars at local, regional, and national levels. Those probably did more to scare me and better equip me to deal with the future of public broadcasting than anything else in my life. In one of those seminars Howard Rosenbloom of Harvard pointed out that technological changes give us two things to consider—threats and opportunities. How we as society, and more specifically as individuals, handle those two challenges is what I believe our experiences in Spokane is all about. It has become necessary for me to approach, in a very calculated manner, that notion of threat versus opportunity. I've been developing a planning procedure based on the principles of marketing. Marketing television demands that you look at primary demand, market segmentation, channels of distribution, competition, possible governmental control, the product's life span, and the effect of technological innovation on these. The prospect that faced those of us in Spokane several years ago was using cable for an alternative delivery system of instructional materials to schools. What was the primary demand for the school system in combination with the television station? Our primary obligations at channel 7, because we are licensed to school district 81, are to school teachers and students. The demands that we perceived from the teaching staff of this rather large, urban school district were to (1) find material that would assist the classroom teachers in instructing their students, (2) deliver it at a time of year to fit the curriculum, (3) deliver it at the time of day when students and teachers are in the same place, and (4) deliver it more than once for adjustment to (or bill) schedules at individual schools. As these questions were being answered, we should have realized that some of the same technology could have been used to establish similar benefits to the rest of the community.

The first market segment was the seventh through twelfth grades, divided into about equal numbers occupying an equal number of classrooms. Again, if we were Monday morning quarterbacks, we would have realized that many other segmentations could have been served with the same techniques of cable transmission. We didn't deal with the life span of the product, but then none of us thought through the life cycle of over-the-air television either. If we had, and had calculated the cost of pupil viewer-hours with the relatively bad programming available in those dark days, any sensible person would not have put a noncommercial station on the air, for instructional television. None of us can afford any longer to be afraid of looking to the future and what the life cycle of certain technologies will be.

Which technologies should be used, or should we use electronic technologies at all? Was instructional television fixed services (ITFS) the answer? Was cable the answer? Was cable going to be available in sufficient numbers under sufficient control of ours to use it properly? What governmental authorities would we be dealing with? It became obvious early that our first hurdle would be to convince the Spokane city fathers that education could indeed use more than the one channel that the Federal Communications Commission was allowing at that time. The question of multiple channels was a real bone of contention with the city council. That conversation, which took nearly a year, finished with the city fathers agreeing to write into the ordinance the requirement that five channels be set aside for educational use. The language of that ordinance in section 21 is

> That a grantee shall reserve an initial five (5) channels, or such number as the F.C.C. shall allow, solely for education on a development basis. Upon completion of the basic trunk line, if it is found that 80% of the week days from September through May, both months inclusive, during any consecutive three (3) hour period for six (6) consecutive weeks, the number of these channels may be reduced to the extent necessary to meet this less than 80% level of usage. If it is also found that 80% of these channels are being used 80% of the week days from September through May, both months inclusive, during any consecutive three (3) hour period for six (6) weeks, additional channels may be requested to meet this more than 80% level of usage. These channels will all remain available to the schools without charge for a period of five (5) years beginning from the completion of the basic trunk line of the system, provided, however, that said channels shall be furnished without charge subsequent to said five-year period if F.C.C. rules and regulations allow.

After the ordinance was passed, we entered into discussions with each of the cable companies bidding for the franchise. The language that finally ended up in Cox Cable's bid reveals that a great deal of politicking took place (probably on both sides). The next governmental agency to come into play was, of course, the FCC. There is no question that some cable companies practiced the technique of promising the moon when they felt that the FCC, would probably deny any "moon shots." The use-it-or-lose it language written into the ordinance, which we were told was unique, is what prompted the FCC to allow that to be maintained.

Last, we come to changes in the technology. We know that changes in technology occur at relatively predictable rates. Widespread adoption of technology occurs at a relatively predictable lag behind the time that something is invented. How could we as public broadcasters respond to this? We must be looking for new changes of technology, and we must be clear about our goals and try to keep the two in balance.

After we dealt with all these questions, something new began to emerge. Whether there had always been a need and it had not been recognized, possibly because it was felt there was no answer, or whether the introduction and discovery of a new technology provided the impetus for it, we at least realized that it was becoming possible to provide individualized instructional packages. As public broadcasters we realized that we could deal, even more than we had in the past, with smaller segments of specialized audiences.

The transmission of video through a wire was available some time ago. The first cable system was initiated in 1948, in Astoria, Oregon. It was originally used to bring existing television signals from distant stations into homes that couldn't get it any other way. It is possible that even in the short time from 1948 to now, cable is an idea that may have already come and gone. However, the fact remains that we must get from point A to point B.

Now comes admission time. Quite by accident, we started the marketing process to see how to solve a problem. I have already dwelled on the primary demands, market segmentation, and governmental interference. Almost every one of the marketing procedures we did in one way or another was an accident. Even more accidental was being in the right place at the right time. We had actually spent most of the year in 1970 and 1971 drawing up plans for a closed circuit school-district-run system within our city. That was quite expensive. Cable, as it was then being presented in 1972-1973, was in fact a step toward the individualized information packages that we knew were going to be necessary.

Spokane is in the center of an area known as the Inland Pacific Northwest. There are 285,000 total television viewing homes in the United States within that viewing area. Canada provides another 120,000, (which includes Calgary and Edmonton on cable along with other smaller cable installations). Spokane county's metropolitan household count is 108,600, with certainly 98 percent being television households. Within that county lies the city of Spokane, the major portion of the cable franchise. Cox has also acquired the franchise to the surrounding area. The school system to which channel 7 is licensed has roughly thirty thousand youngsters and seventeen hundred teaching stations located in fifty-one cities. Over the past ten years it has invested well over $2 million in plant facilities for channel 7, more recently roughly $100,000 in cable origination facility has been invested. They have backed that up on the receiving end by purchasing over the past twenty years fifteen hundred television sets and completely wired eight of our secondary schools. We are in the process now of wiring the remainder of the school buildings and purchasing enough new color sets, roughly 670, to replace most of the tired black and white sets and put sets where there were none before, an investment of well over $600,000. This is the result of the success of our cable study program.

We are building a system within each school building in district 81 so that the five channels coming from the telecommunications center can be received along with the city government channel 5, the three commercial stations, channel 7, and one internal orgination channel. All this progress, of course, did not take place without research and study; we had some excellent help. The study was done with fine cooperation of Philip Rubin and Brian Brightly of the Corporation for Public Broadcasting, with a combination of funds from CPB and school district 81, with Herbert Dordick as our consultant.

The CPB telecommunications study, still in progress, was intended to cover the waterfront, so to speak. The director of the study set about uncovering all the possible users of cable and other forms of telecommunication. She was also to uncover all the potential funding for cable use and all possible means of using public radio and television. Again a chronological accident took place. A group of citizens in the city of Spokane were quietly working to put a public radio station on the air.

Maureen Hathaway, our telecommunications director at the same time uncovered a group of people involved in the medical areas who wanted to take advantage of the cable channel that had been set aside for them but did not exactly know how to go about it. She also began to meet regularly with the consortium of educational institutions in the area, made up of the colleges, universities, Catholic schools, and school district 81, for the composite use of the five educational cable channels. As you can well imagine, those who want to use and those who can afford to fund are not necessarily the same.

At the same time the telecommunications study covering all aspects of telecommunications was going on, school district 81 was beginning to invest funds to bring along their end of the investigation. After all with over thirty thousand children in school, it still represented the largest single block of potential users. Our instructional television coordinator drew up a plan for a school district cable implementation study to run parallel with the CPB study in December 1976. The school district study then started in January 1977, while we were awaiting approval of the CPB request.

The mere existence of a multitude of channels on a cable system will not constitute immediate utilization by either school people or the general public. We have brainwashed the public over the past thirty years to expect tremendous promotional campaigns, ads in newspapers, *TV Guide*, encouraging people to watch a particular television show. Teachers, students, parents, and general public not connected with the school, are just human beings. They are used to that kind of promotion, and when these new channels came into being we soon discovered that they were not accepted in the same way that those established, existing channels, either commercial or public, were. The challenge of selling this service was before us. We narrowed our problems to three: (1) Software: where could we get *good* programming with appropriate rights? (2) Flexibility: How flexible could our schedule be, not only by the time of the year a program might be aired, but the time of day and how many times? (3) Dependability: Nearly all of the people in the schools that had been wired were used to a rather badly run closed-circuit system with tape machines and tapes that were not compatible. We had some problems to overcome and some new things to offer. The experiment finally took fruition, in April 1977 on channels 8 and 10, using sixty teachers stationed in four schools equipped with new color televisions.

Our engineering department with the cooperation of Cox Cable came up with a neat solution to a sticky problem. In Spokane the commercial stations and our public station are channels 2, 4, 6, and 7; leaving channels 8, 9, 10, 11, 12, and 13 open. Engineering discovered that channels 8, 9, 10, 11, and 12 could convert nicely down from 15, 16, 17, 18, and 19 on the cable system. Those were the channels assigned to education. They developed a block converter that would bring these channels into the head end of the schools so teachers merely continued to use the channel selector on the television set and did not require a converter in each room. As a matter of fact, the cable company had no intention of putting converters into every room. They assumed that by the end of the first day half of them would be in someone's house.

We started the programming procedure using the only technique we knew. We drew teachers from various disciplines, selected programs, got the best consensus we could on which program, which day, what frequency, what sequence, and then scheduled them. In the short period of time from April 1, 1977, to the end of the school year in mid June, we found our utilization climbed the first two or three weeks and then gradually dwindled, until by the end of the school year we were disheartened and quite concerned about using cable at all. We had not reckoned with some environmental conditions that neither the teachers or we could have anticipated. Unscheduled state testing programs, flu epidemics, and schools going to state championships began to make a shambles of our prescheduled plans. Teachers who had helped put the schedule together were the first ones to call and report that they were so far behind that they would never be able to use it. We did some soul searching in the summer, and when we started again in the fall, we feared that it might happen again. It did. By the first of December we could see the same attrition taking place. It was then that our television coordinator unscheduled one of the channels completely. She scheduled one in a tight schedule and left the open one for call-ins. It took some reeducation, but within a few days, that call-in channel was full. Now we are operaing on three channels and by the end of this week on four channels. It is our intent to schedule only one and leave three on a call-in basis. With that change in scheduling or nonscheduling, a cooperative effort between Maureen Hathaway, our telecommunications director, and Velma Churchill, our instructional television (ITV) coordinator, began. They set about finding suppliers of software that would be agreeable to a ridiculous set of rights demands. We wanted fourteen months to go from

April of one school year to the end of next year; we wanted unlimited plays. We used the same previewing technique again, bringing in teachers who sat through hours and hours of different programs. We noticed, however, that their enthusiasm was far greater when they knew there would be maximum flexibility in scheduling.

We were only part way through that part of the experiment when some of the brighter children in the classes started turning it on at home and telling their parents about the flexible ways their teacher had of drawing material from a videotape library at KSPS-TV. About the same time we started asking, why couldn't the same process be developed for other than school-time viewing? Publish a catalog, which each teacher now has, to be picked up at the Spokane public libraries or published in the newspaper. Applying the same call-in tactic to the five educational channels after 3 p.m. and on weekends. The idea is being explored now with the cable consortium made up of the educational entities. Some of the colleges are scheduling telecourses in the evening hours, but there are many times when the cable capacity is not being used to its fullest. We now believe that proper promotion, again process of educating the general citizenry, will lead us toward an individualized program offering to the general public. You can be certain that the cable company will help us promote that as a new service.

It was only five years ago that our chief engineer and I were arguing with the city fathers that we could use five channels. The language in our ordinance indicates that we can ask for additional channels if we can keep those channels busy under those percentage restrictions. Last week that process was started. We don't have all the manpower and equipment and share of support figured out yet, but for once it seems that a demand for the product is preceding the ability to deliver it.

Five years ago the director of engineering and I made a quiet promise that we would in fact develop and pull all the pieces together for the most sophisticated telecommunications center we could have.

The basic segments of the center are in place now. Our board of directors has now accepted a proposal from KSPS-TV staff to cooperate with Spokane public radio on a capital fund drive to establish a radio studio in an expanded facility and to expand our cable origination space as well. We are, to say the least, excited. We have already found great interest by the general public in such a center.

And now we come to new technologies that we know are practically in our lap. We know that people are experimenting with fiber optics, and we are aware of its tremendous capacity. We know we will soon be able to receive more than one signal from our national source of programs through the satellite. We know there is still tremendous divergence of opinion on which format we are going to settle on for videotape. We know that direct satellite to home use is somewhere on the horizon. So take the title of this chapter and check one of the boxes. If everyone checked the same one, I would be very surprised. If 40 percent checked the same one, I would be somewhat surprised. If the vote were split three ways I would not be surprised at all.

Cable Television: An idea whose time has ☐ Come? ☐ Gone? ☐ Not yet come?

16 Another Way to Put It: Alternative Distribution Systems for Public Broadcasters

Robert D. Smith

For more than a decade some people in noncommercial broadcasting have though that the time would come when the single open-circuit broadcast channel simply would not be sufficient to the opportunity or the responsibility that we have to serve our constituency.

This has not been a pressing problem to date because almost all our money, our attention, and our energy have gone into making just that one open-circuit channel both publicly useful and economically viable. The underfinanced circumstance of public television and radio has focused our attention on making one service excellent and discouraged notions of widespread diversification into other distribution technologies. At this very time many persons in the public broadcast enterprise are concerned about the possibility that limited capital funds, now reserved for the construction and improvement of noncommercial educational television and radio stations, may soon be dispersed to all manner of telecommunications institutions and projects, diluting the impact of those limited annual allocations and shutting down the already constrained expansion and improvement of public broadcasting.

If we accept the proposition that in America what can happen probably will, then we face an eventual diversification of audio and video distribution systems. It has already started, of course, and is taken for granted in communities where it exists. WVIZ-TV in Cleveland has been operating instructional television fixed services (ITFS) channels for years, delivering an expanded instructional television service to schools and providing educational services to hospitals and health institutions. WVIZ also operates a large videocassette lending library for schools and other institutions. WVIZ has already diversified from a public television station to a community telecommunications agency. The South Carolina ETV Commission operates one of the largest and most comprehensive public telecommunications systems to be found. The idea of a community or state public telecommunications center is not new; it is just not yet very widespread in fact.

I have been asked to tell you about a plan to expand the Toledo public television and radio station system to a first phase of a telecommunications center.

First, some background. WGTE-TV-FM is a community-type licensee, governed by a board of trustees representing the thirteen thousand contributing members to the stations, the member schools and other nonprofit institutions, and the community at large. WGTE is not one of the major producers of national radio or television programs. WGTE spends an enormous amount of time and energy raising local funds, to qualify for federal funds, to keep a staff of fifty-seven persons going, to pay the rent and electric bills, to maintain and operate a minimum complement of equipment to keep the transmitters on the air and the one-at-a-time program production capacity engaged.

I stress the marginal nature of our medium-market operations so that our planned expansion into public telecommunications services will be understood in the context of reality.

In the early seventies WGTE-TV packed up its cameras every Monday afternoon and drove two blacks from the studio to set them up again in the Toledo City Council chambers. When the council convened on Tuesday mornings, our cameras recorded the proceedings, from beginning to adjournment. The three to seven hours of council meeting were broadcast in the evening by videotape delay. The televised adjournment often occurred well into the early morning hours.

The weekly videotape recording of the city council consumed nearly 25 percent of our television production capacity. A little later, we started to record and delay-broadcast only the first two hours of the council proceedings. The agenda shifted subtly and interestingly when this happened, and soon some issues seemed to be saved for after the time when the cameras capped off.

A few years ago the Toledo City Council reorganized itself from a committee-of-the-whole structure to three committees meeting at various times throughout the week and coming together in a weekly council meeting. The real deliberations and decisions took place throughout the week in committee meetings, when members of the public were permitted to address council members, and weekly council meetings became larely technical and procedural occasions when members of the public were not permitted to address the council.

It was at this time that we discontinued the television broadcast of council meetings—and heard complaints from people who regarded the broadcasts as important public service.

When we put our new WGTE-FM on the air, we began the weekly radio broadcast of the city council that continues today. Now people in Detroit and sixty or seventy miles away from Toledo in all other directions can hear what happens in Toledo's City Council meeting every Tuesday morning. For many of them, for most of our radio audience perhaps, this broadcast of the city council proceedings is an unwelcome interruption of our classical music service.

Which makes the main point of this paper: public broadcasting is asked to be many different things to many different audiences, often concurrently. With one channel, we cannot do many different things for many different audiences simultaneously.

How then, do we deliver the city council proceedings, routine health education, credit or noncredit courses, community information calendars, professional and vocational training, and many other kinds of useful, if not entertaining, program material?

Alternative distribution systems are cable, cassettes, ITFS (instructional television fixed services), SCA (subsidiary communications authorizations).

The Toledo plan for the city council is very simple. We will install three wall-mounted color cameras in the chamber. Just outside the chamber a director and a technician will pan-tilt-zoom-focus the three cameras by remote control. The picture and sounds of the city council will be broadcast throughout the day by ITFS and simultaneously recorded on videocassette for complete playback on ITFS during the evening hours. The cameras and microphones will be omnipresent in the chamber. Any public meeting taking place will automatically be live-broadcast and delay-broadcast the same evening. Between meetings, a variety of printed information will appear on the screen, such as future meeting agendas, community announcements, and other printed information coming from many sources.

The ITFS signal, in the original project, will be received in five of the ten city high schools, at the five head ends of the local cable system and thence by cable, to 25 percent of the homes in the community, at the downtown branch of the public library, at WGTE, and in a mobile display unit which moves from one shopping mall to another on a monthly rotation. Others who may wish to invest $1,500 to receive the city council telecast might be commercial radio and television stations (with unlimited rights to use the video and audio for their own newscasts), the daily newspaper, and even the private citizen who does not happen to have cable at this time and cannot live without the presence of the council at home.

This is a three-year project, completely funded without city funds. If the city cannot be without it by the end of three years, then the city will have to begin financing the annual operating costs.

There is an evaluation component too, based on the following hypotheses:

1. Those viewing the government-access channel would be more informed about governmental operations and current issues.
2. Those viewing the government-access channel would be more likely to become involved in traditional ways such as voting.
3. Those viewing the government-access channel would be more likely to engage in more active types of governmental involvement, such as contacting city officials.
4. If the first three hypotheses are supported, those viewing the government-access channel would also be expected to have higher levels of trust in government, lower levels of political alienation, and higher feelings of political efficacy.
5. There would be a diffusion effect which should be measurable. More

specifically, those who live in the area where the government-access channel is offered on the cable system but do not subscribe to the cable system or subscribe to the cable system but do not view the government-access channel would still be expected to exhibit the trends listed in parts 1-4—albeit to a lesser extent—as a result of their more frequent interaction with those who have been exposed to the government-access channel.

The evaluation component will be carried out by professionals in the political science department of the University of Toledo.

Altogether, the project will cost $975,000, one-third to purchase the camera equipment, videocassette recorders, and ITFS transmitters; one-third for three years of daily operation; and one-third for the three-year evaluation program.

If the service operates five days a week throughout three years, the hourly operating costs of this information service will be $51.

This then is the cost of reaching a relatively small audience with a special service through an alternative distribution medium. It is one very specialized example, but it could lead to other uses of still other alternative distribution media—once we break away from the large-audience broadcast attitude that governs our being.

In addition to testing these hypotheses, the project should raise some interesting policy issues on the use of telecommunications coverage of public events and the use of information systems by local governments.

How significantly will the omnipresence of incumbent council members on cable television favor those incumbents during successive elections? During successive political campaigns, should those opposing the incumbents have access to the system. Should there be an equal-time requirement? Since access to the system is limited to official happenings in the council chamber, and printed information presumably approved and processed by the city's staff information people, should there be any kind of fairness doctrine or principal of balance and objectivity?

If the local public broadcaster operates the system, should it be required to enforce the policies that govern its public broadcast operations, namely balance and objectivity in matters of public controversy, the Fairness Doctrine, equal-time requirements during political campaigns, control of indecent, vulgar, or profane language? Can the local public broadcaster property function as merely a technical operator of the system and leave the programming entirely to the local government?

We believe that these are fundamental policy questions that are raised, particularly in the type of project we are planning. We believe that alternative distribution systems are the only ways that multiple services are going to reach multiple audience both in real time and audience convenience time, and that it is a proper role of the public broadcaster to be among those who pioneer such new applications.

17 Providing Specialized Services Through Alternative Broadcast and Nonbroadcast Technologies

Paul J. Steen

I am reporting on a project that is in its early stages and will therefore be reporting more about what will be, and what might be, than what is. But the project is, it seems to me, central to the focus of the role of public broadcasting stations in providing specialized services for their communities through alternative broadcast and nonbroadcast technologies.

I would like first to give you a little background on the project. San Diego State University is the licensee for KPBS-FM and KPBS-TV, fairly typical full-service FM and UHF public radio and television stations. The Office of Special Services is headed by Donald Wylie, who is not a regular member of the staff of KPBS but rather a fairly typical professor in the Department of Telecommunications and Film at San Diego State, teaching courses in public broadcasting, instructional television, and television production.

A year ago I asked Dr. Wylie for ideas on how the KPBS stations could best use a three-channel instructional television fixed services (ITFS) system that was being installed on campus. My thought was that the possible uses for the ITFS system would depend on how other systems of distribution might be used by KPBS–methods such as community cable television, cassette duplication and distribution, satellites, subcarriers. And, to see the whole picture, one needed also to look at incoming distribution capabilities, especially the PBS/NPR (Public Broadcasting Service/National Public Radio) satellite feeds, and the materials available from libraries, agencies, and producers. We decided to submit a proposal to the Corporation for Public Broadcasting, "Specialized Services for Specialized Audiences in San Diego." The CPB was receptive to the idea since it touched on one of its concerns, the final-mile problem, and in October we were officially informed that the project was approved and funded. Because the fall semester had started by the time we received that information, we divided up the funds (with CPB approval) so that Dr. Wylie was on the project one-quarter time during the fall, and three-quarters time during the spring semester.

One might wonder why the study of potential uses for an ITFS system (and other distribution alternatives) came after the installation of the system. Why was the system built in the first place? Although it is a long story, involving a complete reorganization of media services on our campus, I will simply note that there were four principal motivations for building the system:

1. ITFS had been approved as a way to reach community cable television antenna sites, and San Diego has one of the largest collection of cable-equipped homes in the nation.
2. ITFS is being successfully used to provide management courses to businesses and industrial plants in the Los Angeles and San Francisco areas, and it was thought that the same thing could be tried in San Diego.
3. San Diego State has a branch campus in Imperial Valley, about ninety miles away, and is planning to build a campus in North San Diego County. Although it will take repeater stations to reach them, the system provides the first leg of the interconnection with the branch campuses.
4. Matching money was available from the Department of Health, Education, and Welfare.

Because the ITFS system was operating to some extent when the project began, we divided our activities into three broad categories: (1) conducting limited demonstration experiments with ITFS and other distribution techniques; (2) writing grant proposals for larger demonstration projects; and (3) assessing needs and potentials for long-term, ongoing narrow-casting services in the San Diego area. Obviously these activities overlap, but the categories have been helpful in organizing our planning.

I would like to describe very briefly some of the activities of the project in each of these areas.

For four months we fed college courses through the ITFS system to Mission Cable Television during hours that the Los Angeles NBC station duplicated the programming of the local NBC affiliate. During the demonstration there were very few problems of a technical or administrative nature. We did not attempt to enroll students in the courses, although viewers could make individual arrangements with academic departments. We are now seeking professors and departments interested in offering advanced or special interest courses to be offered by ITFS and cable.

In January we moved our one ITFS receiver from the cable television head end across town to the University of California, San Diego (UCSD), medical school to demonstrate the relay of satellite programming. This experiment utilized a five-foot satellite antenna that we evaluated for NHK, the Japanese public television network. With the permission of the Veterans' Administration, we relayed one of the VETSAT telecasts to UCSD by ITFS, using the program also at San Diego State.

We think that we have the distinction of being the first ones to use the WESTAR satellite for nonbroadcast public service use, on April 5, 1978. We asked Congressman Lionel Van Deerlin to be the opening speaker for a conference that we were holding in San Diego. Congressman Van Deerlin went to the WETA-TV television studio during his lunch hour (which was 9:00 a.m. San Diego time). His remarks were fed to the Western Union uplink which will transmit to WESTAR I for reception in San Diego. At our end things were a bit more experimental than we originally bargained for. We thought we would be able to receive Mr. Van Deerlin on our thirty-three-foot PBS antenna, which was installed in March. But then we found that it would not be aimed and tuned until later in the spring, so we obtained permission to use an experimental fourteen-foot antenna across town, with a microwave relay to the mountain where our ITFS transmitter is located, finally receiving the signal at a downtown high-rise office building where the conference is being held. As I have indicated, these are experiments in solving the final-mile problem!

Some other demonstrations scheduled during April and May include a two-way slow-scan television demonstration with the PEACESAT project in Hawaii, experiments with the ATS-1 satellite using voice and slow-scan television, and an experiment to feed hidden captions to a second cable television channel. We also are discussing with Comsat General an idea for using excess time on Marisat (a computer-controlled message and data service for ships) to offer computer-based credit courses to merchant and U.S. Navy ships. Our ability to demonstrate ITFS relay was greatly enhanced in early April 1978 when two long-awaited ITFS antennas and receivers finally arrived.

In the area of long-term demonstrations, we have already submitted a proposal for a project to caption local and national programs with expressive captions, using different sizes and styles of type to represent different characters and emotions for viewers who have lost their hearing or are becoming hearing impaired. We have completed another proposal to be submitted to the Bureau for the Education of the Handicapped for captioning courses of instruction, both credit and noncredit, for this same population. On campus we submitted a proposal to take over a computerized message system, which has been approved. As soon as we complete the reprogramming of the computer, we will begin captioning local public affairs and cultural programs.

We have prepared a proposal for supplying a variety of telecommunications services for a suburban library to see whether the public library can serve as a satellite community telecommunications center when interest and facilities at the library permit this.

We are developing a proposal in cooperation with the San Diego TV College Consortium that will test the idea of aggregating small groups of students in smaller industrial firms by two-way ITFS to permit specialized college courses for middle-management personnel. The consortium consists of the community colleges in the greater San Diego area, plus University of California Extension, San Diego, and San Diego State University. The consortium produces and obtains lower-division college courses and schedules them for broadcast by KPBS-TV and for transmission on cable television, using the county schools ITFS system to reach the head ends of the cable systems. Our project has attempted to organize activities in cooperation with such existing groups whenever possible, since a community telecommunications center, as we believe many public stations will become in the future, should be functioning in many cases as an interface—a broker or a facilitator—between a community organization and the specialized audience which that organization wishes to reach.

We have completed a proposal for extending the ITFS system to Imperial Valley, and the Office of the Chancellor of the State University and Colleges will assist in extending the system to Imperial Valley.

We are developing a proposal to offer television courses to navy men aboard deployed ships. The videotaped courses will be transmitted on the shipboard television systems to receivers in selected locations throughout the ship.

And we are developing a proposal in cooperation with the San Diego Medical Association to offer medical school grand rounds to hospitals and physicians late at night by scrambled broadcasts. The broadcasts would be received in the home or hospital lounge, unscrambled, and recorded by a low-cost home-market recorder, which would be started and stopped by its own internal timer. This system will be similar to the CADAVRS system (computer-assisted, dial-access, video-retrieval system) tried in the late 1960s.

Although the Office of Special Services has not become involved in the operation of the KPBS-FM SCA services, this activity deserves mention. KPBS has one of the most extensive reading services for the blind in the nation, and it is probably the only SCA reading service fully integrated with the public radio station that provides the main channel signal. The morning-to-night services include reading the local newspapers and the *Los Angeles Times*, reading ads from the paper, and reading portions of a novel that the listeners have selected for reading. A portion of the day is devoted to special audio services for the essentially homebound handicapped, and just recently courses by SCA radio for both blind and handicapped adults have been added to the schedule. Our work with the SCA service has been limited thus far to discussions of serving jails and prisons with SCA programming, the possibility of offering audiocassette courses for commuters, extending the SCA services to Imperial Valley through cable television, experimenting with slow-scan television through SCA, and extending SCA programming through satellite.

In the third broad area of activity, determining specialized audiences and specialized needs, we are doing the following:

In April 1978 we mailed a survey that attempted to identify in precisely what services and programs might be offered through ITFS and other distribution systems—a catalog of the programs and services too specialized for the main-channel television schedule yet needed by some group in the area.

We are designing a special survey for those in health-care professions—doctors, nurses, nursing home personnel, and administrators—to determine what kinds of services, what distribution techniques, might best be undertaken by KPBS and San Diego State University, probably in cooperation with the University of California, San Diego, Medical School's Office of Continuing Medical Education.

A third survey is being developed to determine the best way to serve the needs of shipboard navy personnel for television credit courses.

We are also developing lists of available programming for specialized audiences, especially for those in nursing homes and law enforcement work.

Finally, in what was one of the most important single activities of the project, we organized a conference last April 5, 1978, in cooperation with the San Diego Community Video Center (the local cable television public-access organization) and with partial underwriting from the Higher Education Association of San Diego, which brought together all those in the community who provide radio, audio, and television programs and services on the one hand, with all those public agencies and organizations which do or might utilize programs and services on the other. (It was this conference that Congressman Van Deerlin opened by satellite.) The last item on the agenda at this conference was whether there should be a community telecommunications consortium which would continue the dialogue between telecommunications disseminators and telecommunications consumers.

These, then, are the major activities of the KPBS/CPB special services projects. It is too soon to come up with project conclusions or recommendations, but I would like to make some observations which obviously will be based in part on the activities of the project.

First, it is obvious to me that at least some public broadcasting stations should strive to become community telecommunications centers and that all stations should be concerned to some extent with supplying specialized, nonbroadcast services for specialized groups in the community. The opportunities are there; of that I have no doubts. And I believe the obligation is there for many stations. We have recommended that the Carnegie Commission consider ways that pilot national public services programs for nonbroadcast transmission can be organized. I do not recommend that the opportunities for national public service activities by public broadcasters be further investigated. The studies of the past two years seem adequate.

Second, I think it important that the Carnegie Commission and others become concerned with the entire matter of national instructional services. It is amazing and frustrating to me that one of the models for public broadcasting in the United States, the BBC, has considered education a high priority, while our national public broadcasting organizations treat formal education as a very low concern (despite valiant efforts by the individuals in the education offices of these organizations). Even the commercial broadcasters in England encourage and support use of their programs in schools, including the recording of them for later use. This is a far cry from the situation in the United States, where at least one commercial network wanted to license schools to record newscasts! It is our feeling that all contracts for public programming (such as the agreement originally signed for the CPB/BBC Shakespeare series) should have few if any restrictions for use of the programs in public education.

Finally, we recommend further investigation of ways to encourage ITFS use by public television stations. There are problems—for example, the cost of equipment. In some ways the situation today in developing ITFS is similar to the situation in the early 1950s for noncommercial television—the potential is there, but the capitalization for individual systems and the national coordinating service is a major problem. I do not think the answer is expanding the provisions of the educational broadcasting facilities program to permit purchase of ITFS equipment, since there is never enough money in that program for technical support of the regular public broadcasting services of the stations, and further dilution of funds would weaken that program and divide the public broadcasting community. The answer, I think, is to create special legislation for developing nonbroadcast distribution systems for community services that will provide matching grants for both the technical and human costs involved in establishing the ITFS systems. The proposal of the ITFS Council of the Central Education Network that instructional service grants and an instructional telecommunications facilities program deserve serious consideration. The Central Educational Network ITV Council has offered five suggestions for action to the Carnegie Commission, all of which merit consideration:

1. Establish a national council for ITV to give this function a base of support.
2. Establish a national fund for ITV programming development.
3. Fund a research and development center for ITV to look into production and distribution methods and evaluative techniques.
4. Establish an instructional telecommunications facilities program similar to the one that proved so valuable for broadcasting facilities (restricted to state agencies and public television stations providing instructional television services).
5. Set up instructional service grants similar to community service grants given by the CPB to stations, for the improvement of local instructional television services. (*ETV Newsletter*, February 27, 1978, p. 5).

And finally, we believe ways must be found to encourage more flexibility in using ITFS channels for community services beyond the classroom. This obviously will require new policies and rules by the Federal Communications Commission. We are certain that this idea will not be universally embraced. As I have discussed our project with school administrators, some have perceived the idea of using ITFS to reach industry, hospitals, libraries, and the like as a threat to expanded use of ITFS in schools and the ownership of an ITFS system by public television and radio stations as an attempt by public broadcasters to compete for ITFS frequencies, which are becoming scarce in a few communities. I obviously do not see things this way. I believe everyone will benefit if public television and radio stations increase their services to the public, responding to narrow as well as broad needs, with a broad view of the alternatives for disseminating information, education, and cultural enrichment.

Part II
Computer-Communications Networks and the Emerging Information Economy

Introduction

Herbert S. Dordick

The postindustrial society, the knowledge society, the age of uncertainty—all are manifestations of the emerging information economy. That this nation is becoming a nation of information workers cannot be denied. But the definitions of information workers, products, services, and industries are still being debated. While debates occupy the theoreticians, new communications services are being developed and cautiously offered to the market, generating new sets of policy issues to be resolved.

This part explores the ramifications of the information economy, reviews the emerging networks and network services that evidence this emerging information economy, and poses policy issues for which research is required.

A.J. Walderhaug, of the Bureau of Economic Analysis of the Department of Commerce, reviews the concept of national economic accounting with specific reference to Porat's study of an information sector within the U.S. economy. Pointing out that to date the descriptions of the sector are essentially statistical, Walderhaug suggests possible applications and calls for the formulation of some real problems for testing by Porat's system.

Tyler accepts the information economy as suggested by Machlup, Porat, and others as becoming an almost everyday reality and raises the question of just what are information products, services, and markets. His paper addresses one telecommunications development that may influence the answers to these issues, the emergence of versatile technologies that allow telecommunications systems to deliver textual information and simple graphics at low cost to a wide range of users—wholly electronic publishing. The technology and economics of viewdata and other teletext systems are reviewed, and the policy question raised by their introduction is defined. Such innovations in the United States may be significantly constrained by regulatory factors and industry structures rather than market forces.

Johansen adds another dimension to the information economy with a discussion of computer mail and computer conferencing or interpersonal communications through computers. His paper provides an introduction to current technology, but its primary focus is the concept of interpersonal communications through computers. The new services are defined with emphasis on benefits, costs, and available services.

The personal computer appears to have emerged in the United States with the introduction of the first hobbyist computer kit. But the concept of the personal computer, as Caulkins points out, has gone considerably beyond the kit. Among other ramifications is the network of personal computers described in this chapter. What emerges is that the most exciting use of personal computers, regardless of how defined in terms of size and capacity, may very well be in networks. And this raises significant issues in telephone use, standards, technologies, and rates. Caulkins also points out that personal computer networks using radio create another set of policy issues comparable to those raised by the network wired to the telephone. There are significant societal issues, too.

Advances in communications technology, the emergence of the information economy, and the availability of universal telephone service has given us the luxury of dealing with the specialized use of communications by those who are essentially denied the telephone. Krauss discusses the provision of electronic message services for the deaf. For the more than 13 million people in the United States with impaired hearing, the telephone is a constant frustration. Reviewing the current means for electronic communications available to the deaf (teletype systems), Krauss suggests that the first public electronic message service is likely to be a service designed for the deaf. The federal government, Krauss states, could play an important role in stimulating the industry to serve this large audience. It is appropriate at the present time to examine emerging policy issues such as those concerning standards, protocols, and the role of the federal government in the provision of these services.

Nielson perceptively points out that a large part of the communications services needed for the deaf community anticipates precisely the needs of the business community. Policy issues evolving from the wide use of small inexpensive computer terminals on the telephone network might largely be avoided, Nielson holds, if the potential services become widely available to the general public. He describes the characteristics of just such a terminal.

Grundfest and Baer address the regulatory issues arising from the increased utilization of computer communications in the home—the emerging home information center that could evolve from the linking of the family television set to computer data banks by existing telephone lines. Their paper examines potential barriers such as the separation of communications and information processing, regulatory jurisdiction, communications pricing, and information pricing and billing. Can these issues be resolved so that information services to the home can be offered in response to an unknown and unexplored market demand?

Availability of data bases does not necessarily mean immediate usage of these data bases. Lave, Leinhardt, and Ravin address the rate of adoption or usage of the MEDLINE system of the National Library of Medicine—the successor to MEDLARS, the first large-scale attempt to computer-search bibliographic references. Using a diffusion model the authors review usage growth, the patterns of use, and the degree to which access determines usage. Critical to achieving the desired economies of scale is making people aware of its existence. While a diffusion model can explain MEDLINE's usage growth, the process is inherently more complicated because of the system economics. Once potential users have been made aware of the service—a not inconsiderable task for which government subsidy ought not to be discounted—the pricing of services depending on time of day influences usage. Lave, Leinhardt, and Ravin address the issues of information as a product, a matter of growing interest in our emerging information economy.

Concluding this section are two papers examining the long-range policy implications of the introduction of electronic message services (EMS) to the nation. These papers provide valuable insight into the broader issue of new technology and its impact on existing and well-established national institutions. The emerging information economy threatens to seriously alter the structure of several national institutions, including the U.S. Postal Service, the banks, schools and universities, and the very nature of business and industry. Hence Lyon's paper reviews the impact of electronic telecommunications on the U.S. Postal Service. Lyons points out that the U.S. Postal Service has been experimenting with and studying electronic message delivery and other potential applications of telecommunications to their business since 1969, as several regulated carriers as well as private firms have been seeking to capture a portion of the attractive business of first-class mail service. While suggesting that policy decisions must be made that would better define fair and reasonable roles for both the public and private sectors in the EMS business, Lyons argues that the Postal Service may be required to play a significant role in the international field.

18 The Information Economy: Issues of Relevance

Albert J. Walderhaug

The delineation of an information sector within the U.S. economy, as developed in the study by Marc Porat, *The Information Economy* [1], is aimed at providing a tool that can be used by policymakers and interested researchers in fields related to information technology to analyze problems of an economic nature such as those caused by the introduction and diffusion of new technologies. These problems may have impacts that range across a number of market and nonmarket interests and are often wider in scope than the usual focus of individual policymakers. The comprehensive view of information activities displayed across economywide business and nonbusiness sectors in the broad framework of the national economic accounts can help place information activities within the context of the whole economy, and expose intersectoral relationships, and, therefore, coordinate actions needed to resolve problems that have horizontal dimensions affecting several interest groups.

The National Accounts

Objective of National Economic Accounting

The objective of national economic accounting is the creation of an information framework that is suitable for analyzing the operation of the economic system. (A comprehensive treatment of economic accounts can be found in [2] and [3].) A modern system is characterized by an increasingly complex network of transactions. There are three main transactors in this network. *Individuals* receive income, spend, and save. *Businesses* invest, buy materials, parts, supplies, and other services; hire labor; and sell their products on the market. *Governments* collect taxes from individuals and businesses, hire labor, buy goods and services, and provide variety of public services and transfer payments. The flow of data on the myriad transactions is the information on which the economist must rely. Organizing it is his task.

The ideas and conclusions expressed in this paper are the personal views of the author and should not be interpreted as representing the views of the Bureau of Economic Analysis.

Market transactions generally provide the basis for the national economic accounting. In some instances nonmarket transactions need to be identified and evaluated where, they represent an important element in the economic system—for example, home production and consumption of food, barter transactions or payment for services in-kind, the flow of services from homeownership, depreciation recorded on the books of a company.

The task of the national accountant is to organize the data so that the major parts of the economic system can be exposed and the structural changes that are taking place over time can be observed and analyzed. To be useful in economic models of the interactions among different parts of the economy, the framework of the accounts must provide a suitable level of disaggregation and deconsolidation so that the interrelationships can be observed and measured.

While the classification or arrangement of the accounts may imply an economic model, it is not necessary or desirable that a specific economic model underlie the design of the accounts. The accounting system should be one of general purpose, providing information for major classes of models and capable of modification or adaptation to specific analysis. Studies and analyses carried out in a national accounting framework are macro in character and must be oriented or conducted in terms of their setting in the economic system as a whole rather than focusing on the response of the individual decision maker or a narrow slice of the economy.

Structure of the U.S. National Accounts

The Porat study sets the information economy in the context of the U.S. national accounts. This section describes the accounts, to set the scene for an explanation of how they have been adapted to the information economy format.

The national income and product account is a consolidation of production accounts of individual producers. The production account for a firm would show revenues from the sale of its products as a credit and expenses plus earnings as a debit. In measuring the output of the nation as a whole, the sales of one firm to another for use in further production are netted out in order to avoid double-counting goods used in further production. The measure of the output of an economy is the sum of the flow of goods and services to final users or, alternatively, the claims against these goods represented by the income generated in their production. Figure 18-1 shows the relationship in T-account form.

Receipts from final product flows (credits) consist of sales to consumers (personal consumption expenditures), sales to business on capital account and change in business inventories (gross private domestic investment), sales to government (government purchases of goods and services), and net sales to foreigners (net exports). The sum of these final product flows equals gross national product (GNP). The same total can be derived by summing the incomes generated in its production. This is essentially the value added by each producer. It

1. Compensation of employees	10. Personal consumption expenditures
2. Proprietors' income	11. Gross private domestic investment
3. Rental income of persons	12. Net export of goods and services
4. Corporate profits and inventory valuation adjustment	13. Government purchases of goods and services
5. Net interest	
6. Business transfer payments	
7. Indirect business tax and nontax liability	
8. Less: Subsidies less current surplus of government enterprises	
9. Capital consumption allowances	
Gross National Product	Gross National Product

Figure 18-1. The Gross National Product: National Income and Product Account

is the difference between the value of his output and what he pays to other producers for the input of goods and services required for its production. This value added is shown on the debit side of the T-account, and the items can be grouped into payments to employees (employee compensation), payments to government (indirect business taxes), and the return to owners of business and capital (the remaining debit items).

In figure 18-2 the components of GNP have been arranged in an input-output format—a matrix containing data arranged in rows and columns. The row labeled "producers" shows the sales of these producers to the final markets; the column headed "producers" shows the income types originating from production. The black box represents producer-to-producer sales of the goods and services used in production which have been omitted in the calculation of GNP to avoid duplication in the measurement of output.

However, the sales of one producer to another are important in revealing the structure of the economy and understanding the way industries interact in producing the GNP. Figure 18-3 expands the black box to display separately the industries producing and consuming raw and semifinished products and intermediate services. In addition, the sales to final markets are disaggregated to show their industry composition; similarly, the value added is broken down to show the industry in which it originated.

The Information Industry Accounts

The Porat study contains three approaches to describing and measuring the information economy. The trade-offs in each approach are span of time versus detail. The three sets of data are (1) workforce and occupations: 1860-1980 (projected), (2) national income: 1929-1974, and (3) an input-output table: 1967.

	Producers	Persons	Investors	Foreigners	Government	
Producers		Personal consumption expenditures item 10	Gross private domestic investment item 11	Net exports of goods and services item 12	Government purchases of goods and services item 13	Gross national product
Employees	Employee compensation item 1					
Owners of business and capital	Profit-type income and capital consumption allowances items 2,3,4,5,6,8,9					
Government	Indirect business taxes item 7					
	Gross national product					

Figure 18-2. The Gross National Product: In Input-Output Format

		Producers								Final Markets			
		Agriculture	Mining	Construction	Manufacturing	Trade	Transportation	Services	Other	Persons	Investors	Foreigners	Government
Producers	Agriculture									Personal consumption expenditures	Gross private domestic investment	Net exports of goods and services	Government purchases of goods and services
	Mining												
	Construction												
	Manufacturing												
	Trade												
	Transportation												
	Services												
	Other												
Value Added	Employees	Employee compensation								Gross National Product			
	Owners of business and capital	Profit-type income and capital consumption allowances											
	Government	Indirect business taxes											

Figure 18-3. Input-Output Flow Table

Information Occupations

In defining the information sector of the economy, it has been necessary to classify occupations into the information domain. The definition has to be somewhat judgmental because a certain amount of knowledge is required to do any job. The information occupation is taken as being primarily engaged in the production, processing, or distribution of information products or services.

Of the 422 occupations used by the Bureau of the Census in compiling data on occupation by industry in 1970 [4], 188 have been identified as information workers. They have been further classified according to broad functional categories: (1) the *knowledge producers*, which include the scientific and technical professions as well as the consultative occupations such as lawyers, accountants, special counselors, who provide private information services; (2) the *knowledge distributors,* who are the educators, the librarians and archivists who manage information resources, and the media people, authors, critics, editors, artists, radio, and television announcers; (3) the *market search* and *coordination specialists* including the information gatherers, ranging from investigators to meter readers, the people operating in markets to bring buyers and sellers together, planning and control workers such as the administrators and managers as well as supervisors who control processes; (4) the *information processors*, including secretaries, clerks of all kinds, tellers, bookkeepers, cashiers; (5) *information machine workers* such as office machine workers, operators of printing machinery, computer operators, telecommunication workers, and repairers of the related equipment.

The nature and content of jobs may change over time, but the main functions remain, and their relative importance can be traced over time through the data in the decennial censuses of population. These data were used to separate occupations into information and noninformation categories. Figure 18-4 traces the changing relationship of the information workforce—first to the noninformation workforce as a whole and then to the latter broken down into occupations associated with the broad sectors of the economy: agriculture, industry, and services. The information worker versus noninformation worker detail shows two alternative estimates—one using a more restrictive definition of information worker and the other a more inclusive classification. The more inclusive classification shows the two workforces almost equal by 1970.

The four-sector aggregation of jobs (figure 18-5) shows the startling shift in occupations among broad producing sectors. Industry becomes dominant over agriculture sometime between 1900 and 1910. The share of industrial occupations starts its relative decline about 1950, and they are exceeded by the information occupations before 1960. Service occupations are projected to top the industrial occupations before 1980. It is not hard to understand where the characterization of the U.S. economy as postindustrial came from.

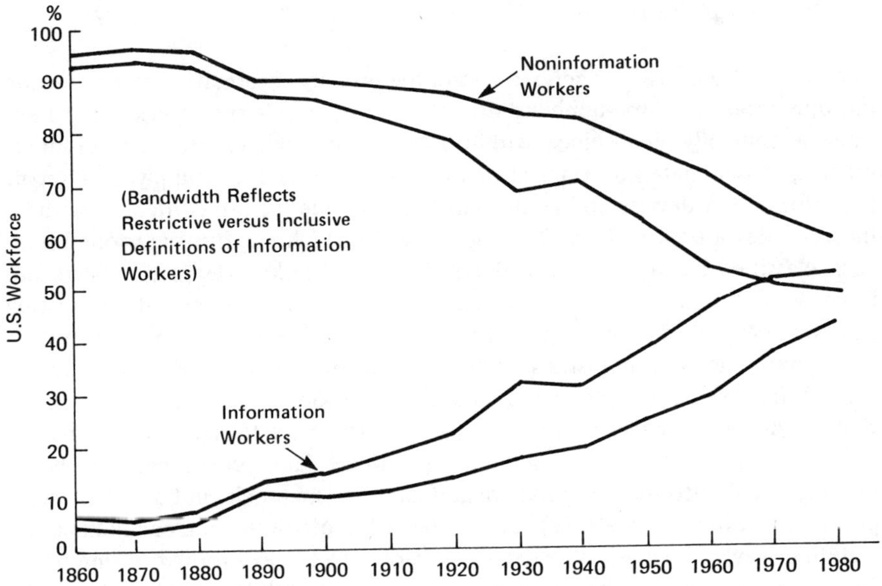

Source: Marc Porat, *The Information Economy* (Washington, D.C.: U.S. Department of Commerce, 1977).
Note: 1980 projections supplied by the Bureau of Labor Statistics (unpublished)

Figure 18-4. Time Series of U.S. Labor Force (1860-1980) Two-Sector Aggregate, by Percentage

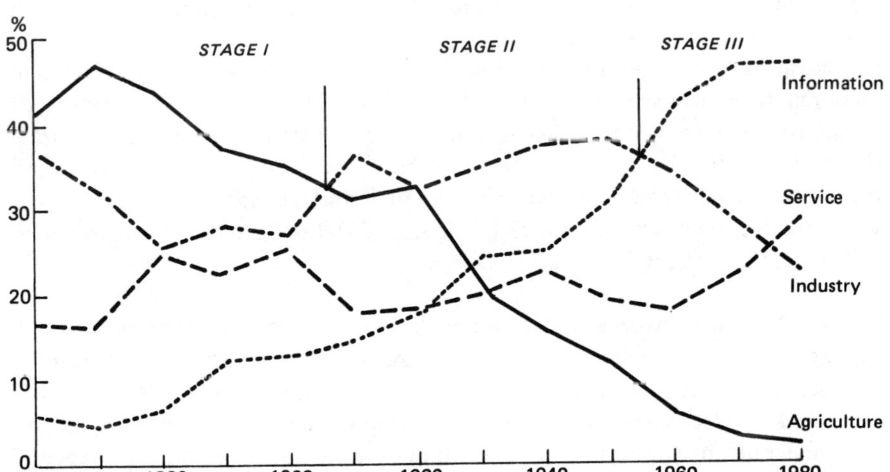

Source: Marc Porat, *The Information Economy* (Washington, D.C.: U.S. Department of Commerce, 1977).

Figure 18-5. Four-Sector Aggregation of the U.S. Workforce, 1860-1980, by Percentage (using median estimates of information workers)

The Primary Information Sector

The information activities defined within the primary information sector include the operation of firms supplying the information goods and services, and their value is generally determined within a market context. These firms produce, process, disseminate, or transmit knowledge or messages, supply and repair the information devices and equipment, and provide the infrastructure in which the activities are carried on. The industries in which the firms are found have been classified according to broad function: (1) the knowledge producers and inventive industries such as commercial research and development and testing laboratories, professional and consultative services; (2) information distribution and communication industries: education, information services, the media; (3) risk management: insurance, financial institutions; (4) search and coordination industries: exchanges, brokerages, agencies, advertising; (5) information processing and transmission services: printing, data processing, telephone, telegraph; (6) information goods manufacturers: materials and supplies, components, investment goods; (7) support facilities: office, school, communication facilities, new or rented. The detailed typology of the information industries defined as primary is shown in table 18-1 identified by standard industrial classification (SIC) number. The list is long and comprehensive, and while one might argue about the inclusion or omission of one or another item, the general magnitude or trend of the sector as a whole should not be affected. The study also identifies and includes in the primary information sector the portions of the government activities that provide primary information services; that is, operations that have counterparts in the private business sector such as the Government Printing Office.

Isolating and exposing these industries in the framework of the national accounts is shown schematically in figure 18-6. Depending on the nature of an industry's output, it is assigned in whole or in part to the primary information sector. The diagram indicates that both the input detail (columns) and the output detail (rows) of the information industries are separated from the sector where they are classified. In figure 18-7 they are shown reassembled and called the primary information sector.

Primary National Income and Product. The output of the primary information sector is sold to both intermediate and final users. The contribution of information output to GNP is measured by the sales to final demand. It is difficult to isolate these sales from the accounts. The detail needed is usually compiled only in the years for which input-output tables have been prepared. The estimates of primary information final demand were made only for the year 1967. Figure 18-7 shows that information products were \$175 billion of a total GNP of \$795 billion in 1967—about 22 percent. Further detail on sales of primary information goods to final purchasers is given in table 18-2.

Table 18-1
The Primary Information Sector at the 507-Order Level

		Related SIC or Census Codes
Construction: Information buildings		
11	Information buildings: Office, education, communication	
	110202 New office buildings	pt. 15, pt. 17
	110207 New education buildings	pt. 15, pt. 17
	110301 New telephone & telegraph facilities	pt. 16, pt. 17
	110504 New oil & gas exploration	pt. 138
	110505 New military facilities	pt. 15, pt. 16, pt. 17
12	Maintenance & repair on information buildings	
	120201 Maintenance & repair on other nonfarm buildings	pt. 15, pt. 17
	120204 Maintenance & repair on telephone and telegraph facilities	pt. 16, pt. 17
Manufacturing: Nondurable information goods		
24	Paper: Printing (Excluding Boxes)	
	240200 Paper mills, excluding building paper	2621
	240400 Envelopes	2642
	240701 Paper coating & glazing	2641
26	Printing & publishing	
	260100 Newspapers	2711
	260200 Periodicals	2721
	260301 Book publishing	2731
	260302 Book printing	2732
	260400 Miscellaneous publishing	2741
	260500 Commercial printing	2751, 2752
	260601 Manifold business forms	2761
	260602 Blankbooks & looseleaf binders	2782
	260700 Greeting card publishing	2771
	260801 Engraving & plate printing	2753
	260802 Bookbinding & related work	2789
	260803 Typesetting	2791
	260804 Photoengraving	2793
	260805 Electrotyping & stereotyping	2794
Manufacturing: Durable information goods		
23	Office furniture & equipment	
	230100 Wood office furniture	2521
	230200 Metal office furniture	2522
	230300 Public building furniture	2531
48	Printing & paper machinery	
	480400 Paper industries machinery	3554
	480500 Printing trades machinery	3555
51	Computers, calculators, office equipment	
	510101 Electronic computing equipment	3573
	510102 Calculating & accounting machines	3574
	510200 Typewriters	3572
	510300 Scales & balances	3576
	510400 Office machines	3579

Table 18-1 (Continued)

		Related SIC or Census Codes
53	Electronic measuring instruments	
	530100 Electric measuring instruments	3611
56	Radio, television, communication equipment	
	560100 Radio & television receiving sets	3651
	560200 Phonograph records	3652
	560300 Telephone & telegraph apparatus	3661
	560400 Radio & television communication equipment	3662
57	Electronic components	
	570100 Electron tubes	3671, 3672, 3673
	570200 Semiconductors	3674
	570300 Electronic components	3679
58	Miscellaneous electronic instruments	
	580300 X-ray apparatus & tubes	3693
62	Mech Measuring & control instruments	
	620100 Engineering & scientific instruments	3811
	620200 Mechanical measuring devices	3821
	620300 Automatic temperature controls	3822
	620701 Watches & clocks	3871
63	Photographic & related equipment	
	630100 Optical instruments & lenses	3831
	630300 Photographic equipment & supplies	3861
64	Advertising Signs & Displays	
	640501 Pens & mechanical pencils	3951
	640502 Lead pencils & art goods	3952
	640503 Marking devices	3953
	640504 Carton paper & inked ribbons	3955
	641100 Signs & advertising display	3993
Information services		
66	Telecommunications, excluding radio & television	
	660000 Communications, excluding radio & television	48 (excluding 483)
67	Radio & television broadcast & CATV	
	670000 Radio & television broadcasting	483
69	Trade Margin on information goods	
	690100 Wholesale trade 690200 Retail trade	50, 52, 53, 54, 55, 56, 57, 58, 59, 7396 p. 8099
70	Finance & Insurance: Information transactions	
	700100 Banking	60
	700200 Credit agencies	61, 67
	700300 Security & commodity brokers	62
	700400 Insurance carriers	63
	700500 Insurance agents & brokers	64
71	Real Estate: Fees, royalty, office rentals	
	710200 Real estate	65 (excluding 6561)

Table 18-1 (Continued)

		Related SIC or Census Codes
72	Repair: Radio, television equipment	
	720200 Personal & repair services, excluding auto repair, barber, beauty shops	72 (excluding 723, 724), 76 (excluding 7694, 7692 & pt. 7699)
73	Miscellaneous Business information services	
	730100 Miscellaneous business services	73 (excluding 731, 7396) 7694, 7692, pt. 7699
	730200 Advertising	731
	730300 Miscellaneous professional services	81, 89 (excluding 8921)
76	Motion pictures, theater	
	760100 Motion pictures	78
	760200 Amusement & recreation services	79
77	Medical, education, nonprofit	
	770100 Physician's offices	801, 802, 803, 804
	700300 Other medical & health services	0722, 807, 809, excluding pt. 8099
	700400 Educational services	82
	770500 Nonprofit organizations	84, 86, 8921
78	Postal Service	
	780100 Post office	
82	Office Supplies	
	820000 Office supplies	
84	Federal Government: Information workers	

Personal consumption accounted for 48 percent of the total. Government purchases were next, with 38 percent. Twelve percent were investment goods and 2 percent, net exports. Within personal consumption expenditures, information goods made up 17 percent of the total. In investment their share was slightly higher, 18 percent. Their share in government purchases was 37 percent.

Sales of information products to final demand, however, are not a true measure of the contribution of the information sector to GNP. Part of its contribution has been absorbed as inputs in the making of noninformation products through the intermediate sales to other producing industries. These sales lose their identity as information products as they are incorporated into other goods and services. A truer measure of the primary information sector's contribution to GNP is the value it *adds* to the materials and services it purchases from other industries and uses in the production process. Again, in 1967 the value added in the primary information industry is estimated at $200 billion, or 25 percent of GNP. Table 18-2 also shows the detail for the components of value added.

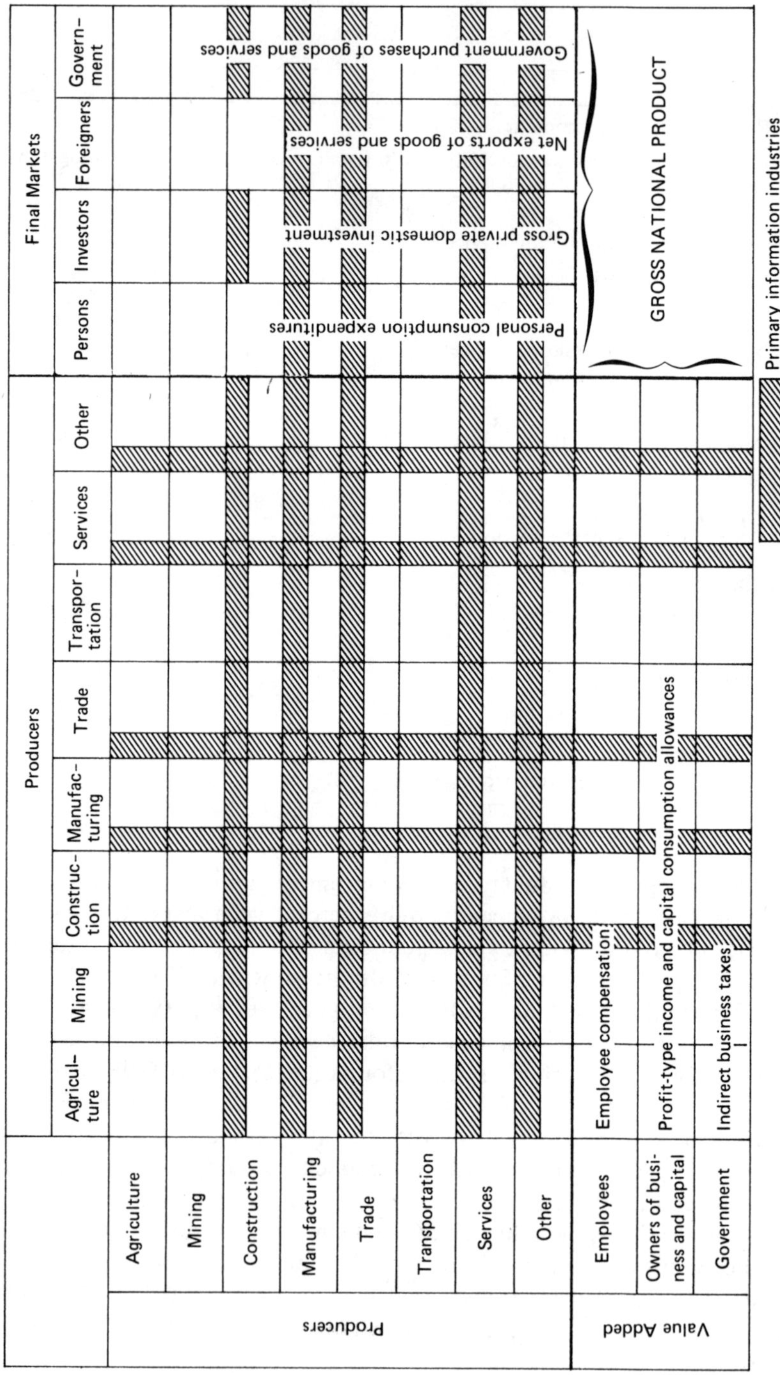

Source: U.S. Department of Commerce, Bureau of Economic Analysis.

Figure 18-6. Primary Information Sector in the National Accounts

Producers	Intermediate Consumers			Final Demand	Percentage of GNP
	Primary	Secondary	Noninformation		
Primary information sector	$69,754	$78,917	0	$174,585	21.9
Secondary information sector	0	616	227,169	27,440	3.4
Noninformation sector	59,538	0	511,965	593,363	74.6
Value added	200,025	168,073	427,290	GNP = $795,388	
Percentage of GNP	25.1	21.2	53.8		

Source: Marc Porat, *The Information Economy* (Washington, D.C.: U.S. Department of Commerce, 1977).

Figure 18-7. The Structure of the Information Economy, 1967 (millions of dollars)

Table 18-2
Summary of Gross National Product and Income Originating in the Primary and Secondary Information Sectors: 1967
(millions of dollars)

	Total	*Percentage*	*Primary*	*Percentage*	*Secondary*	*Percentage*	*Noninformation*	*Percentage*
Final demand								
Total GNP	795,388	100.0	174,585	100.0	27,440	100.0	593,363	100.0
Percentage	100.0		21.9		3.4		74.6	
Personal consumption expenditures	490,358	61.7	83,742	48.0	–	–	406,616	68.5
Percentage	100.0		17.1				82.9	
Gross private domestic investment	120,829	15.2	21,583	12.4	–	–	99,246	16.7
Percentage	100.0		17.9				82.1	
Net exports	4,937	1.0	2,942	1.7	1,586	5.8	409	0.1
Percentage	100.0		59.6		32.1		8.3	
Government purchases	180,188	22.7	66,318	38.0	25,854	94.2	88,016	14.8
Percentage	100.0		36.8		14.3		48.9	
Statistical adjustment	–924	–0.6	–		–		–924	–0.2
Gross product originating								
Total GNP	795,388	100.0	200,025	100.0	168,073	100.0	427,290	100.0
Percentage	100.0		25.1		21.1		53.8	
Employee compensation	467,240	58.7	136,488	68.2	139,405	82.9	191,347	44.8
Percentage	100.0		29.2		29.8		41.0	
Profit-type income	188,377	23.7	37,447	18.7	22,848	13.6	128,082	30.0
Percentage	100.0		19.9		12.7		68.0	
National income	655,617	82.4	173,935	86.9	162,253	96.5	319,429	74.8
Percentage	100.0		26.5		24.7		48.8	
Capital consumption allowance	68,895	8.7	12,151	7.0	5,820	3.5	50,924	11.9
Percentage	100.0		17.6		8.4		74.0	
Indirect business tax and business transfer payments	70,876	8.9	13,939	6.1	–	–	56,937	13.3
Percentage	100.0		19.7		–		80.3	

Source: Marc Porat, *The Information Economy* (Washington, D.C.: U.S. Department of Commerce, 1977).

Note: Estimates in the study used data from the national accounts at different stages during the benchmark revision, and therefore the totals do not reflect the latest published figures in all cases.

To observe the growth of the primary information component in a national accounts context, estimates were made to match the national income series spanning the period from 1929 to 1974. National income includes employee compensation and the returns to investors in property–profits, interest, rents, and so on. It excludes measures of capital consumption (depreciation) and indirect business taxes paid which are included in the full measure of value added. The estimates of national income originating in primary information industries were made at the two-digit SIC level of detail so that their growth can be observed relative to the primary information industry as a whole as well as to the non-information component of the industries from which they were extracted.

Table 18-3 summarizes the growth of the information share of national income originating in several broad categories of industries and relates it to the changing relative importance of the industry group itself as contributors to national income. The primary information share of national income rises from 18 percent of total national income in 1929 to almost 28 percent by 1972. Which industries are mainly responsible for this growth?

We get a clearer picture if we focus on the relatively undisturbed postwar period starting in 1948, when the primary information share of national income was about the same as in 1929. The industries in which the relative size of their contribution to total national income increased *and* in which the relative importance of the primary information component of the national income they originate is growing include finance, insurance, and real estate; services; and state and local government. The finance group's share of total national income rose from 8 to 12 percent, and the importance of its primary information component grew from 37 to 49 percent. The service industry's contribution to total national income rose from 9 to 13 percent, and the importance of its primary information component increased from 39 to 48 percent. National income originating from state and local government rose from 4 to 10 percent of the total, and its primary information component grew from 43 to over 50 percent.

Producers whose contribution to total national income remained relatively constant but who show an increasing share of their national income coming from the primary information component include durable goods manufacturing and the federal government. The national income share of manufacturers of durable goods remained fairly constant, between 16 and 18 percent, during the period. The primary information income share of the income they originate rose from 9 to 15 percent. The federal government's contribution to national income was about 6 percent through the period, and its primary information component rose from 35 to 40 percent.

In the nondurable goods manufacturing industries and in wholesale and retail trade, increases in the primary information component of the income they originate are offset by declines in their share of total national income. Construction and communications show relatively constant shares in both components during the period.

Table 18-3
Percentage of National Income (NI) in Major Industry Groups and Percentage of Primary and Secondary Information National Income in Each Group: 1929-1972

	1929	*1933*	*1939*	*1948*	*1954*	*1958*	*1968*	*1967*	*1972*
All industries									
Industry NI/total NI	100.0	100.0	100.0	100.0	100.0	100.0	100.0	100.0	100.0
Information NI/industry NI	31.4	31.9	31.8	37.9	42.8	46.0	49.2	51.2	52.8
Primary information NI/industry NI	18.3	22.8	19.4	16.8	20.6	22.9	24.5	26.5	27.8
Secondary information NI/industry NI	13.1	9.1	12.4	21.1	22.2	23.1	24.7	24.7	25.0
Construction									
Industry NI/total NI	4.4	2.0	3.2	4.5	5.1	5.1	5.1	5.2	5.5
Primary information NI/industry NI	14.8	15.2	15.9	15.9	18.7	19.3	18.1	18.0	17.3
Secondary information NI/industry NI	–	–	–	–	–	–	–	34.0	–
Manufacturing									
Industry NI/total NI	25.3	19.1	24.9	30.4	31.0	29.1	29.6	29.5	26.3
Primary information NI/industry NI	10.0	12.7	10.2	10.0	11.4	12.4	13.4	15.2	14.9
Secondary information NI/industry NI	–	–	–	–	–	–	–	28.2	–
Nondurable goods									
Industry NI/total NI	13.2	12.3	12.5	14.6	13.0	12.3	11.8	11.4	10.2
Primary information NI/industry NI	14.3	16.0	13.8	11.2	13.0	13.7	14.1	14.8	15.1
Secondary information NI/industry NI	–	–	–	–	–	–	–	28.6	–
Durable goods									
Industry NI/total NI	13.0	6.8	12.4	15.8	18.0	16.7	17.8	18.1	16.1
Primary information NI/industry NI	5.9	6.8	6.8	8.9	10.3	11.4	12.9	15.5	14.8
Secondary information NI/industry NI	–	–	–	–	–	–	–	28.0	–
Communications									
Industry NI/total NI	1.3	1.7	1.5	1.2	1.7	1.9	2.0	2.0	2.1
Primary information NI/industry NI	100.0	100.0	100.0	100.0	100.0	100.0	100.0	100.0	100.0
Secondary information NI/industry NI	–	–	–	–	–	–	–	–	–

Wholesale and retail trade									
Industry NI/total NI	15.6	14.0	17.4	17.7	15.8	15.7	15.1	15.0	15.1
Primary information NI/industry NI	13.0	14.5	12.5	10.0	11.8	12.4	12.8	13.4	13.7
Secondary information NI/industry NI	–	–	–	–	–	–	–	42.9	–
Finance insurance and real estate									
Industry NI/total NI	14.7	14.6	11.0	8.4	10.8	11.6	11.2	11.4	12.0
Primary information NI/industry NI	34.7	42.4	38.1	36.5	39.6	41.1	40.9	42.8	49.5
Secondary information NI/industry NI	–	–	–	–	–	–	–	3.3	–
Services									
Industry NI/total NI	10.2	12.8	10.4	8.8	9.1	10.3	11.2	12.1	12.8
Primary information NI/industry	34.7	42.4	38.1	39.4	42.6	44.8	46.5	48.6	48.1
Secondary information NI/industry NI	–	–	–	–	–	–	–	24.4	–
Government									
Industry NI/total NI	5.9	13.2	11.7	8.7	11.8	12.6	13.4	14.4	15.9
Primary information NI/industry NI	43.5	43.1	39.1	38.9	37.6	40.3	43.0	43.8	46.7
Secondary information NI/industry NI	–	–	–	–	–	–	–	19.6	–
Federal government									
Industry NI/total NI	1.7	4.1	5.7	4.7	6.6	6.4	6.1	6.3	6.2
Primary information NI/industry NI	47.8	43.6	34.7	35.0	32.5	35.8	36.9	35.9	40.3
Secondary information NI/industry NI	–	–	–	–	–	–	–	19.0	–
State and local government									
Industry NI/total NI	4.2	9.1	6.0	4.0	5.2	6.2	7.2	8.0	9.7
Primary information NI/industry NI	41.9	42.8	43.3	43.5	44.0	44.9	48.1	50.4	50.8
Secondary information NI/industry NI	–	–	–	–	–	–	–	20.0	–

Source: Marc Porat, *The Information Economy* (Washington, D.C.: U.S. Department of Commerce, 1977).

Some insight into the methodology used to create the primary information sector national income will give a better appreciation of that sector's significance. The BEA national income series are shown by industry at the two-digit level of detail in the Standard Industrial Classification (SIC). If a whole two-digit industry is assigned to the primary information sector, no further estimating is required to derive the primary information allocation. In 1929, 40 percent of the primary information national income required no splits. These industries include communications, most of banking and finance, the motion picture industry, education, and legal and miscellaneous professional services. By 1972 the share of these industries dropped to just over 30 percent of primary information national income. In 1929 industries representing about 55 percent of national income were judged to have no primary information content. By 1972 the share of these industries dropped to 30 percent. Thus over the period of the estimates an increasing proportion of national income (16 to 40 percent) was subject to allocation between information and noninformation categories.

The general method of making the allocation was proration of the industry's national income based on a related measure of the information component. Availability of data made it necessary to use different measures for different industries. The use of the prorate implies that the national income share of the allocator is constant for all the industries within the two-digit classification.

Input-Output Table with Primary Information Industries. In setting the primary information sector in an input-output (I-O) format, the transactions detail is increased to show the interrelationships among the information and noninformation industries in the economy as well as the sale of information goods to final demand and value added in their production. Table 18-4 begins to reveal the technology underlying the functioning of the economy.

The primary information sector is shown as part of a 108-industry input-output table which has been adapted from the U.S. table for 1967 [5]. Information components have been extracted from 26 of the 82 industries in the U.S. table and set up as additional producing and consuming industries. In splitting out the primary information industries, data from the 484-industry U.S. table was used. Of these 484 industries, 76 producing industries were assigned in whole or in part to the primary information sector. Of these, 59 are allocated in total and account for over 60 percent of the primary information sector. The remaining 17 industries were split based on the composition of their products in 1967. The inputs were split apart in proportion to the production of information and noninformation products. The assumption in making the split is that an industry uses the same input technology or input structure for all its products, both information and noninformation.

Table 18-4
Data Base in *The Information Economy*

	Available Detail	Time Period
I. Labor force		
A. Information occupations	Total	1860-1980
B. Noninformation occupation by sector	3	1860-1980
C. Employee compensation		
1. Information occupation	188	1967
Function	13	1967
2. Occupation	422	1967-1970
Industry	108	1967-1970
II. Primary information sector		
A. Gross national product		
1. Final demand by category of purchaser	5	1967
Supplying industry	2-digit SIC	1967
2. Value added by industry	2-digit SIC	1967
VA component	5	1967
B. National income, total	2-digit SIC	1929-1974
C. Input-output matrices		
Interindustry transactions	108 industries	1967
Direct requirements	108 industries	1967
Total (direct and indirect) requirements	108 industries	1967
III. Secondary information sector		
A. Gross national product		
1. Final demand by category of purchaser	2	1967
Supplying industry	20 groups	1967
2. Value added by industry	20 groups	1967
VA component	3	1967
B. National income	Total	1929-1974
C. Input-output matrices		
Interindustry transactions	190 industries	1967
Direct requirements	190 industries	1967
Total (direct and indirect) requirements	190 industries	1967
IV Other		
A. Detailed I-O transactions	507 industries	1967

In splitting the rows of the industries producing both information and noninformation goods, it was frequently possible to distribute the products separately to using industries so the output was not split proportionally between all users.

The Secondary Information Sector

If the concept of information as an activity or a definable segment of the economy were limited to the information goods and services exchanged in the marketplace, a substantial share of information activities would be missed.

All the information services that are produced for internal consumption by noninformation firms and government would not be covered. These are the services supporting all the planning, coordinating, management, and communication functions that are essential to controlling the operation of any enterprise or organization. Together they have been characterized in this study as bureaucracies—public if they are in government, and private if they are in business. Although essential to the operation of the firm, they are not directly involved in fabricating the product or performing the service, so they are in some sense ancillary, or secondary.

In defining a secondary information sector, the aim is to measure these self-performed information services in some systematic and useful way. Many of these services have counterparts in the market economy: business management, accounting, and legal services can be hired; advertising, duplicating and printing services, and electronic data processing have market alternatives. In fact, the self-performed functions may be in competition with their commercial counterparts. Hence for analytical purposes, it would be very desirable to set up the various internal information functions as sort of quasi firms if information were available.

The estimates of secondary information activity as a whole had to be made under the simplifying assumption that the sum of the functions is equivalent to the information resources purchased by the firm: (1) the labor services of all the information workers (employee compensation and labor income of proprietors); (2) the capital services of the information machines (measured as the capital consumption allowances of buildings and equipment identified as products of the primary information sector); and (3) a portion of the intermediate inputs associated with the performance of the information activity. These intermediate inputs have been measured by the purchases of information goods and services from the primary information industries.

Indirect business taxes and profit-type income, the remaining components of value added, have been allocated entirely to the noninformation part of the industry. Theoretical and data problems precluded a meaningful allocation of these items to the two parts of an industry, and the essentially conservative decision was made to leave the taxes and profits with the principal activity of the firm. Value added in the secondary information sector, then, is equal to compensation of information workers (employees, proprietors, and unpaid family workers) plus capital consumption allowances on information structures and equipment. National income originating in the secondary information sector therefore reflects just the compensation figure.

The Secondary Input-Output Matrix. The input-output table, developed to isolate the primary information sector, provides a convenient basis for isolating the secondary information sectors as defined. Figure 18-8 shows schematically how it was extracted from the I-O table. The primary information goods and services and the information labor and depreciation on information capital

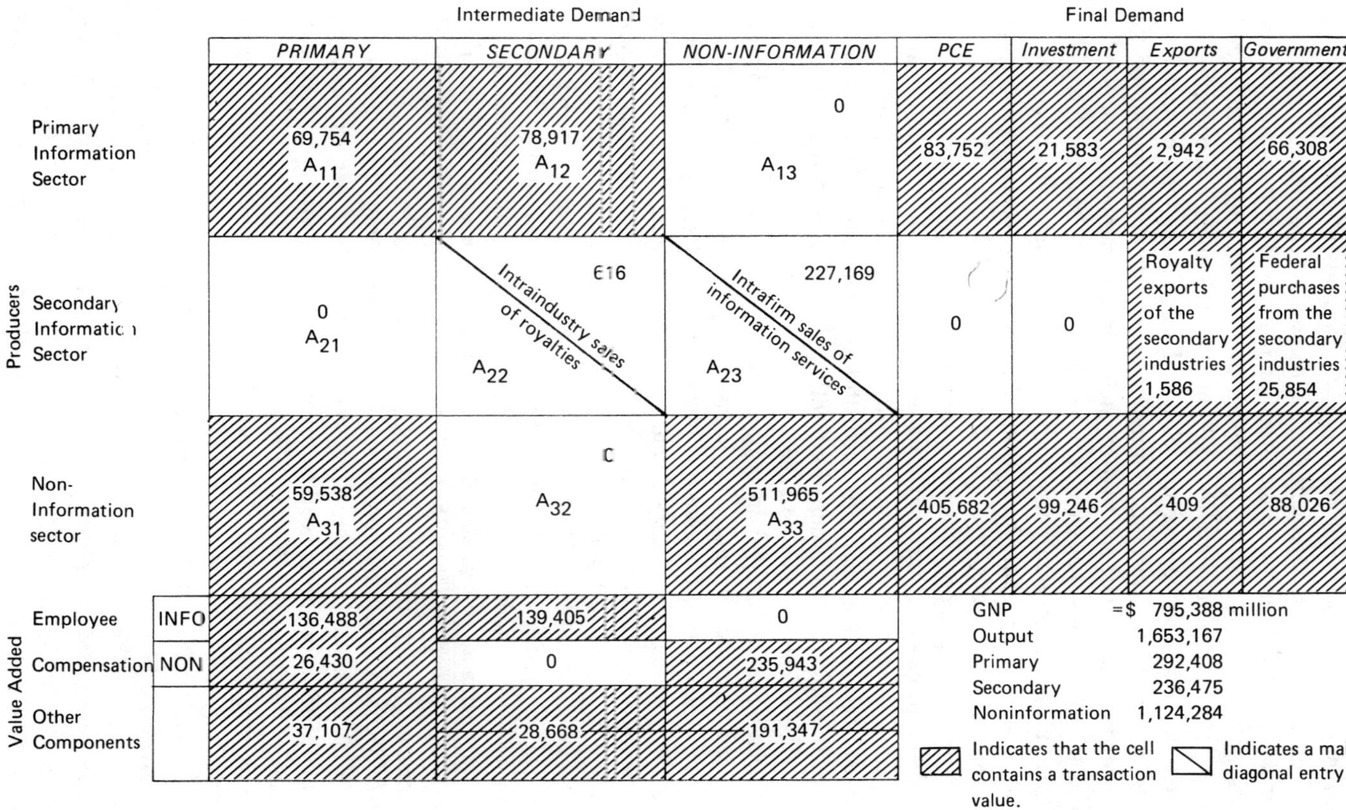

Source: Marc Porat, *The Information Economy* (Washington, D.C. U.S. Department of Commerce, 1977).

Figure 18-8. Input-Output Schematic Diagram Showing the Secondary Information Sector

components of value added are removed from the column (inputs) of the noninformation industries. The real output of the secondary sector representing receipts from royalties, management fees, and research and development sold to others is separated from the noninformation industry. Royalties and management fees are sold intraindustry or as exports to foreigners. Research and Development is sold to government. In addition, the employee compensation of government information workers not classified as primary (the bureaucracy) is sold to government. The remaining amounts, the imputed output of the quasi-information industries, are sold to their principal noninformation industries to replace the inputs that were removed in setting up the secondary industries. The result is an input-output table with 190 rows and columns consisting of the 26 primary information industries, 82 secondary information industries, and 82 noninformation industries.

The estimates of labor income (employee compensation and the labor component of proprietors' income) were developed using detailed information on occupation by industry, to break down the labor income component of value added for the industries into occupational detail. The compensation of workers in the information occupations within the noninformation industries could then be extracted and assigned to the secondary information sector.

Depreciation on investment in information capital in the noninformation industries was estimated in rather crude fashion by first prorating capital consumption from the national income and product account (NIPA) industry detail (two-digit SIC) to the I-O industry detail. Then for each noninformation industry the capital consumption total was apportioned between information and noninformation capital using data from the 1967 capital flow matrix which shows the investment by each industry in new structures and equipment by type [6]. The detailed investment inputs were separated into information and noninformation goods yielding the weights to break down the depreciation.

The final-demand and value-added components of the secondary information sector are shown in table 18-2, which summarizes the estimates of GNP and income originating in both the primary and secondary information sectors for 1967. Secondary information products constituted only about 3 percent of total final demand. However, they were 32 percent of net exports, and compensation of information workers in government constituted 14 percent of government purchases.

The secondary information sector accounted for 21 percent of GNP compared with 25 percent for the primary information sector. Together they make up 46 percent of GNP in 1967.

Employee compensation in the secondary information sector is roughly equal in amount to that in the primary information sector, and together they make up about 60 percent of total employee compensation. Profit-type income originating in the secondary information sector is less than in the primary information sector—13 percent versus 20 percent of the all-industry total. However, profit-type income in the second sector is limited to labor income of proprietors, whereas the primary sector includes return on investments as well the measure of profit-type income. National income originating in both information sectors was about 25 percent each, accounting for just over 50 percent of total national income in 1967.

In commenting on the definition and methodology for constructing the secondary information sector, I would like to make a couple of points. Imputing the value of information services performed by the quasi-information firms as equal to the input of information products plus the labor income of information workers and consumption of information capital may either overstate of understate individual industries.

The possibility for understatement is due to the exclusion of inputs from noninformation industries in valuing the output of secondary information services. The input structure of their market counterparts in the primary information sector includes purchases from both information and noninformation industries. About 25 percent of intermediate inputs into information industries such as finance and insurance and the business service industry are from the noninformation industries. They include such items as petroleum products and utilities for heat and light, expenditures for business travel and entertainment, and other noninformation materials and services.

On the other hand, it is likely that all information material inputs are not consumed in performing information services. Some of the materials are components installed in the noninformation products of the noninformation industries. For example, mechanical measuring and controlling instruments (information products) include speedometers installed in automobiles (noninformation product). Similarly, electronic components are inputs to a wide variety of industries such as ordnance, industrial machinery, electrical machinery, and transportation equipment.

The latter item, particularly, may have differing impacts on the industries in the economy and needs to be recognized and dealt with systematically in defining an information activity and searching for functional relationships within the economy.

Time Series of the Secondary Information Sector. To assess the growth of the secondary information sector, the study provides estimates of national income originating in the sector for the period 1929 to 1974. These estimates match the estimates of national income for the primary information sector and together they provide a measure of the growth of national income in the information sector as a whole. Table 18-3 shows the growing relative importance of the information sectors in the economy as a whole in the 1929 to 1972 period. National income in the secondary information sector increased from 13 percent to 25 percent of total national income. The primary information sector increased from 18 percent to 28 percent; together, representing the information sector, they increased from 31 percent to 53 percent of national income.

The task of estimating national income originating in the secondary information sector is difficult inasmuch as no data regarding these activities are available directly from the national accounts on which to base the estimates. They rest primarily on a time series of information workers derived from occupational information in the decennial censuses. To estimate the series for the years between the decade years, it was necessary to use the annual rate of increase in the proportion of information workers in the workforce to allocate the total of persons engaged in production between information and noninformation workers.

The methodology used to derive the secondary national income has explicit and implicit assumptions that lead to estimates that I fear may be questionable. First, it is assumed that national income generated by a worker in the primary information sector is equal to national income generated by a information worker in the secondary information sector. However, the workforce from which national income in the primary information sector is generated consists of both the information workers and the noninformation workers who fabricate the information devices and support the information services produced in the sector. Further, primary information national income includes the return on invested capital (profit-type income), which is excluded by definition from the secondary information sector. Differences in wage rates between information and noninformation workers and variability in profit-type return over the time period would result in different amounts of national income per worker in the two sectors.

The second assumption, not stated explicitly, but implied in the formula used to calculate secondary income is that national income per worker is the same in the information and noninformation sectors of the economy except for a constant factor applicable to the secondary national income estimate.

$$N^s = [N^t (L^i/L^t) - N^p]\lambda$$

where

N^s = secondary information national income

N^t = total national income

N^p = primary national income

L^i = primary and secondary information labor force

λ = correction factor to link with estimate of N^s for 1967

The resulting estimates of total information national income are proportional to changes in the information share of the workforce. This can be seen graphically in figure 18-9 showing the increase in the information economy's share of national income. The smooth upward trend reflects the increase in the proportion of the information workers in the total workforce.

The estimates of primary information national income, however, were not based on information workers but were factored out of national income on the basis of the relative importance of information products in the industry as indicated; they contain income attributable to noninformation workers as well as information workers and also profit-type income. To subtract these estimates from a series of total information national income that is not derived in a consistent manner results in a residual estimate of secondary information national income that is difficult to interpret. Figure 18-9 shows that the secondary information share of national income mirrors the movements of the primary information national income yielding trends that do not appear to be a reasonable reflection of expected development of the secondary information sector.

Questions of Relevance

So far, this paper has dealt with the conceptual framework for defining an information activity and the major effort to measure it in an objective way within the conventional system of national accounts. The study by Porat, aside from being a quantum step forward in dealing with the phenomenon of information in the economy in a systematic fashion, has left a legacy in the form of an extensive set of data that has been made available by the Office of Telecommunications, now the National Telecommunication and Information Administration (NTIA), of the U.S. Department of Commerce.

The data are intended for use in the kinds of research needed to advance the understanding of developments occurring with increasing intensity in information technology, to assess their possible impacts on the economy, and to assist in the development and evaluation of policy. The sets of data available from the study and data in the files of the NTIA are summarized in table 18-4. In the following discussion the components of *The Information Economy* data base are examined in turn to assess their use or relevance for these purposes.

Information Labor

The two-way and four-way breakdowns of information and noninformation labor for the years 1860-1980 provide an interesting description of how the demand for skills has shifted as the nature of production has changed over the years. The categories are too broad, however, to have any direct application from a policy standpoint. The aggregates obscure developments that affect the component skills that have more immediate functional connections.

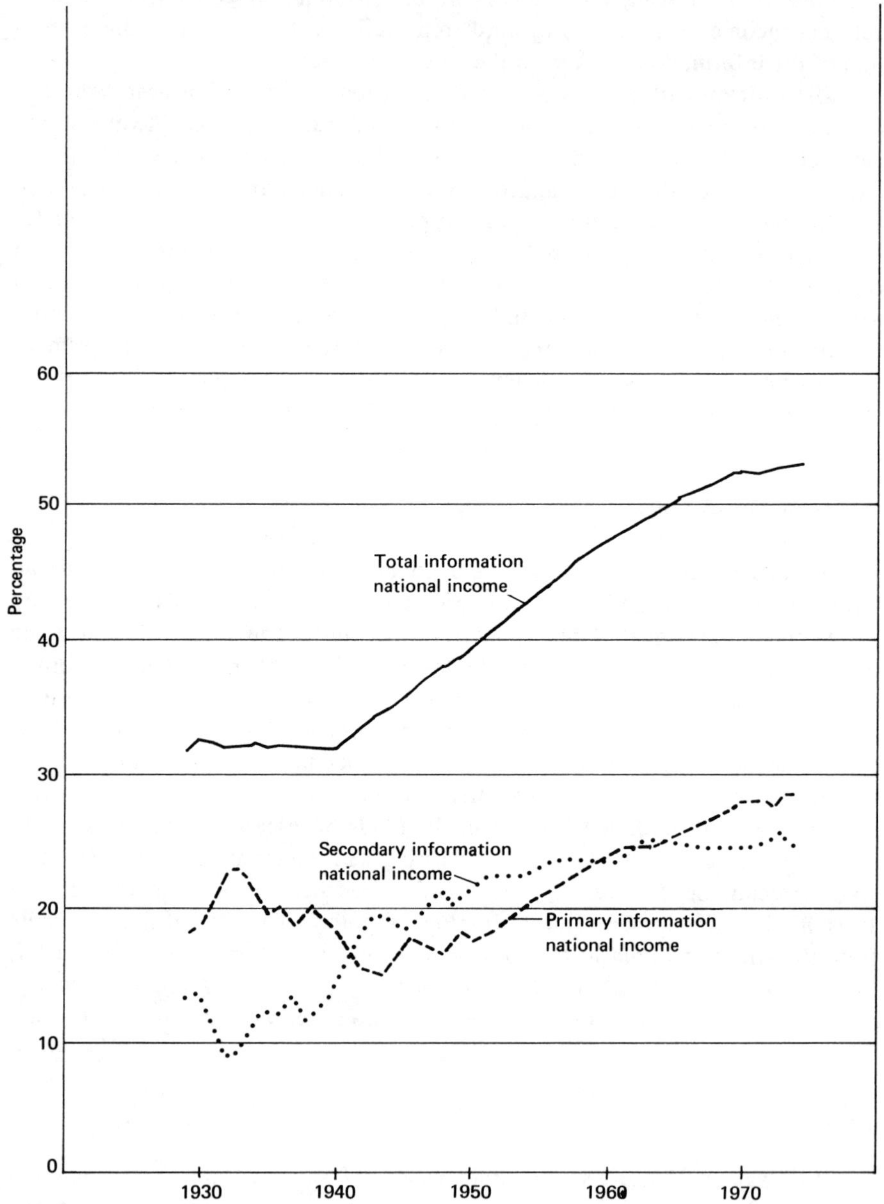

Source: Marc Porat, *The Information Economy* (Washington, D.C.: U.S. Department of Commerce, 1977).

Figure 18-9. Percentage of National Income Originating in the Information Sectors

A useful extension of the data set would be to include a more detailed classification of the census data on information occupations according to functional breakdowns, such as the one discussed in the third section. For example, the trend of scientific and technical workers may be directly related to research and development expenditures; the relative growth of administrative, managerial, and supervisory occupations would indicate a change in the institutional framework of production toward a progressive hierarchic and bureaucratic form; the relative growth in the use of equipment requiring specialist personnel may mean increasing need for educational services.

Observed trends may indicate that simple extrapolations are sufficient to provide the background for decisions. In more dynamic situations regression analyses can reveal the variable factors to which the growth of skills is related. The classification of information occupations can be extended to occupation-by-industry detail from Census Bureau data. This added dimension to the classification becomes more significant if one can project or forecast output levels by industry.

Information GNP and National Income

This study provides data on GNP in the primary and secondary information sectors for a single year, 1967. Detail on the product side is given for purchases by final users and the industrial composition of the goods and services they purchase. On the income side, the detail includes the components of value added in total and for two-digit SIC industries.

National income in the primary and secondary information sectors has been estimated for the period 1929 to 1974, starting at the same time as the regular national income and product series with detail by industry at the two-digit SIC. The choice of national income to provide the time series was no doubt the desire to get the most industry detail and as broad a time span as possible within the framework of the national accounts. Gross-product-originating (value-added) data, published in less industry detail, are available starting in 1947 but have the advantage that they are available in both current and constant dollars which makes them useful to analyze real growth. It would be possible to derive the primary and secondary information components of gross product using the same assumptions as those used to derive the national income estimates, and a thirty-year time span for gross-product-originating data would provide useful information.

Models for projecting GNP yield only aggregate estimates. The information share of GNP can be broken out as a proportion of the total based on historical relationships. But the kind of detail available from this method is not likely to be relevant to the immediate concerns of a policymaker dealing with information technology whose problems are likely to result from shocks to the system that would have sharp and differential impacts on the course of components of the model. A question of continuing concern, however, is what effect the growing information sector is having on the conventional measure of growth, the GNP in constant dollars. The arguments tend to get colored by conceptual issues related to welfare. However, in dealing with the problem of externalities, such as whether to include the cost of cleaning up pollution as a contribution to GNP, it would seem best not to attempt to incorporate every issue into one measure of output. If externalities are perceived as a policy problem, they should be formulated and entered into the analyses as an explicit calculation.

Concern regarding the impact of the growing information sector, and particularly the nonmarket or "bureaucratic" aspect of information activity on real output, however, is a matter of productivity. The demand for information services competes in the marketplace for scarce resources. As the activity becomes an increasing share of national income, it would be well to know whether it contributes more or less than its cost.

The question is addressed in the Porat study, and the analysis is done in terms of a calculated ratio of real output to information overhead. Real output is measured conceptually as GNP less information overhead inputs. The information overhead inputs consist of the factor cost (value added) of the self-performed information activities in both the information and noninformation industries:

$$\frac{\text{GNP} - \text{information overhead}}{\text{information overhead}}$$

The ratio, dollars of real output per dollar of overhead, shows a dramatic decline from \$11.65 in 1933 to \$2.78 in 1974 broken only occasionally and briefly during the period [1, vol. 1, table 9.11].

The calculations were done in current dollars, however, which means that relative price changes have affected the input and the output measures, obscuring the actual relationships. Some questions were raised in the third section regarding the measure of overhead expense (secondary information activities), so judgment should be reserved on the actual numbers. However, the trend is sufficiently spectacular to warrant considering the possibility of undertaking a more conventional analysis of productivity of the resources devoted to the information activity in terms of real output and real input. For this reason, it would be desirable to complete the time series of gross product originating in the information sectors in current and constant dollars.

The measure of productivity in conventional terms is the comparison between the quantity of goods and services produced and the quantity of resources (factors of production) employed in turning them out. Real output is measured by GNP in constant dollars. The work of Denison [7] in sorting out the sources of economic growth has developed a framework and methodology for making measures that represent inputs of factors in physical terms to measure their relative contribution to real output. The measure of labor input is employment, but it is modified by a number of characteristics to reflect quantity and quality of labor input. Shorter hours reduce labor input; but if the shorter workday increases the output per hour, the reduction is partially offset. Educational attainment is a measure of the quality of labor input; as are experience, reflected in the age of the worker; and the number of women in the workforce. These characteristics are all measurable and have been found to explain wage differentials that in an equilibrium situation would reflect the marginal contribution of workers in each category to output. Thus wages can be used to weight the individual indexes to develop an index of labor input in real terms.

The contribution of capital service is measured by input indexes of five kinds of capital goods including nonfarm residence, other structures and equipment, inventories, U.S.-owned assets abroad, and foreign assets in their country. Changes in capital stock are weighted by rates of return in the base period to develop the index of capital inputs.

The difference between the growth rate of real national income in constant dollars and the growth rate for the combined labor and capital inputs is the measure of the productivity of the factor inputs. Denison has examined this residual and has attempted to factor out some explanatory variables such as improved resource allocation, economies of scale, and advances in knowledge. These variables are undoubtedly linked to the application of information, and it would be useful to try to sort out the relationships.

It is probably not yet feasible to deal with productivity in the information sector in this kind of detail. However, it should be possible to get a start in measuring the productivity of labor in the information component, for the decennial years at least, using the detailed demographic data that are available in the population censuses.

Input-Output Analysis and the Information Sector

The information-oriented interindustry transactions table described in the third section shows the flow of goods and services among 190 producing industries specially designed to expose those flows related to information goods and services—as well as sales to final users of information and noninformation goods and services. Twenty-six of these industries are the primary information industries, 82 are the secondary information industries, and the remaining 82 are the producers of noninformation goods and services. The interindustry relationship described in the transactions table serves as the basis for a model that can be used to describe the effects that changes introduced into the system at various points can have on the producing industries.

The so-called technology matrix–direct requirements per dollar of output–is derived from the interindustry transactions table by dividing the entries in each column–the inputs to a producing industry–by the total output of the industry. The technology represented in such a table is an average of the mix of technologies in use when the table was constructed. If one assumes that the inputs required to produce an industry's product are a fixed proportion of the output of the industry, then a column of the technology matrix can be used to calculate the inputs required from each of the industries that supply the producing industry to produce a given level of output. This is the direct impact of the industry's production requirements on the economy.

The effects do not stop there, however. The requirements placed on the supplying industries in turn require them to acquire inputs to produce their output, and their suppliers need inputs from their suppliers, and so on. Thus the effects of production in the initial industry ripple through the production system. The total of the direct and indirect effects are calculated mathematically and form a third table in the input-output system. (The mathematical formulation of the model is shown in *The Information Economy* as well as in standard texts in the field and will not be repeated here.) It appears in Porat's study as the total-effects matrix and shows the production required directly and indirectly from all industries per dollar of a given producing industry's product sold to a final user; that is, the product is not involved in further production and therefore generates no further input requirements. Although the basic input-output model abstracts from problems of substitution of inputs and nonlinearity in input requirements, it does provide a useful framework for assessing the systemwide impact of changes in demand, in technology, and so on.

In simplest terms, the total-effects matrix is a model that translates a final demand variable into its direct and indirect impact on the output of all the producing industries. Before it can be used, however, the problem under investigation must be specified in a form that is compatible with the model. It has to be stated as a set of initial impacts on industries using the same classification as the model. It also must be stated in the same prices as reflected in the model.

Several kinds of problems can be examined in an input-output format. The most commonly asked questions are about the effects of changes in the size and structure of final demand. As the economy grows, it is not only the size of GNP that changes but also the relative importance of purchases by the final-demand components–consumers, investors, government, foreigners. Further, the industrial composition of the purchases by the components change. Projections of GNP require a methodology, an adjunct model, to estimate these changes by category of final demand and in the industrial detail that is compatible with the basic input-output table. General-purpose models have been developed to perform this task, but they employ conservative assumptions regarding the adaptiveness of the economy to change; that is the institutional framework of the economy will not change radically; economic, social, technological, and scientific trends will continue; no war or long-lasting energy shortages, and so forth. (An excellent description of the Bureau of Labor Statistics (BLS) growth model is given in [8]; [9] and [10] contain the latest BLS projections to 1985.)

The portion of the economy defined as the information sector in the study is in the vanguard of technical and institutional change. This is the purpose of isolating it for more intensive study. Therefore, special investigations of the response of the final consumers to rapidly changing development in information technology may be needed before the problem can be stated in a form that can be used in input-output systems to assess the impacts on other producing industries in the economy. What impact will pay television have on the spending patterns of consumers? What type of investment is required to establish a new communications technology? How fast will it be introduced into service? These questions require studies at a micro level of detail. They may be quite narrowly focused, and the answers they arrive at may in themselves be sufficient for decision making without a further assessment of their impact on the other producing industries in the economy.

The relative importance of imports and exports of information goods and services and their contribution to the balance of payments is of considerable concern. the technologies transfer among countries quite readily and can affect the balance of imports and exports and, therefore, production in domestic industries fairly quickly. International division of labor may also be affected. Current data on imports and exports need to be organized to bring out the information-related components for observation and evaluation.

Input-output methodology not only involves specification of the exogenous (final-demand) variable but also requires that the coefficient matrices be updated or projected to reflect current technology or that expected to be in practice at the time. The techniques used to project the coefficients are usually based on observed trends from earlier input-output studies. New developments in technology require specific modification of the existing table, in the input pattern of the producing industries to reflect the requirements to make the new products and in the input coefficients of the consuming industries to reflect the costs associated with using the new technology. Here again the information will have to come from studies of the relevant technology at the micro level which is the only level at which the factors affecting demand can be evaluated. Once the job is done, the resulting relationships can be inserted into the framework of the national model and the conventional macro analysis can proceed to determine the impact on the levels of production in the other industries.

It is important, however, that the micro studies and the macro model be reasonably consistent in structure so that information from the micro studies can be incorporated into the input-output table. This requirement makes it desirable that the form of the micro study not be preempted by the format of the macro model. The structural relationships discovered in the micro studies should have a role in defining the interrelationships shown in the macro input-output model of the information economy.

The impact of new developments in information technology on labor requirements and skills can be evaluated in the input-output framework. Once new output levels are projected for the producing industries, they can be translated into jobs by ratios of output per employee for each of the industries. The jobs for information workers can then be broken down using the occupation-by-industry matrices developed from data in the decennial census.

The policy implications of shifts in demand for labor skills would reflect problems of structural unemployment and retraining due to displacement of skills and, conversely, shortages of skills and the need for providing special education programs and facilities.

Another problem amenable to analysis in the input-output format is the requirement for capital investment associated with changed patterns of output, for replacement of capital used up in the production process and for new capacity to accommodate expansion of output. The state of the art in developing capital stock data and capital-to-output ratios, as well as the related capital consumption on an industry basis, is not well established. But that should not preclude investigation and experimentation to develop and incorporate into the system data that will permit estimates of the amount and kinds of capital that are associated with changes in the levels and methods of production in the information sector.

Conclusion

In his study *The Information Economy* Porat has provided a most comprehensive analysis of the information sector in the U.S. economy for a recent period. He documents very convincingly that the sector has emerged as a significant part of the economy. Yet the description of the sector presented in the study is essentially statistical. Illustrations of possible applications are provided. But I cannot point to a case of its use to analyze an actual problem.

The present need is to do some evaluation of the information sector accounts as they are now constituted. It is time that some real problems be formulated and brought to the system as tests of its capability, in its present form, to yield estimates that can be demonstrated effective as a tool for policy analysis. These tests must also show that the additional information detail provided by the model gives a clearer understanding of the forces working in the economy that are subject to manipulation by action of the policymakers.

The experience in using the accounts in the context of real problems may yield new insights regarding their present structure and suggest modifications that will make them more usable. It is too early at this point to say that the present structure of the accounts should be formalized.

Porat sees the information policy issues as stemming from "the combined effects of information technologies (computers and telecommunications) on market and nonmarket events." Developments occurring in these technologies are among the most dynamic in the economy. This would seem to indicate that an industry-oriented or micro approach is better suited to the investigation of their impacts on the economy at least in the initial stages of investigation.

The data base presented in the study is a foundation that can be built on. It needs to be updated, and I have indicated where I think the components can be usefully extended or the estimates improved. The most important are the estimates of the secondary information sector. The concept and measurement of the private bureaucracy should be extended to include the self-performed information services in the primary information sector. Intermediate inputs (information goods and services) into the secondary information sector should be cleaned up regarding their use, whether they support the information activity or are components of the product made in the industries of the non-information sectors. An attempt should be made to estimate value added originating in the secondary information sector directly, rather than as a residual of total national income in the information sector and national income in the primary information sector. If the estimates of secondary information national income are made directly, it is likely that industry detail can be provided to match that for the primary information sector.

The functional classification of information occupations by industry would be a useful addition to the data set. And estimates of value added in the information industries in current and constant dollars are required for assessment of real growth.

References

1. Porat, Marc U., *The Information Economy*. Vol. 1: *Definition and Measurement*; vol. 2: *Sources and Methods for Measuring the Primary Information Sector*; vol. 3: *Interindustry Transactions Matrices* (1967); vol. 4: *The Technology Matrices* (1967); vol. 5: *The Total Effect Matrices* (1967); vol. 6: *The Labor Income-by-Industry Matrix of Employee Compensation* (1967); vol. 7: *The Labor Income by Industry Matrix of Employment Compensation* (1970); vol. 8: *National Income, Work Force, and Input-Output Accounts;* vol. 9: *User's Guide to the Complete Data Base* (Washington, D.C.: U.S. Department of Commerce, Office of Telecommunications, 1977).

2. Kendrick, John M., *Economic Accounts and Their Uses* (New York: McGraw-Hill, 1972).

3. Ruggles, Nancy, and Ruggles, Richard, *The Design of Economic Accounts* (New York: National Bureau of Economic Research, 1970).

4. U.S. Department of Commerce, Bureau of the Census, *Census of Population: 1970, Occupation by Industry, Final Report PC(2)-7C.*

5. "The Input-Output Structure of the U.S. Economy: 1967, "*Survey of Current Business*, February 1974 (eighty-five-industry classification); "The Input-Output Structure of the U.S. Economy: 1967," a supplement to the *Survey of Current Business*, 1974 (367-industry classification), vol. 1: "Transactions Data for Detailed Industries"; vol. 2: "Direct Requirements for Detailed Industries"; vol. 3: "Total Requirements for Detailed Industries."

6. "Interindustry Transactions in New Structures and Equipment, 1963 and 1967," a supplement to the *Survey of Current Business*, vol. 1.

7. Denison, E.F., *Accounting for United States Economic Growth, 1929-1969* (Washington, D.C.: Brookings Institution, 1974).

8. "The Structure of the U.S. Economy in 1980 and 1985," U.S. Department of Labor, Bureau of Labor Statistics, Bulletin No. 1831, 1975.

9. Kutscher, Ronald E., "Revised BLS Projections to 1980 and 1985: An Overview," *Monthly Labor Review*, March 1976.

10. Bowman, Charles T., and Morlan, Terry H., "Revised Projections of the U.S. Economy to 1980 and 1985," *Monthly Labor Review*, March 1976.

19 New Media in the Information Economy: Prospects and Problems for Viewdata and Electronic Publishing

R. Michael Tyler

The information economy mapped out by Machlup (1962), Porat (1976), and others, with its emphasis on know-how and communication as the framework of all human activity, is rapidly becoming an everyday reality. We can already identify in broad outline not only its econmic parameters but also its characteristic technologies. Just as the energy-intensive boom of the 1950s and 1960s was associated with mass production in manufacturing and mass ownership of automobiles and domestic consumer durables, so the rise of the information economy will rely to a large extent on telecommunications, microelectronics, and the digital computer. But what can we yet say about more specific kinds of products, services, and markets?

Any comprehensive attempt to answer this question would have to range across such disparate issues as the automation of manufacturing, the role of information technology in education, health, and other social services, or the impact of the office data-handling technologies often known collectively as the office of the future. Moreover there are certain general questions about the information economy's technological infrastructure that clearly affect most or all of these sectors, for example:

> Will the present reliance on transport of paper (such as the postal service, check clearing) continue or be extensively replaced by electronic media?
>
> What will be the role of printed media, including books, in the future information economy, and how will it relate to electronic media of text transmission?
>
> Will the present relationship between business activity, travel (including commuting), and the location of homes and workplaces continue or, as some observers have suggested, be radically altered?

This chapter addresses just one telecommunications development that will influence all three of these questions and will consequently shape events

---

I am indebted to Charles Jonscher and Raymond R. Panko for much indispensable advice and assistance on the subject of this paper.

throughout the information economy—the emergence of versatile technologies that allow telecommunications systems to deliver textual information and simple graphics at low cost to a wide range of users—in other words, to support totally electronic forms of publishing. This paper deals only with wholly electronic publishing, that is, systems that deliver information to the ultimate user by electronic means such as a printing terminal, a television screen, or other cathode-ray tube display.

Among such systems, increasingly known generally as teletext, the United Kingdom's initiatives with broadcast teletext and with viewdata have attracted particular attention. Both involve the adaptation of television receivers as low-cost display terminals for text and graphics. Information is transmitted either in the blanking period of the broadcast television signal (as in Ceefax or Oracle) or through the telephone system (as in viewdata). But other forms of electronic publishing for the general user must also be considered close potential competitors. One is based on the use of frame grabbing to display text and graphics selected by the user from a constant stream of frames carried by a wide-band cable television channel. Another alternative, often overlooked in this context, is the use of information prerecorded on storage media such as videodisks.

Electronic publishing for the specialist user is not new, of course; systems such as LEXIS, MEDLINE, or the Information Bank at the *New York Times* are well established. What remains in question is the next step, a cheap, easily-understood Model T version of electronic publishing suitable for a much wider market among unsophisticated users. This paper analyzes that next step. It surveys developments so far and introduces some of the evidence already available, technical feasibility and costs, market potential, and the far-reaching range of policy implications.

It begins with a brief and nonspecialist sketch of the teletext technologies and the services they make possible; and a general international review of the present state of development of teletext services. This provides the starting point for a discussion of policy issues, concentrating on the interests of the many organizations with a stake in the future of teletext, and the effects of industry structure and regulation on its likely development.

The Technology and Its Service Potential

Four distinct kinds of teletext systems now dominate the field. Each has a different technological basis, and the differing cost limitations are likely to shape a distinct role for each of them in future. For brevity, we shall call these by the following terms:

Narrow-band interactive teletext, often known as *wired teletext*, using the telephone system or similar networks to distribute data under the user's control at relatively slow speeds—not more than 2,400 bits/sec. This is the viewdata concept originated by the Post Office Telecommunications in the United Kingdom and now marketed under the trade name Prestel.

Broadcast Teletext, in which textual and graphic information is inserted into the redundant intervals in broadcast television signals, creating a stream of television frames of information from which the user can make his or her selection. Ceefax and Oracle in the United Kingdom are each examples of broadcast teletext. This is inherently a one-way service, where the user can select the information frame but does not have a general two-way communication capability at his disposal.

Wide-band broadcast or cabled tetetext, employing the same principles as Ceefax and Oracle but achieving much greater capacity by allocating all (or much) of a complete video channel to transmitting the alphanumeric or graphic characters. In principle an off-air broadcast channel could be used in this way, but a much more likely early application is to cable television.

Wide-band two-way teletext, in principle would be ideal. It can be achieved on sophisticated switched cable television networks such as that used for the TICCIT experiment in Reston, Virginia; but costs are likely to remain prohibitively high for a considerable period.

In many ways the most distinctive feature of teletext is not so much its individual technological content as the assembly of a total system comprising information guides, an editorial and indexing structure, and a billing system. A serious deficiency of broadcast teletext systems is that except where they form part of a television package, it is virtually impossible to bill for service. It can therefore yield commercial advantage only by attracting subscribers and viewers to an overall program package of which it forms a part.

Some key technical points of examples of each of this kind of service are summarized in table 19-1. In a viewdata system the distinctive data-base organization and method of search are both a strength and, as I shall suggest later, a limitation.

The data base is organized in a tree structure. If the user knows the page number of the information he requires, he can access it directly. His alternative is to start from the index page and use his push-button key pad to input selection digits in response to a menu of defined choices that are presented sequentially, so that he can pick his way through the tree structure from, say, travel to package holidays to package holidays in Spain, until he arrives at the page he is interested in. Although this method, illustrated in figure 19-1 is a little slower, it allows a very easy method of browsing through the data base in search of interesting information and is simple for untrained users to understand. Unlike broadcast teletext, viewdata allows the user to specify the page he requires rather than presenting him with the complete data base. The system's storage capacity can be very large: the Post Office expects, even for its initial public trial, to be offering more than a hundred thousand frames of viewdata information, each containing up to twenty-four lines of forty

Table 19-1
Comparison of Four Teletext Systems

	Ceefax, Oracle	*Reuter Monitor*	*Prestel (Viewdata)*	*Mitre TICCIT*
Channel	Broadcast One-way	Cable One-way	Telephone wire Two-way	Cable Two-way
Access time per page	20 secs typical (up to 192 secs if 800 pages provided)	1-2 secs	11 secs (maximum)	1-2 secs
Information per page (characters)	960	768	960	Limited by screen resolution only
Character set	Uppercase and lowercase Graphic elements	Uppercase	Uppercase and lowercase Graphic elements	Unlimited
Display quality	Color Characters and limited graphics	Black and white No graphics	Color Characters and limited graphics	Color Analogue (characters and full graphics)
Information base size (frames)	150 typical 800 maximum	2,000	70,000 initially (unlimited capacity)	– (unlimited capacity)

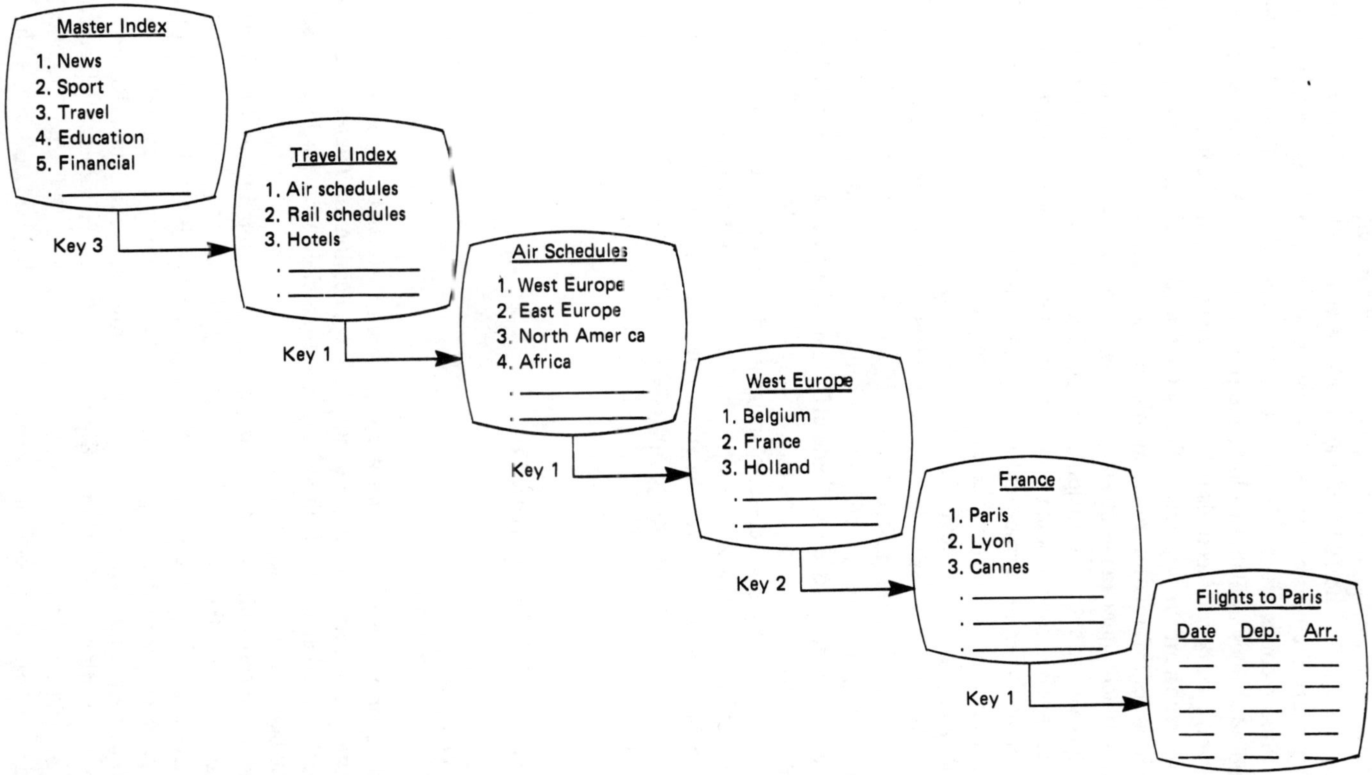

Figure 19-1. An Example of the Use of a Tree-Search Procedure: Locating Information on Scheduled Flight Times to Paris

characters, and there is a distinct possibility that this will be increased to over 1 million frames for the start of the public service in 1979. Even with a data base of this size the user can expect to start receiving the frame he selects only two or three seconds after he inputs the last of his selection digits. With only one-hundred frames available to him, the present broadcast teletext user could have to wait twenty-four seconds before he receives the page he wants.

It is important to appreciate that an interactive teletext system, though originated as an information retrieval machine, provides the essential infrastructure for a much wider range of services. The provision of message services, for example, is an attractive evolutionary step, although its implementation on a large scale involves solving some problems of computer-system architecture and dimensioning. Computer games and instruction—either remotely or by means of teleloaded software called from the central computer by an intelligent terminals—are another very attractive field of service development.

The inherent constraints on broadcast teletext are best understood in terms of the trade-off diagram in figure 19-2. If a one-way system is to allow rapid access to a large data base, a wide-band channel must be provided. This may be contrasted with the different trade-offs that characterize the relationships between the transmission parameters in a two-way system. For example, if a teletext system is to allow access to an information base of ten-thousand pages of about six-hundred characters each, then to achieve a maximum delay time of ten seconds would require a channel transmission rate of about 5 million bits/sec in a one-way system compared with 600 bits/sec in an interactive wired system.

Teletext in Practice: Current Activities

United Kingdom

It seems appropriate to begin an account of current teletext operations with the United Kingdom because the practical application of teletext has gone considerably further in the United Kingdom than elsewhere, partly for reasons of public policy and industry structure.

The broadcast teletext idea was developed by the British Broadcasting Corporation under the name Ceefax and by the Independent Broadcasting Authority under the name Oracle.[1] Broadcast teletext is fast becoming a consumer product: estimates suggest that about five-hundred teletext-equipped television sets have so far been sold in the United Kingdom, but BREMA, the industry association for the United Kingdom's set manufacturers, has predicted that there will be 12,000 such sets in use by the end of 1978 and 250,000 by the end of 1979, when they are expected to cost about $1,000 in the United Kingdom, about $300 more than a basic television set. Ceefax signals are now broadcast on a regular basis, with the data base constantly updated. Topics

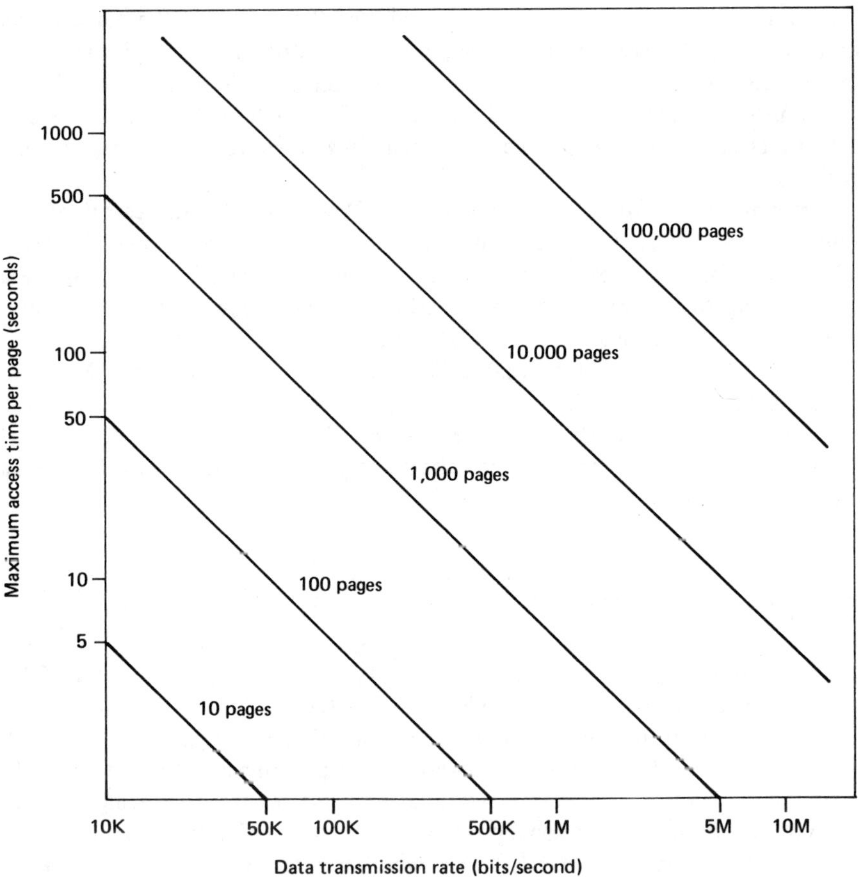

Note: Curves shown are for a 600-character page and 8 bits per character coding.

Figure 19-2. Relationship between Information Base Size, Data Transmission Rate and Maximum Access Time per Page for One-Way Systems

covered include news, weather forecasts, consumer information such as food prices, sports events and results, road traffic conditions, and cooking. Perhaps significantly, the IBA's parallel service, Oracle, has been off the air since October 1977 owing to a union-management dispute over policies toward the staffing of the new activity.

The British Post Office's viewdata system, known since April 1978 by the registered name Prestel, is now operating on a pilot scale, and intensive preparations are under way for a market trial to run from June 1978 to September 1979. Fifteen hundred specially equipped television sets will be in use for the trial. Of these, 850 will be in a sample of homes in London, Birmingham, and Norwich;[2] 300 will be in a random sample of businesses, and 350 will be in businesses selected for their special relevance to the market study.

The Post Office is not, however, waiting for the results of the market trial before making its basic decision to press ahead with a full public viewdata service. That decision was announced on 28 February 1978. The Post Office is committed to spending £23 million (about $42 million, U.S.) in 1978 and 1979 to launch a public service, with ten Prestel viewdata centers, starting early in 1979.

Perhaps the most striking feature of Prestel is the successful effort, mounted at an early stage in the development of the service, to involve a wide variety of information provider organizations in the establishment of the data base. So far, over seventy-five information providers have agreed to publish their information through Prestel. They range from the weekly *Exchange and Mart* classified advertising magazine to the government's Central Office of Information. Other notable examples include the Consumer's Association with their 'Which?' reports and the *Financial Times* newspaper. Even for the initial market trial, over one-hundred thousand frames of viewdata information will be offered to users.

Other European Countries

The brief summary of current European developments given in table 19-2 is sufficient to demonstrate the widespread activity in the field of teletext development. Interest is intense and experimentation is widespread. It is likely to be some years before the longer-term pattern of development is clear. In France the two separate public sector teletext activities are now being rationalized within a single system concept, Titan. In Germany the diversity of systems reflects fundamental disagreement over who should provide teletext services—broadcasters, newspapers, or telecommunications carriers. A similar debate is surfacing in several other countries, notably the Netherlands.

Japan

There is known to be keen interest in teletext in Japan, and there can be little doubt that Japanese electronics manufacturers will be quick to offer teletext-equipped television receivers in the marketplace. However, the future of teletext within Japan itself remains problematic. The broadcasting authority, Nippon Hoso Kyokai (NHK) has developed its own broadcast teletext system, similar in principle to Ceefax and Oracle; but the costs of such a system dimensioned to deal with the very large number of Japanese characters may make the system uneconomic for many years to come. Similar considerations would of course apply to a system of the viewdata type.

Table 19-2
European Teletext Initiatives as of March 1978

Country	*System Name*	*Type X*	*Transmission Organization*	*Other Information*
France	Tictac	V	CNET (Centre Nationale des Etudes de Telecommunications)	Undergoing trials trials with PTT administration
France	Antiope	V	CCETT (Centre Common d'Etudes de Televisions et Telecommunications)	Undergoing trials with TDF (Broadcast monopoly transmission agency)
Germany (Federal Republic)	Bildschirmtext	V	Deutsche Bundespost (PTT Admintration)	Uses U.K. viewdata software and U.K. GEC 4080 computer. Demonstrated at 1977 Berlin Radio Fair Service trial 1980
Germany (Federal Republic)	Videotext	B	ARD–first German television channel	Also demonstrated at 1977 Berlin Fair
Germany (Federal Republic)	Videotext	B	ZDF–second German television channel	Also demonstrated at 1977 Berlin Fair
Germany (Federal Republic)	Bildschirmeitong	B	BDZV–newspaper publishers association	Also demonstrated at 1977 Berlin Fair
Sweden	Text TV	B	Sveriges Radio	Experimental operations since 1976
Sweden	?	V	Televerket (PTT administration)	Known to be under development by Televerket/Philips/Swedish publishers
Finland	Telset	V	PTT/Sanoma publishing/Oy Nokia	Demonstrated in March 1977; pilot trial during 1978

V: Viewdata-type system
B: Broadcast teletext

United States and Canada

Although no fully developed Teletext systems such as viewdata or Ceefax are now found in the United States and Canada, there have already been three kinds of developments of relevance to the future of teletext in the United States market:

1. The Public Broadcasting Service has two limited broadcast teletext capabilities, the HRI Add-On System developed by Haseltine Research and used to provide captioning for people with hearing impairments; and TV Time added on to the Add-On System by PBS and the National Bureau of Standards to provide accurate time signals for general public use.

2. Specialized broadcast services using the wide-band transmission capabilities of cable television networks, notably the Reuter's Monitor Service. Reuters offer specialized services for financial interests in the lower Manhattan area in a joint venture between Reuters and IDR, the latter being a Reuters subsidiary company set up to develop and assemble the terminals required by the system.

Subscribers can purchase any subset of numerous categories of information, including, for example, metals markets, grains markets, livestock markets, cocoa/coffee/sugar markets, money markets, securities and options.

3. Interactive information services through cable television, ranging from the advanced, experimental–and costly–Mitre TICCIT system operated in Reston, Virginia, until 1977, to the simpler but more commercial range of interactive information and game-playing capabilities offered by Warner Communications' Qube System in Columbus, Ohio.

The high and increasing penetration of cable television in the United States, and more especially in Canada, makes the cable teletext concept attractive. This and the increasing capability for interconnection demonstrated by the program distribution activities of Home Box Office and other companies, suggest that cost-effective teletext service of high capacity using wide-band broadcast methods could be achieved by aggregating markets nationally. Reuter's News-Wire service, now sold to over two-hundred cable television systems in the United States, gives some sense of the possibilities. A brief description of News-Wire in given in Appendix A.

Economics: Costs and Demand

Costs

It is now possible to cost the technical infrastructure of teletext systems with reasonable accuracy; but it remains difficult to estimate the total costs of the service because of the limited experience on the editorial side–information provision. Moreover it is extremely difficult to convert cost estimates into pricing schemes because of large uncertainties about the likely extent and intensity of use.

Nevertheless some broad indications can be given. Consider the example of one-way broadcast and two-way viewdata services serving a user population of 250,000. Using cost estimates by Johnson (1975), adjusted to take account of inflation and U.S. wage levels, U.S. costs for the broadcast system would be about \$3 million annually for the physical infrastructure and about \$5 million for information provision. The cost per user per month would be \$2.70.

Consider a viewdata system serving the same market. Both the nature of the service and the price will be different; the information base will be more extensive and inherently more costly, say \$7 million in the United States at current prices. A physical system capable of accommodating an average of ten minutes of usage per user per day would imply operating costs of perhaps $35 million, including the opportunity cost of using telephone system capacity. Thus the total cost per user per month—additional of course to terminal rentals—would be of the order of $12.

To these figures must be added the costs of terminal rentals. At present, terminals are produced in small quantities and costs are high, as the examples in appendix B show. However, manufacturers are developing terminals for mass production: for Prestel in the United Kingdom, eleven different terminal types are being developed by eight different manufacturers. Most of these terminals are capable of receiving both a normal television picture and Ceefax and Oracle. An indication of the prices that customers and set manufacturers may expect to pay for teletext decoders during the next several years can be obtained by considering the forecast prices of equipment currently under development as shown in table 19-3. Of particular interest is the development, by General Instruments, of a set of five integrated circuits containing the components required for viewdata reception on a standard television screen. The forecast prices of these adapter modules and components are summarized in table 19-4.

These prices cover only the components required for decoding and display control. The costs of the key-pad unit and of the modem for interconnection to the telephone line are not included. The consensus of views on likely future terminal prices arrived at by Communications Studies and Planning after considerable study shows that the extra monthly rental for a teletext-equipped television receiver in 1980 would be about $5 for the viewdata-type and a little less if broadcast teletext alone is provided.

Potential Demand: Evidence and Uncertainties

The growth of the information economy charted by Marc Porat can usefully be thought of as driven by two distinct forces: the growing demand for information goods and services by final consumers (household and government) and the growing purchases of information-related inputs by business

Table 19-3
Forecast Prices of Teletext Adapter Modules and Components

Manufacturer	*Item*	*Forecast Price*
General Instruments	Set of 5 ICs for converting standard television for viewdata reception (modem excluded)	$25 (in 1978)
Texas	TIFAX board (already available see table 5.1); reduced production costs. For broadcast teletext only.	$55 (in 1978-1979)
Texas	Further development of TIFAX board to accept broadcast teletext and Viewdata	$550 (initially) $200 (eventually)

Table 19-4
Forecasts of the Premium Price of a Wired Teletext Terminal (over that of a Standard Television Set) in 1980 and 1985

Item	*Forecast Retail Price 1980*	*Forecast Retail Price 1985*
Adapter module	$100	$45
Modem	45	20
Key pad	5	5
Total	$150	$70

enterprises. Both could give rise to demand for teletext; but what evidence do we have for an evaluation of that demand? Research into the detailed modeling of demand teletext is now under way in the United Kingdom,[3] but since no quantitative results from that work will likely be available before, early 1980, we must take a more informal approach to forming a judgment about the market prospects. One useful way to approach the problem is to pose the fundamental questions:

> Are there significant behavioral barriers to user acceptance of teletext as an information medium?
>
> If not, would a forecast of the adoption of Teletext service on a large scale be compatible with reasonable projections of future expenditure patterns by households or businesses?

While it would not be appropriate to develop each of these questions fully in a paper oriented toward issues of public policy, some brief discussion of each may help give a general impression of the problems and opportunities.

Acceptability Issues: An Illustration. A simple example will illustrate how dangerous it is to adopt too simple a view of what the uses of existing information media are and hence how acceptable any kinds of new electronic media are.

It is useful, for example, to examine how newspapers and magazines and books are used, that is, readers' perusal styles.[4] Teletext may well have to become compatible with some existing perusal styles, although it is possible that behavior will adapt so that some entirely new perusal style appears among teletext users.

In the United States 84 percent of all newspaper pages are seen by the average reader. That browsing is not very selective is indicated by the finding that 80 percent of all women's pages are opened by the average male reader, while the typical female reader opens 70 percent of all sports pages (ANPA, 1973).

In one Canadian survey (Lusty, 1976) 55 percent of the readers were found to habitually start by looking at the front page and read all the way through to the back page. Another 17 percent looked at some specific item first, then read the paper from front to back. Another 25 percent scanned the paper quickly, with 13 percent looking at specific items first, the other 12 percent not. Only 3 percent looked only at preselected parts of the paper seeking specific kinds of information.

U.S. experience is similar. Only 4 percent read only specific information, and 53 percent start on the first page and read through, rather leisurely, through the back of the paper. Another 37 percent turn first to specific items, then go through the rest of the paper (ANPA, 1973).

Even the most enthusiastic devotee of viewdata has to admit that a selective information-retrieval system organized in a logically structured tree system is not well adapted to accommodate patterns of browsing of this kind.

Expenditure and Demand. Given one illustration of the quite extensive evidence that teletext is complementary to existing media, and in many respects not a close substitute, it is interesting to examine some broad but nevertheless interesting economic evidence relevant to the future consumer market for teletext. Despite the onset of the information economy, the share of information goods and services in household expenditure remains modest. Consider the case of the United Kingdom. In 1976 the average British household spent about $8 per month on the printed media; about $7 on telecommunications and postage; about $12 on television and radio receivers (to rent or buy); and license fees; about $5 on tickets to cinemas, theaters and sports events. U.S. and Canadian figures are not dramatically higher. Typical U.S. cable television rentals are not unreasonable in relation to these figures, but there are clear implications for teletext. If Teletext services at a cost of say, $12 per month are to be widely taken up in the marketplace, we can be sure that (1) this level of expenditure cannot be sustained by any plausible (or indeed any) level of diversion of demand for printed media; (2) such a level of expenditure would imply a considerable diversion of expenditure from other information-related items in the household budget; or, if the diversions hypothesized are to be kept within reasonable limits, we must assume some diversion of other kinds of discretionary expenditures (luxury clothing, say, or leisure automobile trips) in favor of Teletext.

Whatever the answers that experience and analysis will eventually provide to these puzzles, there can be no doubt that a great deal is at stake. The total revenues of the top seventy magazines in the United States exceed $1,500 million, and the top ten include some (notably *TV Guide, Time,* and *Newsweek*) where signifcant competition from a teletext information service might be found. In Canada, print media received over $700 million in advertising revenue each year, much of it in the form of classified advertising and informative announcements that could be published in a very cost-effective way using teletext. The stakes involved and the policy issues they imply are the subject of the remainder of this paper.

Problems of Policy and Institutions

The Stakeholders

From the discussion so far, the features of teletext that promise to puzzle any policy analyst are apparent. The organizations and institutions likely to be affected by teletext are astonishingly varied, ranging from newspapers and their groups of employees (as varied as journalists and compositors) to consumer protection groups or manufacturers of telephone receivers. The way in which the packaging of viewdata service cuts across the boundaries of conventionally distinct activities such as telecommunications, publishing, or computing.

Both features create policy problems distinctly different from those that are familiar in connection with other telecommunications media.

It is useful, in trying to understand the policy issues raised by teletext, to identify the organized interests that might gain or lose by its development, and their known and possible future attitudes toward the new technology. They can be divided into five main groups.

1. *Electronics manufacturers, especially manufacturers of television receivers* facing the prospect of future saturation of consumer markets, welcome teletext as a logical next stage in a marketing sequence that leads from monochrome television through color television and teletext to the ultimate intelligent home terminal. In the United Kingdom the industry association BREMA has been a major force behind teletext, promoting, for example, the harmonization of technical standards between viewdata (Prestel), Ceefax, and Oracle.

2. *The carriers*, in the United Kingdom, Post Office Telecommunications and the broadcasting authorities. The objectives of the Post Office as a telecommunications carrier are more apparent than those of the broadcasting authorities. Viewdata provides the Post Office not only with direct revenue in the form of direct billings for usage charges (per-frame charges, under the present system[5]) but also with increased revenue from telephone call charges.

3. *The information provides*, a wide and rapidly growing range of organizations wishing to publish their information (which may range from job advertisements to scientific information to the weather or the jokes of the week) through teletext. For viewdata, with its potentially limitless data base, the information providers are significant force. In the United Kingdom, for example, they include major business organizations such as the *Financial Times* group of companies and are in the process of forming their own industry association.

4. *The print media*, many of whom have hedged their bets by establishing themselves as viewdata information providers. Nevertheless, the interests of

the print media as such may be threatened by competition from teletext. Whether such fears are justified is a matter of controversy. Some observers feel, on the contrary, that there will be synergistic effects stimulating markets for both. It does seem clear, however, that some particular kinds of printed media are particularly at risk; a notable example is specialized magazines and local newspapers, which are heavily dependent on classified advertising.

5. *Unions*, whose reactions to teletext are a particularly sensitive indicator of the kinds of fundamental issues that teletext media can raise. A clear trend can now be discerned, especially in Europe, toward increased union anxiety about the effect that new electronic technologies will have on employment. How far this is justified in the long term, and how far it is a reflection of essentially temporary economic conditions of slow growth in demand, is a profoundly important topic but is beyond the scope of this paper. Given that this anxiety *does* exist, it is clear that teletext will be a focus for it. A few quotations from a recent report by the National Union of Journalists in the United Kingdom will give the flavor of some current union thinking:

> The Annan Committee on the future of broadcasting failed to recognize the dangers inherent in accepting, for the time being, that all three systems should develop without restraints. . . . Members must be alerted to the extent that Teletext already affects our jobs, and its imminent dangers [NUJ, 1977, p. 39]. The threat to newspapers from viewdata is . . . much greater than that from Ceefax and Oracle. . . . If specific information is available through a quick call to viewdata, newspaper sales could be in danger [NUJ, 1977, p. 40].

Of course, while the new teletext services are still on a small scale, no countervailing voice can be heard from those who will ultimately fill the many new jobs they will create, although the telecommunciations unions in the United Kingdom, notably the Post Office Engineering Union, are vociferous in their support of service innovations such as viewdata. So far, no serious attempt has been made to analyze the relative labor intensiveness of teletext and traditional printed media or to model the effects of future growth and interaction on total employment. As the policy debate unfolds, it will become increasingly clear what a serious deficiency this is. If the necessary research is to be available in time to inform policymaking, steps to initiate it will have to be taken quickly.

The Teletext Policy Agenda

The public policy issues generated by the development of teletext stem from the interaction of the interests of the various stakeholders with the initial policies of the teletext entrepreneurs (in the United Kingdom, innovators within the Post Office and the broadcasting authorities) and the existing structure of rules and institutions for telecommunications. Some are relatively independent of special features of the industry structure or regulatory regime in one country, and some are quite specific to a single country. In the following section I discuss in turn issues of general international applicability (most current policy issues in the United Kingdom fall into that category) and the regulatory issues that can be expected to complicate any major introduction of the teletext technology into the United States.

Policy Issues of General Relevance

Policy problems fundamental to teletext and likely to arise under almost any policy context that is not uniformly totalitarian, fall naturally into three groups: problems of editorial policy and control; problems of intellectual property; problems of competition and competitive impact. Each of these phrases summarizes a complex set of issues, and only a few key points can be covered here as a starting point for discussion.

Two essentially different strategies are open to the organization owning and operating the physical infrastructure of a teletext system. The U.K. Post Office, although it is not strictly a common carrier by the terms of the 1969 Post Office Act that set it up as a public corporation, has chosen to maintain an editorial policy of more or less strict neutrality toward information providers on Prestel, subject only to the provisions of generally applicable laws on obscenity incitement to violence, and so on. The Post Office has consciously rejected any attempt to endow it with general editorial responsibility.

Such a stance is not without its problems, however. The Advertising Standards Authority, an unofficial but prestigious advisory body, has criticized the Post Office policy, claiming that such problems as the absence of clear demarcation between editorial content and advertising may violate accepted standards of ethics in advertising. Others have raised complex legal issues about liability for any breaches of law that might occur. On the whole, however, the common carrier stance appears to have weathered the storm of criticism rather successfully.

So far the problem of intellectual property, copyright payment of journalists and so on, has not become acute. But there are clear signs that it is being raised by the National Union of Journalists (NUJ, 1977) and others, as indeed has the related problem of possible rights of particular trade unions to monopolize particular kinds of jobs in the teletext staffs, by analogy with existing union demarcation in the staffing of the printed media.

The issue of competitive impacts has already been touched on. Even at the likely points of maximum conflict—between viewdata and regional classified advertising or between use of Oracle and prime-time viewing figures for television advertising—open conflict over future market shares or substitution of one medium for another is not yet apparent. They can be expected to be a highly visible and contentious feature of the telecommunications policy scene in the longer term, however.

Distinctive Problems of Policy and Institutions for Teletext in the United States

Considerable policy problems and institutional barriers stand in the way of the extensive implementation of teletext in the United States; indeed, these obstacles undoubtedly have a good deal to do with the remarkable lead that European countries, notably the United Kingdom, have obtained in this field.

All three kinds of organizations that could have taken the lead in developing and implementing teletext face peculiar difficulties in seeking to do so within the existing U.S. industry structure and regulatory framework.

Telephone companies are potentially very well placed to provide a teletext service. Not only do they have a distribution network that is virtually universal, but they also possess a versatile, enforceable, and virtually universal billing system. Their participation in teletext systems have been held back, however, by regulatory policies requiring the separation of data processing from communications and telephony from text-message communications. A viewdata system would span both these pairs of categories.

Cable television companies, notably multiple system operators (MSOs) and satellite interconnection operators like Home Box Office are in an increasingly strong position to provide teletext of the one-way wide-band type as individual markets become aggregated into larger groups of users capable of supporting a wider range of services.

Broadcasting organizations, like their European counterparts, could, readily provide a broadcast teletext service provided that no regulatory objection were raised on the grounds that teletext is a means of message transmission. It is hard to see, however, what economic motive the commercial television companies would have for introducing such a service, given the virtual impossibility of charging for it, except perhaps on the debatable argument that the network with the first or best broadcast teletext service would attract improved ratings. It seems most unlikely that this effect would outweigh the direct loss of advertising revenue due to reduced audiences for television advertising.

The Public Broadcasting System might consequently find itself in a particularly favorable position to be the leading innovator in the United States with broadcast teletext services, employing them for purposes of social interest along the lines already pioneered by their work on captioning for the deaf.

Conclusions

At this early stage in the development of Teletext, no firm conclusions can be drawn about its future potential in the marketplace or about the policy implications. Enough is known, however, to indicate a significant potential for the service, indicated by market research surveys, (for example, European Market Surveys, 1976) public interest, and the judgments of information providers and to moderate some of the more extravagant forecasts of early growth in the residential market. A research agenda and specific research projects are emerging to meet the need for better data, quantitative models, and a generally better understanding of the marketplace for such communications media.

The emerging policy issues can be discerned in outline, at least. It seems clear that in the United States two of the three potential sources of teletext innovation may be significantly constrained by regulatory policies and industry structure. Unless public policy changes come about, therefore, the least known of the three teletext technology options—wide-band one-way broadcast teletext—may soon emerge in one form or another as the leading from of teletext in the U.S. marketplace.

Notes

1. The IBA is a public authority with a monopoly on the provision of physical facilities for networking and transmission for the private television companies. It is also the public regulatory body for the private sector of television broadcasting.

2. Half of the residential sample will receive the viewdata service entirely free; the other half will pay £12 to £15 (depending on the type of receiver) for the first six months and £18 subsequently. Usage-related charges will be zero for the first two months and half-rate subsequently.

3. A two-year program funded at Communications Studies and Planning, London, by the Long Range Studies Divisions of Post Office Telecommunications.

4. In discussing this issue I am drawing on work by Raymond R. Panko and Rosemary Panko (private communication, 1977).

5. The viewdata system charges on a per page basis with a Post Office basic charge of ½p, or about one U.S. cent, to which the information providers' charges are added. The information providers select their own frame charges on a range from 0 to 50p (90 cents) per page.

References and Supplementary Documents

ANPA (American Newspaper Publishers Association). 1973. News and editorial content and readership of the daily newspaper. Washington, D.C., April 1973.

BBC, IBA, and BREMA. 1976. Specification of standards for broadcast teletext signals. IEE Colloquium on Teletext, London; January, 1976.

Council for Cultural Cooperation. 1975. Advanced uses of cable systems. CC/DC (75)5, Council of Europe, Strasbourg, March 1975.

Esdale, G. 1976. Demographic and economic trends and the market potential for new consumer electronic products. Presented at a Stanford Research Institute International Seminar, October 1976.

European Marketing Surveys. 1976. Report on a survey to establish Viewdata information requirements. European Marketing Surveys London, June 1976.

Fedida, S. 1976. Viewdata, an interactive information medium for the general public using the telephone network. IEE Colloquium on Broadcast and Wired Teletext, January 1976.

Fedida, S. 1977. Viewdata networks. Paper presented to the NATO Symposium on the Evaluation and Planning of Interpersonal Telecommunications Systems, Bergamo, September 1977.

Industry Week. 1974. Computer-to-computer ordering looms on the horizon. *Industry Week*, 180, no. 4 (January 1974).

Johnson, T. 1975. Teletext: data transmission by television. *Financial Times*, London.

Lusty. 1976. National newspaper readership study, Fall 1975. Conducted for the Canadian Daily Newspaper Publishers Associated. Gordon Lusty Survey Research, Ontario, May 1976.

Mitre Corporation. 1974. An overview of the TICCIT program. M74-1, Mitre Corporation, January 1974.

NUJ (National Union of Journalists, United Kingdom). 1978. *Journalists and New Technology*. London: National Union of Journalists.

New Scientist. 1977. Foot in the door for the home computer. *New Scientist*, 14 April 1977.

Post Office Telecommunications. 1978. Press and Broadcast Notice DW40, February 28, 1978.

Statistics Canada. 1974. Consumer income and expenditure division, Family expenditure section. Unpublished document 62-541.

Wireless World. 1977. Viewdata crushes the Post Office armour. *Wireless World*, July 1977.

Appendix 19A: Reuters News-Wire

Information is transmitted by Reuters to the cable companies over its own worldwide private-circuit network. Two channels are offered: one provides general news throughout the day, the second broadcasts financial news during office hours and sports news at other times. The tranmission rate is very low: 470 bits per second. This data stream carries all the information for both channels. At the cable company a special-purpose convertor translates this data-bit stream into a television display image for analogue transmission over the cable.

The display consists of two parts. The first is a single row at the top of the screen that is continuously moving, like a ticker tape output. The remainder of the display consists of eight rows of thirty-two characters per row (uppercase only). It takes eight seconds to fill a screen and the image is retained for another twelve seconds, so that a new page is transmitted every twenty seconds. A complete cycle of news lasts fifteen minutes, and it is claimed by Manhattan Cable Television that many viewers watch the whole fifteen-minute cycle.

Appendix 19B: Prices of Teletext Decoders and Adaptors

Manufacturer	*Item*	*Approximate Selling Price*
Pye Labgear	Stand-alone adapter for broadcast teletext reception on standard television	$700
Gandalf Data Communications	Stand-alone adapter to allow use of television set and telephone line as general data terminal, suitable for viewdata-like service	$1,500
Videocraft	Component kit for installation into user's own television set to receive broadcast teletext (including charge for installation)	$450
Videocraft	Stand-alone adapter for broadcast teletext reception on standard television	$660
Videocraft	Complete receiver and decoder for broadcast teletext	$850
Texas	TIFAX: printed circuit board for integration in television sets by manufacturer, to allow reception of broadcast teletext	$230 (1 off) $130 (1000 off)

20 Interpersonal Communication through Computers: An Introduction to Computer Mail and Computer Conferencing

Robert Johansen

Both computer mail and computer conferencing are horseless carriage terms, defining a technology of the future in terms of a technology of the past. Such an approach is particularly misleading in this case, since *mail, conferencing*, and even *computer* are all imprecise terms. What is clear, however, is that a new context for interpersonal communications is being created and that this future communications context will have unique characteristics not present in the media of the past.

The goal of this paper is simply definitional, although unfortunately the definitions are not simple. My goal is to step back from the confusing terminology of today's exemplary systems for interpersonal communication through computers and draw some conclusions about basic concepts involved. Policy issues relating to these systems will not be discussed. The paper is intended for nontechnical readers with no exposure to computer mail or computer conferencing but is intended to be useful to those who have been involved as well.

Interpersonal communication through computers means that two or more people are communicating with each other and that the communication just happens to be occurring with a computer as an intermediary. The basic notion is that people are communicating with other people, not with a machine (man-machine communication); nor is a computer communicating with another computer (computer communication). Of course, both man-machine and computer communication are often involved in an episode of interpersonal communication through computers, but these are means toward the goal of interpersonal communication.

Computer Mail

Computer mail[1] is person-to-person communication, similar in kind to that provided by the telephone or the conventional postal system. The similarity with present media, however, is more misleading than helpful. Computer mail is a medium for sending messages from one person to one or more other people.

The computer provides a means for instantly delivering the message (basically doing better what conventional mail already does), but it also provides remarkable facilities for editing, organizing, storing, and otherwise manipulating the message. It is these latter capabilities that move far beyond the capacities of conventional mail and belie the title *computer mail.*

Computer mail involves an individual sending a message, a computer terminal which encodes the message, a transmission system (often a telephone line), a computer, and one or more recipients of the message—again using a computer terminal of some type. Typically, an individual types a message on a typewriter computer terminal. This message is sent to a computer where it is stored until it is claimed by the intended recipients when they check into the computer system using their own terminal.

The type of communication that occurs through computer mail is generally ongoing as opposed to the specific time limitations of a computer conference. For instance, computer mail may be used within a corporate structure to provide a standard administrative communication system. It might also be used to coordinate activities of individuals who are traveling frequently, as was the case during the 1976 election when the Carter campaign staff used Scientific Time Sharing Corporation's Mailbox system. The structure for this person-to-person communication depends on the particular computer mail system being used. All computer mail systems, however, provide some basic alterations in the way interpersonal communication has occurred in the past.

Uhlig, describing the considerable experience of DARCOM (U.S. Army Material Development and Readiness Command) with computer mail, states:

> The biggest single lesson we have learned from our experimental use of computer based message systems is that this is a major new medium of human communication and interaction, with a very positive impact on the way we do business. It is not just a replacement for letters, phone calls, TWXs, Telex, and Telegrams.[2]

Table 20-1 summarizes the benefits of computer mail that have been observed to date. These benefits arise primarily from the storage and manipulation capabilities of the computer, thus allowing even geographically separated working groups to communicate independent of the usual time and space constraints. The paperless office becomes more feasible as information is stored in computers rather than in filing cabinets. Individuals can select the times they want to send or receive messages; the system can even remind them of important dates or provide a structure for organizing complex information. Panko and Panko (1977) speculate about what it might mean to merge the communication capabilities of the computer with some of its recognized capabilities:

> Just as authors have suggested that initiative-taking "robots" will some day clean our households and run our shops, we suggest that

Table 20-1
Benefits of Computer Mail

1.	Permanent, searchable, stored record [of all communications desired]
2.	No simultaneous activity necessary
3.	No meeting schedule necessary
4.	Optimum time for composing, reading, and responding may be selected
5.	Physical co-location [of participants] not necessary
6.	No interruptions
7.	One action for general information distribution
8.	Fast delivery at low costs
9.	Automatic distribution [to designated persons]
10.	Automatic headers [for messages]

Source: From James H. Bair, "Communication in the Office of the Future: When the Real Payoff May Be," *International Conference on Computer Communication*, Kyoto, Japan, August 1978, p. 10.

> large-scale automation systems will eventually evolve into reasonably autonomous "mindservants" that will not only take initiative in pacing tasks but will even be able to program themselves to understand and serve their masters' needs.[3]

Such a development in interpersonal communication through computers would go far beyond the definition of mail, as well as redefine current understandings of interpersonal communication itself. Notions such as the office of the future come closer to describing this activity, but even it only provides a vague feeling for what might be possible, not a precise definition. The current systems for interpersonal communication through computers can thus be viewed as flirtations with a type of computer-aided communication whose form and effects are not yet known.

The British viewdata system provides a taste for a possible evolution of the computer mail concept. Viewdata is a hybrid of computers, television, and the telephone to provide information services to both business users and the general public. While message services will not be present in the initial system, they will certainly be possible if policy decisions are made to implement them. A general message service could be provided through standardized messages for users with limited-function viewdata terminals or full message sending (computer mail) for those with full terminals. Market tests of the system are already under way with a large commercial usage scheduled to begin in early 1979.[4] Of course, the focus of the viewdata system is not currently on mail functions but on providing information services (news, weather, sports, consumer information, business data). Yet the mail function, when residing on a computer system, has broad potential to adopt many features far beyond the scope of postal activities. The current viewdata system is but one example of the possibilities for an extension of early notions of interpersonal communication through computers.

Computer Conferencing

Computer conferencing can be viewed as a modest extension of computer mail, in much the same way as a conference telephone call extends a two-person call. However, research on the social psychology of small groups has indicated that there are sharp differences between dyadic communication and communication involving more than two persons.[5] Similarly, group communication through computers is different in several ways from two-person communication. The current characteristics of computer conferencing, as distinct from computer mail, can be summarized as follows:

> Computer conferences generally have a group task orientation, often for a specific time period.
>
> A group record is kept automatically and can be reviewed by participants as needed.
>
> Current computer conferencing systems are often easier to use for non-technical users; computer jargon is less imposing.
>
> Computer conferences can easily shift into synchronous group meetings, where more than one participant is present simultaneously. In such situations, everyone can be simultaneously typing, with a computer keeping a transcript of the proceedings and sorting the incoming messages.
>
> Computer conferences require group facilitation skills and leadership; organization of the meeting is critical.

Computer conferencing was used as early as 1970, and discussions of the concept took place even earlier. The Office of Emergency Preparedness (now the Federal Preparedness Agency of the General Services Administration) developed a system for computer conferencing, combined with data-base resources, for monitoring and responding to national crises (Turoff, 1971, 1972). In the early and mid 1970s, a series of field tests was organized by the institute for the future, with primary support from the National Science Foundation. National Aeronautics and Space Agency (NASA), U.S. Geophysical Survey (USGS), Earth Resource and Development Agency (ERDA) (now Department of Energy (DOE), the Charles F. Kettering Foundation, and other organizations were involved. These field tests generally involved groups of scientists, most of whom were geographically separated, who were engaged in joint tasks (Vallee et al., 1975; Johansen, DeGrasse, and Wilson, 1978). Field tests of computer conferencing are continuing at the New Jersey Institute of Technology (Turoff, 1977) as well as the University of Michigan (Zinn, 1977), and one commercial firm is offering computer conferencing services (see table 20-2).

Table 20-2
Examples of Currently Active Computer Mail and Computer Conferencing Systems

Computer Mail Systems	*Computer Conferencing Systems*
On Tyme Tymshare, Inc. Cupertino, California	Electronic Information Exchange Systems (EIES) New Jersey Institute of Technology Newark, New Jersey
Hermes Bolt, Beranek, and Newman Cambridge, Massachusetts	Confer University of Michigan Ann Arbor, Michigan
MSG; MS Advanced Research Project Agency Washington, D.C.	Planet Infomedia Corporation Palo Alto, California
HP-2026 Network Hewlett-Packard Palo Alto, California	Mint Nonmedical Use of Drugs Directorate Canadian Federal Government Ottawa, Ontario, Canada
Mailbox Scientific Time Sharing Corporation Washington, D.C.	Forum (Swedish version) Stockholm University Data Center Stockholm, Sweden

Note: These are most of the currently active systems, although no attempt was made to develop a list of all such systems. A more complete description of computer mail systems is contained in Konrad K. Kalba et al., *Electronic Message Systems: The Technological, Marker, and Regulatory Prospects* (Cambridge, Mass.: Kalba Bowen Associates, 1978).

At least six styles of computer conferencing have been identified to date, although there are frequently overlaps among these styles.[6]

The *exchange* is typically carried out over a period of months. The participating groups are usually quite large, ranging in size from twenty to forty.

The *community* implies a qualitative change from the *exchange* toward more cohesiveness as a group and a higher degree of personal interaction. The individuals become committed to the other participants, as well as to the substantive purposes of the group.

The *seminar* is focused on a specifically defined topic. The most common example is the research seminar or open conference which involves asynchronous usage (when participants are not all present simultaneously), usually over a period of one week to one month.

The *assembly* is an extension of the *seminar*, with more participants (perhaps a hundred or more), multiple topics, and a number of separate parts to the proceedings. Rarely observed to date, this style would be a computer conferencing analog to professional society conventions.

The *encounter* is somewhat similar in style to a face-to-face meeting, where participants are all present simultaneously discussing a topic for a short time (usually a few hours or less). The intensity of the encounter is often quite high, since computer conferencing allows all participants to speak at once with the computer program sorting out the order.

The *questionnaire* involves an unlimited number of participants in a structured question-and-response format. Such questionnaire formats are typically used as part of other computer conferencing styles rather than by themselves.

These six styles, however, only document the ways in which computer conferencing has been used to date. Like computer mail, however, there is little reason to assume that the first uses of this medium will be good precursors for future uses. The designation *conferencing* is already inadequate to describe current applications, and it is likely to become even more so. As with computer mail, the major extensions of this medium are likely to come from the computer's capacity for storing, organizing, and manipulating information. For instance, mathematical models might be used as participants in computer conferences to provide quantitative inputs to group discussions of particular substantive areas.[7] Another possibility is to use computer programs to aid in group decision making, perhaps through consensus building or weighted-voting techniques.[8]

Costs and Comparisons

The costs of both computer mail and computer conferencing are difficult to estimate and even more difficult to project for future periods. Panko, however, after doing a number of studies on the costs of computer mail systems, has concluded that these services will become very reasonable by 1980 (compared with existing media) and even more reasonable by 1985.[9]

Costs of computer conferencing are slightly higher but again should be quite attractive by 1980.[10] Direct comparison among media in terms of costs are complex, though, since very different services are provided. The actual costs will vary dramatically depending on the needs of specific users and the value they place on various features. Since this paper is only an introduction to interpersonal communication through computers, detailed cost comparisons will not be made here. The simple point is that costs of these media are already competitive with existing media in some circumstances, and the longer-range outlook is even more economically attractive.

While there are important differences between two-person and group communication through computers, computer mail and computer conferencing have been separated more by historical accident than by logical differences, These media are but early examples of a communications context not yet created: interpersonal communication through computers. A mix of new communications media seems most likely, with eventual discarding of terms such as *mail* or *conference* which no longer describe the phenomenon taking place.

But the development of interpersonal communication through computers may be a slow process. The great majority of current users are either systems developers or members of organizations that are consciously experimenting with these technologies. There are very few examples of actual operating systems for interpersonal communication through computers outside research and development environments. At the present time, even a motivated potential user would have difficulty trying out various systems and would quickly discover that choosing the right system is no easy task. Existing systems are rarely on interconnected computers and almost always have their own idiosyncratic structure of user commands. Current users experience frequent frustration as they are required to check several incompatible mail systems for messages, each with its own subtle (and not-so-subtle) barriers to use. Integrated systems are not yet here, and they could easily be slow in coming.

Conclusion

Interpersonal communication through computers is a frontier. The first few explorers, using crude systems typecast as computer mail and computer conferencing, have returned from the frontier with promising reports about what might be possible. And there are a few settlements on the frontier, each of which is using its own version of interpersonal communication through computers and each of which would like to think of itself as a model for the future. Enough is known to say with some certainty that the domain of interpersonal communication through computers is a frontier. But that is only a beginning; to think it more would be to severely limit the shape of future possibilities.

Notes

1. The term *computer mail* will be used in place of the term *electronic mail*, although the choice is quite arbitrary. The systems being discussed here, however, are directly tied to computer systems. Facsimile systems, for instance, are not discussed, although they could be considered in the general family of electronic mail systems.

2. Ronald P. Uhlig, "Human Factors in Computer Message Systems," *Datamation*, May 1977, p. 126.

3. Raymond R. Panko and Rosemarie U. Panko, "An Introduction to Computers for Human Communication," Proceedings of the National Telecommunications Conference, Los Angeles, December 1-3, 1977, p. 21.

4. For a description of the current state of viewdata activities, see chapter 19 in this volume. For information about viewdata, contact the Telecommunications Systems Strategy Department, Long-Range Studies Division, Post Office, 88 Hills Road, Cambridge CB2 1PE, England.

5. A general review of the small group literature can be found in Paul A. Hare, *A Handbook on Small Group Research*, 2nd ed., (New York: Free Press, 1976).

6. For more detail on computer conferening styles, Johansen, DeGrasse, and Wilson (1978, pp. 30-43).

7. See Jacques Vallee, *Technology Forecasting and Social Change* 10, no. 4 (March 1977). A project is under way at the Institute for the Future, under support from a National Science Foundation grant, to explore the possible hybrids between computer conferencing and mathematical modeling processes. For information, contact Dr. Hubert Lipinski at the Institute.

8. These methods have had limited testing, with mixed results. The major problems seem to be in blending the group dynamics issues with the specific decision-making aids. Computer programs rarely provide easy answers to group decision-making problems, and they can sometimes make them worse. While development of computer programs to aid group decision making is still an important task, it may be more important to explore ways of integrating these programs with the normal processes of small group communication.

9. See Raymond R. Panko, "The Outlook for Computer Mail," *Telecommunications Policy*, June 1977, pp. 242-253. Panko has also done a more recent study, available only in draft form at this writing. He is currently at the College of Business Administration, University of Hawaii, Manoa, Hawaii 96822.

10. An analysis of actual cost of a recent computer conference organized for the Planet System at commercial rates is contained in *Planet News*, Infomedia Corporation, 430 Shreman Avenue, Palo Alto, California 94306.

Bibliography

Bell, Daniel. "Teletext and technology." *Encounter*, June 1977, pp. 9-29.

Crocker, David H. *Framework and Functions of the "MS" Personal Message System*. Rand Corporation, Report No. R-2134-ARPA. Santa Monica, Calif., December 1977.

Edwards, Gwen C. *An Analysis of Usage and Related Perceptions of NLS: A Computer Based Text Processing and Communications System*. Montreal: Bell Canada, October 1977.

Electronic Mail: A Multi-Client Study, A Summary of Survey and Forecasting Methods. London: Communications Studies and Planning, and Luton: Mackintosh Consultants Company, July 1977.

Feinler, Elizabeth J. "The Identification Data Base in a Networking Environment." *Conference Record, National Telecommunications Conference*, vol. 21, section 3, December 1977, pp. 1-5.

Geller, Henry. "*Electronic Mail: Legal and Policy Considerations.*" Aspen Institute, prepared for 1976 Telecommunications Policy Research Conference, Airlie, Virginia, April 21-24, 1976.

Irving, R.H. "Usage of Computer Assisted Communications in an Organizational Environment," University of Waterloo, Ontario, Canada, June 1, 1976.

Jackson, Charles L. "Electronic Mail: What Is It? What Might It Be?" Massachusetts Institute of Technology, prepared for 1976 Telecommunications Policy Research Conference, Airlie, Virginia, April 21-24, 1976.

Johansen, Robert, DeGrasse, Robert, Jr., and Wilson, Thaddeus. Group Communication through Computers. Vol. 5: Effects on Working Patterns. Institute for the Future, Report R-41, February 1978.

Kalba, Konrad K., Sirbu, Marvin A., Jr., de Sola Pool, Ithiel, and Thompson, Janet Taplin, Electronic Message Systems: The Technological, Market, and Regulatory Prospects. Cambridge, Mass.: Kalba Bowen Associates, April 1978.

McLaughlin, John. "Electronic Mail: Defining a Role for the U.S. Postal Service." U.S. Postal Service, prepared for 1976 Telecommunications Policy Research Conference, Airlie, Virginia, April 21-24, 1976.

Myer, T.H., and Vittal, John. "Message Technology in the ARPANET." Conference Record, National Telecommunications Conference, vol. 21, section 2, December 1977, pp. 1-8.

Panko, Raymond, R. "The Outlook for Computer Mail." Telecommunications Policy, June 1977, pp. 242-253.

Panko, Raymond R., and Panko, Rosemarie U. "An Introduction to Computers for Human Communication." Conference Record, National Telecommunications Conference, vol. 21, section 1, December 1977, pp. 1-6.

Pye, Roger, and de Smith, Michael. Automated Communicating Office. Final Report on Feasibility Study, Communications Studies and Planning, London, Report No. AC01/78055/PY, February 1978.

Turoff, Murray. "Delphi Conferencing: Computer-Based Conferencing with Anonymity." *Technological Forecasting and Social Change* 3 (1972): 159-204.

Turoff, Murray. "Delphi and Its Potential Impact on Information Systems." *AFIPS Conference Proceedings*, vol. 39. Montvale, N.J.: AFIPS Press, 1971, pp. 317-326.

Turoff, Murray, and Hiltz, Starr Roxanne. "Meeting through Your Computer," *IEEE Spectrum*, May 1977, pp. 58-64.

Uhlig, Ronald P. "Human Factors in Computer Message Systems." *Datamation*, May 1977, pp. 120-126.

Ulrich, Walter. "Ontyme: A Computer Message System." *Conference Record, National Telecommunications Conference*, vol. 21, section 5, December 1977, pp. 1-8.

Vallee, Jacques. "The Outlook for Computer Conferencing on ARPANET and PLATO." Paper prepared for the Society for General Systems Research Meeting, February 1977.

Vallee, Jacques, Lipinski, Hubert, and Miller, Richard H. *Group Communication through Computers*. Vol. 1: *Design and Use of the FORUM System*, Institute for the Future, Report R-32, July 1974.

Vallee, Jacques, Johansen, Robert, Randolph, Robert H., and Hastings, Arthur C. *Group Communication through Computers*. Vol. 2: *A Study of Social Effects*. Institute for the Future, Report R-33; November 1974.

Vallee, Jacques, Johansen, Robert, Lipinski, Hubert, Spangler, Kathleen, Wilson, Thaddeus, and Hardy, Andrew. *Group Communication through Computers*. Vol. 3: *Pragmatics and Dynamics*; Institute for the Future; Report R-35; October 1975.

Vallee, Jacques, Johansen, Robert, Lipinski, Hubert, and Wilson, Thaddeus. *Group Communication through Computers*. Vol. 4: *Social, Managerial, and Economic Issues*, Institute for the Future, Report R-40, January 1978.

Zinn, Karl L. "Computer Facilitation of Communication within Professional Communities," *Behavioral Research Methods and Instrumentation*, April 1977.

21 Personal Computer Networks

David Caulkins

The Personal Computer Movement

The start of the personal computer (PC) movement in the United States was marked by the introduction of the first hobbyist computer kit, the ALTAIR 8800, by MITS (now a division of PERTEC). Real growth of the movement probably started with the January 1975 *Popular Electronics* article on the ALTAIR 8800. The year 1975 was also a year of expansion in other areas. Many of the largest and most active personal computer clubs were founded during this period in Atlanta, Chicago, Los Angeles, New England, New Jersey, and the San Francisco Bay Area.

There are now at least ten publications dedicated to hobbyist and personal computers, some with monthly press runs exceeding a hundred thousand copies. The oldest of these (*Byte* magazine) published its first issue in 1975.

Widespread distribution of the hobbyist/personal computer was made possible by the development of the microcomputer. This was done initially by one company, Intel Corporation. The first two microcomputers (the 4004 and the 8008) were not cheap, capable, or simple enough in application for widespread personal use. The third was the 8080. This machine enjoyed extraordinary success in a broad range of applications. It is still (three years later) a widely used industry standard. The 8080 served as the basis for the ALTAIR 8800 and then quickly for a host of competing products made by IMSAI, Polymorphic Systems, Southwest Technical Products, Processor Technology, and others.

The total number of personal or hobbyist computers sold rose from less than one hundred before 1975 to approximately fifty thousand in early 1978. This is the same as the total number of computers in the United States in 1968, ten years ago. The year 1977 marked the entry of major manufacturers into the personal computer marketplace, first Commodore Business Machines with the PET (Personal Electronic Transactor), and then Tandy (Radio Shack) with the TRS 80. These are turn-key machines; they need only be taken out of the box and plugged in to function. Their operating system software is in read-only

Many of the ideas in this paper stem from discussions with Walter Baer, Paul Baran, Buffy Caulkins, Phyllis Cole, Ronald Crane, Peter Deutsch, Geoffrey Goodfellow, Robert Maas, John Shoch, Lawrence Fesler, and Michael Wilber. Any errors are my own.

memory (ROM); it cannot be erased by a careless or inexperienced user. The PET and the TRS-80 (like most of the personal and hobby computers) are equipped with the BASIC language.

These machines are not mere toys; the same BASIC benchmark program was run in the PET and the IBM 5110; it ran approximately twice as fast in the PET. (See Appendix 21C for details.) Texas Instruments has announced its intention of entering the personal computer marketplace in full-page ads attempting to recruit engineers and programmers for the TI personal computer project.

The interest of these large companies is one measure of the probable growth in the population of personal computers. Another is provided by marketing surveys that have been made in this area. Figures 21-1 and 21-2 give the results of one such survey, done by Vantage Research of Palo Alto, California. Another indication is that in April 1978 there were eight hundred computer stores, retail outlets specializing in the sale of personal computer equipment.

What are the buyers of personal computers using them for? No one knows for sure. There are, however, strong indications of activity in at least the following areas.

Small Business. Many personal computer systems at the high end of the price spectrum ($3,000-$10,000) are being used for accounts payable and receivable, inventory control, and so forth.

Education. Computer-aided instruction (CAI) programs running in personal computers are in use in some schools. Home computers are being used to amuse and teach children, often quite young ones.

Recreation. Every home computer owner's program library includes games. These range from blackjack, backgammon, and Othello to rather complex econometric and moving-vehicle simulations of real-world systems.

Real-Time Control. A number of products are now being marketed that allow wireless remote computer control of home appliances using high-frequency signaling over ordinary AC power wiring.

Two scenarios have been advanced for the growth of the personal computer population in the United States. (1) *The ham radio scenario*. Personal computer growth parallels that of amateur radio; the personal computer population grows slowly over five or ten years to a total of a few hundred thousand units. Personal computers are too complex and difficult for most people to use; the home computer is a rare novelty rather than an unexceptional household item. (2) *The CB radio scenario*. Computers capture the imagination of a rapidly expanding number of people. Personal computer network-type telecommunications provide a rapidly increasing number of commercial[1] and recreational services to the computer-equipped home. (See appendix 21A for an example of a commercial service already being planned by a U.S. company.) The personal

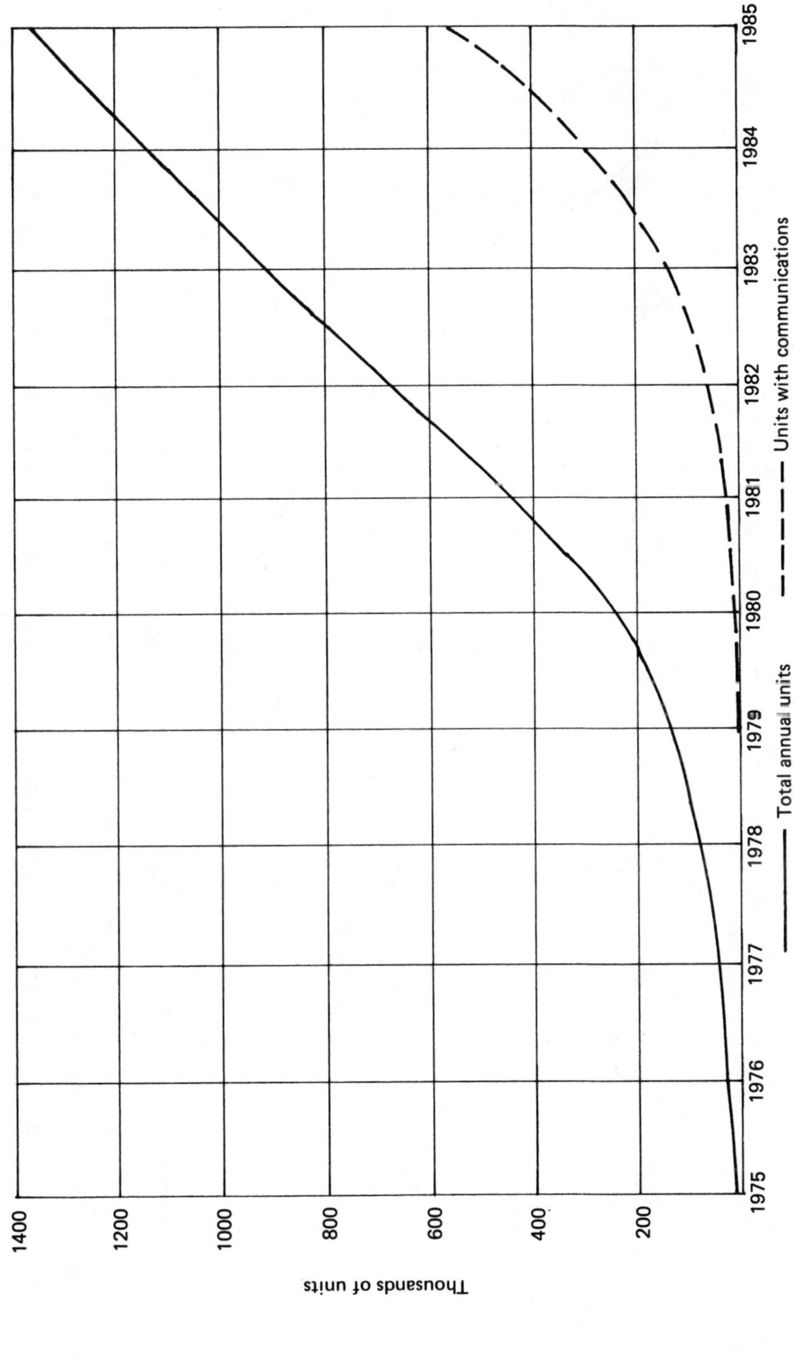

Source: Vantage Research, Palo Alto, California.

Figure 21-1. Annual Unit Shipments of Personal Computers With Communications Capability

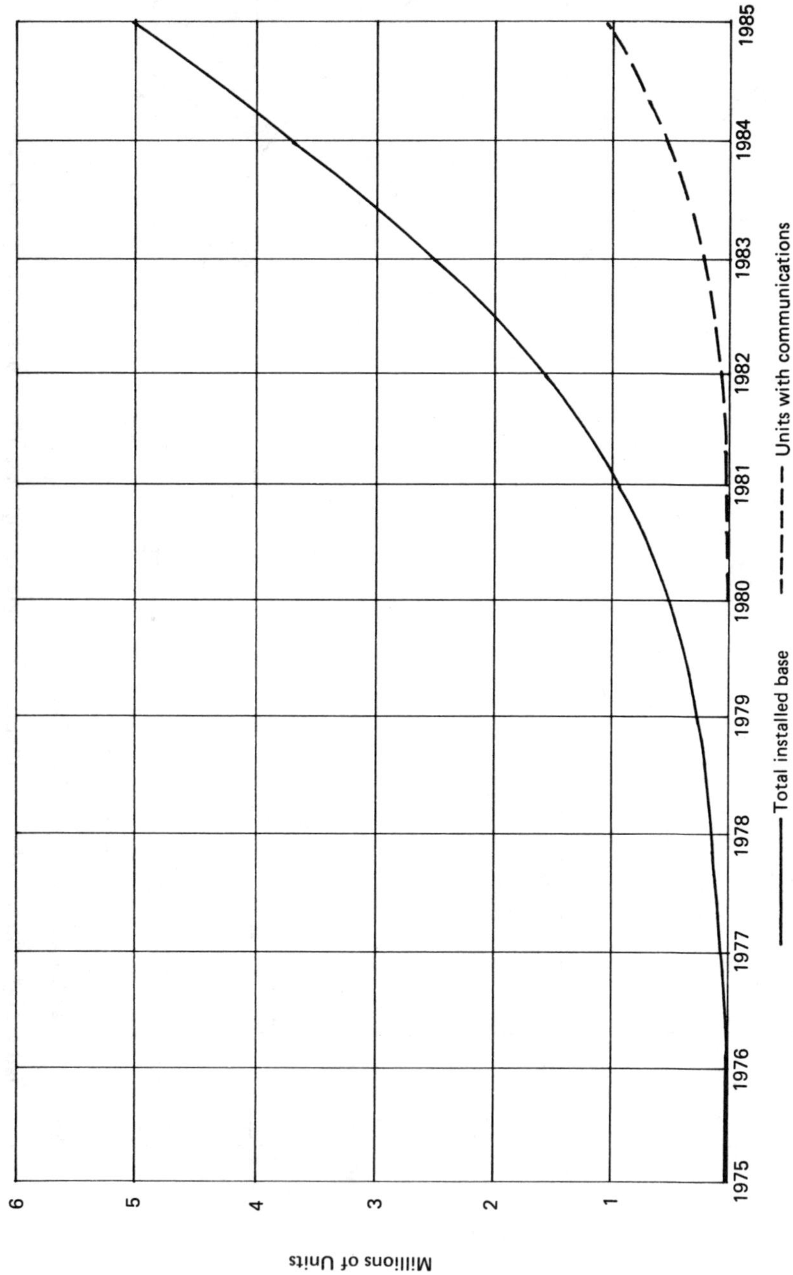

Source: Vantage Research, Palo Alto, California.

Figure 21-2. Cumulative Installed Base of Personal Computers with Communications Capability

computer population grows to several million units in five years. The home computer becomes a common appliance in a sizable fraction of American homes.

I incline toward the second scenario, for two reasons. The first is that technologies enjoying rapidly expanding application are often viewed as threatening by those whose lives are involved in the changes invoked by the new technology. Arthur Luehrmann has suggested that the layman regards the computer in 1977 the way he did the book in 1456—as a device of forbidding complexity usable only by highly trained experts. I believe we are entering an era in which computers can become universally acceptable tools like books, telephones, and wristwatches.

My second reason for favoring the rapid-growth scenario is that many people find computer programming attractive (not to say addictive). Joseph Weizenbaum has vividly described this attraction:

> The computer programmer . . . is a creator of universes for which he alone is the lawgiver. So, of course, is the designer of any game. But universes of virtually unlimited complexity can be created in the form of computer programs. Moreover, and this is a crucial point, systems so formulated and elaborated act out their programmed scripts. They compliantly obey their laws and vividly exhibit their obedient behavior. No playwright, no stage director, no emperor, however powerful, has ever exercised such absolute authority to arrange a stage or field of battle and to command such unswervingly dutiful actors or troops [1, p. 115].

Personal Computer Network Activity

The Personal Computer Network (PCNET) Committee came into existence in May 1977. PCNET activities are carried on by unpaid volunteers and funded by donations. The principal PCNET goal was to provide a reliable, low-cost means of transferring messages or files between personal computers situated in places equipped with a single dial-up voice-grade telephone line. It is surprising how powerful and efficiency-improving such a message exchange facility is. What keeps ordinary message services such as telephone, telegraph, and mail from working as well seems to be a combination of factors: too slow (mail); often hard to catch someone in the office, or pierce the screens of secretaries (phone); hard or time-consuming to use (mail, telegrams); expensive in terms of characters per dollar (telegrams, phone). A computer-based message system overcomes most of these difficulties. People regularly using such a system rapidly develop a whole new communication style. Most messages are brief—one-thousand characters or less. Message text is composed by the sender at his terminal without the filtering and delay of secretarial intermediates, and style tends to be much more informal and direct than conventional media, such as business letters. Message system users move rapidly toward a computer-based, personally oriented file system containing messages, distribution lists, and text files. The difference is one of kind, not just degree.

The PCNET mailing list grew from 15 in May 1977 to approximately 450 in April 1978. The PCNET Committee undertook four tasks: (1) design a set of PCNET protocol (a brief glossary of this and some other computer network terminology is given in appendix 21B); (2) implement these protocols in software; (3) experiment to verify that the software met the design goal; (4) document the software; distribute software packages to interested users. The protocols were judged sufficiently mature in late 1977 to allow software implementation to proceed. The initial experiments with first-cut software began in November 1977. In March 1978 PCNET operation was demonstrated to an audience of two-hundred people; an 8080-based personal computer system in Sunnyvale, California, sent a series of messages to a Commodore PET in San Jose, a distance of about ten miles. Noise was deliberately injected into the phone line to interfere with the transmission. The PCNET software in the two machines successfully detected the damaged packets of information and arranged for their retransmission; complete uncorrupted messages were displayed to the audience in San Jose.

Current PCNET activities involve preparing the PCNET software and associated documentation for general release, finishing a final draft of the PCNET protocol suitable for publication, and system design of PCNET implementations using alternative network link media.

Early in 1978 an interesting variant to the PCNET architecture was developed by members of CACHE, the Chicago Area Computer Hobbyist Club. This system, called a Community Bulletin Board System (CBBS), is intended to provide electronic mail service to an urban community of modest size (probably less than five-hundred people). Each member of the community must have access to a terminal or personal computer equipped with a modem.

Use of the CBBS is best covered by example. Suppose community member Mary wants to send a message to member Tom; she calls the CBBS (which operates unattended twenty-four hours a day). It has been built with good human factors so that even naive users are smoothly guided through usage procedures. Mary follows the instructions, types in the message to Tom, and terminates the call. At some later time (which can be anywhere from minutes to weeks) Tom calls up, reads his list of pending messages, and causes the new message from Mary to be typed on his terminal. If he wishes, he can enter an answering message. All the advantages of electronic mail are present in the CBBS; instant, accurate delivery of text, no need for sender and receiver to be present simultaneously, traffic in machine-readable form. The CACHE CBBS was built in thirty days from concept to operation for less than $3,000 by two talented CACHE members, Ward Christensen and Randy Suess. To try the CBBS, a call must be made with any 110- or 300-baud terminal. When the CBBS answers give it several carriage returns, from which it figures out whether your terminal is 110 or 300-baud. From this point on, the CBBS is self-teaching.

PCNET Link Media Options

The dial telephone network is adequate for many PCNET purposes, but other media offer significant capability enhancements, particularly in increased data rate, and lowered system costs.

A PCNET with radio links (digital block radio, DBR) might be structured as follows. Each personal computer to be used as a PCNET node (network relay point) would be equipped with a radio transceiver. This transceiver would have a useful range of about five miles operating on a single frequency common to all other such transceivers in the United States. (Initial traffic estimates indicate that a single-frequency allocation could handle all the DBR PCNET traffic for several years. See appendix 21D.) The transceiver might have a bandwidth sufficient for a digital data rate of about 50 kilobits per second. The PCNET transceivers could (and probably should) incorporate a microprocessor as a control element. These transceivers could thus be quite "smart," implementing optimal use strategies under firmware control. Because these strategies could be controlled by ROM-based firmware, the PCNET user (or his errant software) could not cause the transceiver to operate in an abusive or channel-hogging way.

This DBR PCNET system could utilize an ALOHA-like method of random, multiaccess broadcast [2-4]. All transceivers simply transmit data packets randomly as required. If two or more packets are transmitted simultaneously from transceivers close enough to interfere with one another, error-detecting codes are used to identify the offending packets. They are subsequently retransmitted. This scheme (an extended description is presented in appendix 21D) requires equipment of minimum complexity, probably costing no more than a CB radio. The potential data rates are significantly larger than those feasible using the dial telephone network, and the costs are considerably lower. A radio-based PCNET of this type would need inexpensive transceivers that could be easily installed and used by people with little or no technical expertise. Frequency bands that have been suggested for DBR PCNET use include the following: 160-170, 220, 230-420, and 900 MHz. This sort of new radio service has significant policy implications.

Demand allocation of communication bandwidth (see [3] and [4] for a detailed discussion) offers many attractive alternatives for PCNET operation. PCNET nodes contain one or more microcomputers; consequently, communication protocols and strategies of considerable sophistication can be implemented.

This simple DBR PCNET system postulates dedication of radio spectrum bandwidth to PCNET use. As suggested in [3], this need not be the case. The "smart" nature of PCNET nodes allows them to gracefully share bandwidth with other systems, utilizing dead air that would otherwise be wasted.

One such possibility is the sharing of a land mobile or other local voice frequency. PCNET nodes would monitor the shared frequency and transmit only when there was no carrier. Nodes could transmit one packet in 0.2 seconds; after each such transmission, they would wait for several seconds and then check for carrier again before transmitting another packet. Even in the worst case, voice users of the channel would see no interference after the first 0.2 second of their transmission. PCNET nodes could get useful bandwidth even in channels judged by voice users as being very crowded. As with the telephone system, PCNET nodes could effectively utilize increased channel capacity during off-peak hours.

Policy Issues

Telephone-Based PCNET

Analysis and experiment by the PCNET Committee indicates that a dial telephone-based PCNET is feasible now. The equipment for a PCNET node (a personal computer with removable storage medium and an auto answer-originate auto dial modem) can be purchased for $1,115. If the curves of figures 21-1 and 21-2 are any guide, it will be some time before even substantial growth in PCNET activity can affect important telephone system statistics on significant parameters such as call holding times. At present the telephone system is essentially idle from midnight until 8 a.m. PCNET has been designed for unattended operation; messages put into the system can easily be marked for automatic transmission during these hours when rates are attractively low.

A packet-switched service for residential customers may prove attracrive; it will certainly be so if it can be competitive with the approximately 120,000 bits per dollar now possible at early morning and weekend rates over dial-up circuits.

Radio-Based PCNET

Digital data transmission over PCNET link distances by radio for nongovernment or nonbusiness use is now sharply restricted. Part 97 of the FCC Rules limits digital radio transmission to the obsolete Baudot Code and to a data rate less than 75 bits per second (BPS).

If unmodified, these restrictions would prevent development of a radio-link-based PCNET; it could not compete with a telephone-based system operating at 300 BPS. Even if these restrictions were removed, the detailed technical examination required for a radio amateur license would sharply reduce the number of people willing to experiment with a radio-linked PCNET. Potential PCNET users bear the same relationship to radio amateurs as television viewers do to television repairmen; PCNET users will want to use the technology but will have no interest in the tedious minutiae of radio transmission and reception. Transceivers for PCNET use can be made smart enough to allow full PCNET capability while preventing unfair or abusive use of the PCNET radio channel by naive or malicious PCNET transceiver operators.

The licensing for CB radio operation is much better designed for easy entry; unfortunately Part 95 of the FCC Rules does not authorize CB band modulation suitable for digital transmission. To foster the growth of radio-based PCNET activity and capture the societal benefits, it seems desirable to make PCNET radio operator licensing no more stringent than that for CB radio.

Societal Consequences of Personal Computer Networks

If the CB radio scenario for rapid growth in personal computer telecommunications is accurate, the ability of geographically distributed groups to interact rapidly will markedly increase. This seems to be a good thing; any increase in the bandwidth available for communication between people is intrinsically desirable.

The power of a modern nation is chiefly the result of its capabilities in four areas: economic, industrial, technological, and military. The digital computer (especially the microcomputer) plays an ever more important role in all four areas. Successful large-scale application of digital computer technology requires a population with a high percentage of "computer literates"—people comfortable with computer systems and able to use them easily and effectively. Recent work in economics [5-7] indicates that the processing of information is a central and expanding part of modern society. Here again, the computer is a vital tool, and computer literacy is an important and productivity-enhancing attribute for the members of such societies.

Personal computer networks of the PCNET type can be a powerful force in the spread of computer literacy, especially in the United States. Americans love gadgets; the personal computer comes close to being the ultimate gadget. Americans love to communicate; encourage the growth of electronic mail networks by personal computers, and the combination should prove irresistible. In the pursuit of this new recreation, a sizable fraction of the U.S. population could become computer literate.

References

1. Weizenbaum, J. *Computer Power and Human Reason*. San Francisco: W.H. Freeman and Co., 1976.

2. Abramson, N., et al. "ALOHA Packet Broadcasting: A Retrospective." Proceedings of the National Computer Conference, 1975.

3. Kleinrock, L., and Tobagi, F. "Random Access Techniques for Data Transmission over Packet-Switched Radio Channels." Proceedings of the National Computer Conference, 1975.

4. Cerf, V., et al. "The Future of Computer Communications." *In Proceedings of the Federal Communications Commission Planning Conference.* Washington, D.C.: Federal Communications Commission, November, 1976.

5. Porat, M.U. "The Information Economy: Definition and Measurement." OT Special Publication 77-12(1), U.S. Department of Commerce, Office of Telecommunications, May 1977.

6. Oettinger, A.G., et al. *High and Low Politics: Information Resources* for the 80's. Cambridge, Mass.: Ballinger Publishing Co., 1977.

7. Parker, E.B. "Telecommunication Policy and Information Services." Presentation to the 1977 ASIS annual meeting.

Appendix 21A
Excerpt from a 1977 JS&A Catalog

Dial-A-Bargain®. You can use a home computer to transmit digital data over the telephone to another home computer. By purchasing JS&A's entire system, you'll eventually be able to play chess, checkers, or other games of skill over the telephone with someone else using his home computer, and you'll be able to see both sides of the play action on your TV screen. To do this you'll need a phone coupler which will be available later this year.

Want products from our catalog? We'll have them ready to ship to you in less than an hour with our program called "Dial-A-Bargain." It works like this. You first insert a special JS&A ordering cartridge into your home computer. On your TV screen will appear our order form. You fill out the form using your keyboard and then place your phone in a phone coupler attached to your home computer. You dial a special toll-free number which is answered by our main computer. You then press a button on your unit, and your order, which is entered directly into our computer, is confirmed and the scheduled shipping data is flashed on your TV screen. If we are back ordered on an item, it will advise you and give you the option of canceling your order.

Meanwhile, your order is processed and a label is printed out at our warehouse. Our warehouse employees will respond the minute a label is printed, even during our night shift, so it is possible that your order will be ready to ship within a few minutes. All JS&A customers who own JS&A computers will be eligible for this truly space-age program.

Appendix 21B Glossary

Circuit switched network A network in which a direct circuit connection is set up between communicating nodes

Connectivity The number of links terminating at a particular node

Demand access Allocation of transmission bandwidth on demand

Distributed computer network A collection of geographically distributed and autonomous computers sharing common transmission and switching media

Hop Transmission of a packet from one network node to the next

Link The communication channel between two adjacent nodes

Node A computer system capable of receiving, testing, storing, and transmitting packets using one or more links to other nodes

Packet A short block of data used to carry information through a distributed computer network. Packets typically contain three kinds of information: the data or message to be transmitted, addressing and control information, and data integrity checking codes

Packet-switched network A network in which packets are routed from the originating node through intermediate nodes to the destination node

Protocols Sets of conventions (formats, control procedures) that facilitate all levels of intercomputer communication. Includes electrical inerface conventions, line-control procedures, digital communication network interfaces, interprocess communication conventions, and application level (file transfer, data base retrieval) standards.

Appendix 21C
Source Code: BASIC

The BASIC benchmark program was written by Greg Dolkes; its source code is as follows:

```
10 for I = 1000001 to 1000003 STEP 2
20 for D = 3 to SQR (I) STEP 2
30 IF I/D = INT (I/D) GOTO 60
40 NEXT D
50 PRINT I
60 NEXT I
```

Benchmark running times were as follows:

Alpha Micro 100 (compile time 0.26 sec)	3.86 sec
PET	7.0 sec
IBM 5110	12.0 sec
IBM 5100	23.0 sec

(The Alpha Micro 100 and PET are hobbyist or personal computer-type machines.) The benchmark was rewritten in FORTRAN and run on a CDC 7600; running time was 0.0013 sec.

Appendix 21D
A Design for a Digital Block Radio PCNET

There are too many unknowns to permit a detailed and explict design at this time. The regulatory issues may be resolved in a number of ways; these would strongly influence the technical decisions that would be made in an actual DBR PCNET design. The design presented here is intended as an existence proof to indicate that a DBR PCNET is both feasible and cost-effective with today's technology and will become more attractive with the continuing decrease in cost of digital electronics.

The design is for a DBR PCNET as it might exist in the 1982-1985 period. It is predicated on the following assumptions.

1. Of the total population, 2.5 percent participates in the DBR PCNET; there are 5×10^6 DBR PCNET transceivers geographically distributed in proportion to population density.

2. There is a single DBR PCNET frequency for the entire United States. Enough bandwidth is allocated to permit a data rate of 50 kilobits per second. The DBR transceivers are designed to communicate over ranges of five miles or less. The retail price of these transceivers in mass production is targeted to be less than $300.

3. The transmission protocol would be Slotted ALOHA with Carrier Sense Multiple Access. Channel capacities with this kind of protocol can range from 37 percent to 100 percent of total channel capacity [3] when total capacity is that available when there are no conflicts for channel use. I will use the 37 percent figure as a conservative value.

Demographic considerations and assumption 1 yield table 21D-1. The amount of data available in a twenty-four hour period to each transceiver in the high population density case is

$$\frac{0.37 \times 24 \times 3600 \times 50 \times 10^3}{511} = 3.13 \times 10^6 \text{ bits}$$

This is equivalent to 391,000 characters per day, or 98 pages of 3,960 characters each. This seems more than adequate.

Three critical design parameters have been rather arbitrarily selected: a single DBR operating frequency with sufficient bandwidth for a data rate of 50 kilobits per second; transceiver range of five miles; and slotted ALOHA CSMA protocol. A much more detailed analysis of the technical, cost, performance, and spectrum allocation trade-offs must be made before a real design can be carried out.

Table 21D-1
Demography and Assumptions

Population Density	*People per Square Mile*	*PCNET Transceivers per Square Mile*	*PCNET Transceivers in a five-Mile Radius*
High	260	6.5	511
Moderate	130	3.25	255
Low	25	0.62	49
Sparse	3	0.075	6

22 Electronic Message Services for the Deaf

Jeffrey Krauss

It is ironic that Alexander Graham Bell invented the telephone while searching for ways to facilitate communications among deaf people, yet the telephone system that resulted has been virtually inaccessible to deaf people. Indeed, in recent years the telephone has become such a common means of communication that an individual who cannot use it may be isolated or otherwise disadvantaged in modern society. For the more than 13 million people with impaired hearing in the United States, the inaccessibility of the telephone and other two-way telecommunications services may well impose barriers beyond the hearing loss. However, new technological developments in telecommunications, electronics, and computers may make it possible to provide for the deaf many of the advantages of telecommunications not heretofore possible. The implementation of electronic message services for the deaf will enhance their ability to communicate with one another and with the hearing, but it may also raise some important policy questions.

Electronic message services (EMS) are growing in importance as an adjunct of business communications service. The concept of office automation, which is likely to have a drastic impact on the structure of job responsibilities and on productivity in the modern office, relies heavily on the deployment of electronic message services. However, relatively little effort has been expended in examining the marketplace for residential electronic message services. The important policy significance of an EMS for the deaf is that it is potentially the first non-business, mass-market EMS service. Consequences of this may include the economies of large-scale semiconductor production, which could make the home terminal a mass-market product, and the impact on first-class mail volume. Other important policy issues, which may require federal action or may simply work themselves out in the marketplace, include questions of standards, protocols, and interconnection. Finally, with respect to both services for the deaf and services for the general public, the issue of industry structure—who should provide the service—is also of importance.

Existing TTY System

Currently, the primary means of electronic communications available to a deaf individual is the TTY system. The abbreviation TTY has traditionally stood for teletypewriter, and the TTY system employs a class of such terminals that operate on a dial-up basis over the public switched telephone network. The TTY system was developed initially by members of the deaf community because the telephone companies did not offer a service to meet their needs. An essential feature of this system was the development of a modem that allowed TTY terminals to be acoustically and inductively coupled to the network. The modem was designed in 1964; by 1966 extensive use had developed in St. Louis and other cities. (Since this was prior to the Federal Communication Commission's 1968 *Carterfone* decision [13 FCC 2d 420 (1968), recon. denied 14 FCC 2d 571 (1968)] the acoustic and inductive coupling was evidently contrary to telephone company tariffs, but there is no evidence of telephone company opposition.) Initially TTY terminals had been almost exclusively surplus or obsolete teletypewriters, donated by the Bell System, Western Union Telegraph Company, and the International Record Carriers. More recently a wider range of terminals incorporating modern technology has been developed and marketed for use by the deaf. It is estimated that fifteen thousand to thirty thousand TTY terminals are currently in use, although a lower bound for the potential market size can be estimated to be the roughly 2 million profoundly deaf individuals in the United States.

Even with the new technology incorporated into the TTY system, it has two significant drawbacks. First, the terminal equipment used by the deaf is incompatible with standard computer equipment used in the provision of data communications services. Second, because it results in the transmission of data over the network at very low bit rates, the system makes very inefficient use of the telephone network. While a detailed discussion of the technical incompatibilities is beyond the scope of this paper, a brief examination shows that they result from the different character representations used in Baudot and ASCII codes, and from the nonstandard frequencies used in the modem designed for the TTY system. The inefficient use of the telephone network may or may not be a burden on the network, but it does seem to be a burden on some deaf users because of high message-unit charges. When these high charges occur, it is because the telephone connection is maintained between two TTY terminals while the users slowly type out their messages.

Electronic Message System

In contrast, an electronic message system, based on packet switching or statistical multiplexing transmission techniques, could make much more efficient use of transmission lines. Moreover, charges to the customer could be based on the number of characters transmitted rather than the connect time. Equally important, a computer-based EMS can overcome the limitation caused by the incompatibilities between the TTY and the telephone. In particular, the conversion between Baudot and ASCII representations can be handled by a computer switch, as illustrated by the conversions between TWX and Telex standards now routinely handled by Western Union and the International Record Carriers.

This conversion capability will remain important as long as any deaf users retain the Baudot-based TTY terminals, although it is likely that most users will eventually have ASCII terminals because of the manufacturing economies. In the meanwhile, the conversion capability is absolutely essential if a transition is to occur.

An EMS for the deaf will bring them additional benefits. For example, store and forward message service will be available. Under the current TTY system, only a conversational mode is available, which requires that both users be at the terminal after the connection is made. If a caller encounters a busy signal, or if there is no answer, then no conversation can take place. An EMS offers the capability to store messages when a connection cannot be completed and to deliver the messages when the recipient requests delivery. In addition, text editing will be routinely available.

By making use of the computational power of computer switches, additional services of a noncommunications nature could be added to the EMS service. For example, a deaf person looking for employment could dial into an automated job-listing service on a computer. Another possible service could correct the user's spelling and grammar, as an adjunct to a text-editing capability. Undoubtedly, experts in the field of education of the handicapped and rehabilitation services will devise many other service delivery programs based on the capabilities of the electronic message service.

Recent Activity

Relatively little activity in the development and provision of electronic message services for the deaf has come to my attention. I understand that the Deaf Community Center of Boston, in cooperation with the consulting firm of Bolt, Beranek, and Newman, has begun an experimental project funded by the Department of Health, Education, and Welfare (HEW). Another HEW activity has just begun, in the form of a request for proposals for an experimental electronic message service for the deaf, coupled with a thorough evaluation of the developing technology. This project is funded for a substantial sum, about $400,000, and important results can be anticipated in one to two years. Finally, the Federal Communications Commission has instituted an inquiry into the telecommunications needs of the deaf, which focuses particularly on the range of service capabilities that might be available through the use of computer communications technologies. The conclusion is that there is little funding and little activity in this area.

Because of this apparent dearth of activity, it is reasonable to ask whether policy analysts should devote much time to the topic. On the other hand, all long-range analysis and evaluation is subject to some risk and uncertainty. In view of the rapid development of technology, the emerging interest in applying this technology to the needs of the handicapped, and the general public fascination with the notion of electronic mail, it seems worthwhile to at least begin to examine the policy considerations, in spite of the uncertain long-term prospects.

Policy Issues

The incompatibility of the existing TTY system used by the deaf and the equipment used for data communications is probably symptomatic of a larger question of standards, codes, and protocols for EMS services. While computers can be programmed relatively easily to convert between Baudot and ASCII representation, this conversion capability becomes more difficult when communicating word-processing terminals are used as terminals in an EMS because terminals with built-in text-editing capabilities often have nonstandard representations for various functions and operations. Similarly, with respect to modem operation, there are already at least two incompatible standard modem designs for operation at 1,200 bits per second. The important policy question is to determine what role the marketplace should play in establishing standards, as opposed to government or other institutional intervention to mandate standards. And if it seems likely that technical standards should be established, what is the appropriate body to handle that responsibility? With respect to services for the deaf, one can conceive that HEW, the FCC, the National Bureau of Standards, the American National Standards Institute will all have an interest.

Interesting policy questions arise in the consideration of industry structure. What sort of organization should provide an EMS service for the deaf? Should it be operated on a nonprofit basis, with subsidies or other financial support from the government? If it is operated on a nonprofit basis, with communications channels leased from other communications common carriers, then how does the FCC's *Resale and Shared Use* decision apply? In that decision resellers of communications services, operating on a profit-making basis, must be regulated as common carriers. Entities that administer telecommunications services on a nonprofit, shared-use basis are not common carriers and thus are not regulated by the FCC.

A related question deals with preferential rates. Suppose a communications common carrier finds that it can offer an EMS service to the deaf as an incremental addition to an EMS service already in operation. The EMS for the deaf could "piggyback" on the existing facilities, and the incremental cost of the service for the deaf might be less than the cost of establishing a new, totally separate service. Is it in the public interest to require or even to allow the communications carrier to provide the service to the deaf at a price based on incremental costs, lower than a separate supplier could price the service? This really speaks to the question of internal versus explicit subsidies and whether these services should be supported by hidden subsidies within the rate structures of existing common carriers.

Another industry structure question relates to the FCC's computer communications rules. This is pertinent to EMS services in general, and not limited to services for the deaf. Although I have not yet defined precisely what an electronic message service is, let us assume that it acts as both a communications service (sending messages from one party to another) and as a data processing service (text editing, storage and retrieval of messages). The question now before the FCC is whether it is possible to define a service of this type as either communications (thus subject to FCC regulation) or data processing (from which communications common carriers are barred, under the FCC's computer communications rules). The new FCC Computer Inquiry (Docket no. 20828) has thus far failed to focus on the particular problems associated with electronic message services when attempting to establish this jurisdictional boundary.

Another policy question deals with the interconnection of networks and services. There are really two parts to this: the interconnection of various EMS services for the deaf with one another, and the interconnection of a service for the deaf with a service for another group of customers. The case of interconnection among services for the deaf could arise if EMS services are implemented locally, with each service designed to serve a single metropolitan area. The potential problems are similar to those that arose in the telephone industry eighty years ago. More than one service may initially be established in a city; can we expect anticompetitive behavior as each tries to expand its geographical area? Is interconnection needed in this case? Is it appropriate? Even if there is only one EMS per city, how should they be interconnected for intercity service? What role will the intercity service suppliers (Telenet, Tymnet) play in the provision of local EMS services? One can conceive of an intercity service supplier's offering interconnection services only to its own local affiliates and denying interconnection to unaffiliated local service suppliers.

Other problems arise in the interconnection of services for the deaf with services for the hearing. A service for the deaf may serve primarily deaf people, but surely some hearing people will have an interest in using the service to communicate with their deaf friends, relatives, or business associates. Emergency service and public service organizations will also be subscribers. Will the rates be established in such a manner that hearing users have an incentive to subscribe to a service for the deaf? This is a particularly important concern if services for the deaf are subsidized or otherwise supported by government funding. Interconnection of a service for the deaf with other EMS services will make it more attractive for the hearing to subscribe to the service for the deaf; it may also make the service more attractive to deaf people, increasing their telecommunications options. Here again, there is historical precedent for the interconnection of two networks: the Telex and TWX networks are now interconnected, although they were not interconnected prior to Western Union's purchase of TWX from AT&T.

Finally, a thorough public policy evaluation of this topic should examine the role of the federal government in the provision of EMS services for the deaf. The federal government wears many hats: it funds research, it provides message delivery service through the Postal Service, and it is a customer. Arguably, these roles may conflict when EMS is involved.

It is reasonably likely that federal funding will be needed early in the development and implementation of EMS services for the deaf. Because of the perceptions of most service vendors and equipment vendors that this market segment is small and unprofitable, there is an unwillingness by these private sector firms to fund exploratory research and development. Consequently, there may be a need for the federal funding of specialized terminal equipment and EMS systems design and development. Currently, HEW has funding available for these types of projects. Pilot projects or demonstration projects are likely to be needed prior to any large-scale deployment of EMS services for the deaf, and here both HEW and the Department of Commerce may have a funding role. The question of continuing operating subsidies is much more complex, but one would hope that an EMS service for the deaf can be operated without needing direct subsidies on a continuing basis.

The federal government is also a customer for telecommunications services and electronic message services. As a customer, it has an interest in seeing that its rates are kept as low as possible and that it does not subsidize other users. The organization with the responsibility to manage the government's communications needs, the General Services Administration, has virtually no interest in the service delivery responsibilities of the mission-oriented cabinet agencies such as HEW or Commerce. GSA, like any other business customer, tries to minimize communications costs for services provided to the government.

The federal government is also a supplier of message delivery services, through the U.S. Postal Service, and experimentally through its operation of the ARPAnet for the Department of Defense. The design principles embodied in the ARPAnet (namely, packet switching) have been very important as basic building blocks of some of the EMS services now in operation. It is possible that the public availability of EMS services, such as a service for the deaf, will hasten the loss of first-class mail from the Postal Service, and, of course, this raises a whole range of public policy issues beyond the scope of this presentation. In any case, the role of the federal government is, as usual, manifold and internally inconsistent.

Summary

The development of the first public electronic message service is likely to be a service designed for the deaf. An examination of the existing services used by the deaf, as well as an examination of electronic message services generally, discloses a number of interesting and important policy issues. Some, such as those relating to technical issues of standards and protocols, are relatively narrow; others, dealing with the role of the federal government, are representative of broad public policy issues that are likely also to surface in other forums. While the current level of implementation of those services is still very low, it is not too soon to begin examining some of these issues.

23 The Policy Impact of Small, Inexpensive Computer Terminals

Donald L. Nielson

This country's telephone system, probably the most reliable, sophisticated, and ubiquitous communication system anywhere, is a source of frustration to millions of America's deaf. The frustration is one of inaccessibility.

About twenty years ago, using discarded and antiquated teletypewriters, a small segment of the deaf community banded together to create an inroad in their war of frustration. With few resources and much dedication, a network of fifteen thousand to thirty thousand terminals has slowly come into existence, but the need for terminals in this network far outstrips their availability. Recently, a number of more modern terminals have been commercially offered, and new devices and services have been conceived. One such device is the ultraportable, hand-held terminal shown in figure 23-1. It is compatible both with the teletype network and with a standard ASCII/Bell 103 time-sharing computer-access arrangement.

Although the fact may not be obvious, this small terminal carries unusual significance for the present U.S. policy on digital communications. A practical mechanical equivalent of the ear is still a long way off, speech-understanding techniques are costly and of limited vocabulary, and video phones are not generally available, so the deaf are usually constrained either to signing or to what might be termed digital dialogue—that is, communications using alphanumeric characters. This latter technique, when accomplished in or near real time, contains exactly the elements of electronic mail. For this reason, a large part of the communication services needed for the deaf community anticipates precisely the needs of the business community in general. These services include, for example, the integration of text processing and communications, but because of the narrow interpretation of present regulations, this combination of services is not available among members of the business community. A major question is, therefore, Will certain digital services be legitimately provided to the deaf before being available universally? Devices such as the one shown in figure 23-1, although motivated by the needs of the deaf, can satisfy many of the requirements of electronic message services.

Some likely features of the hand-held terminal are listed in table 23-1. These lead to a number of uses both in the deaf community and under general use. A few such uses are mentioned in table 23-2. While the needs of the deaf community are segregated here, most of the uses listed are of some value to the general public, with differences only in degree.

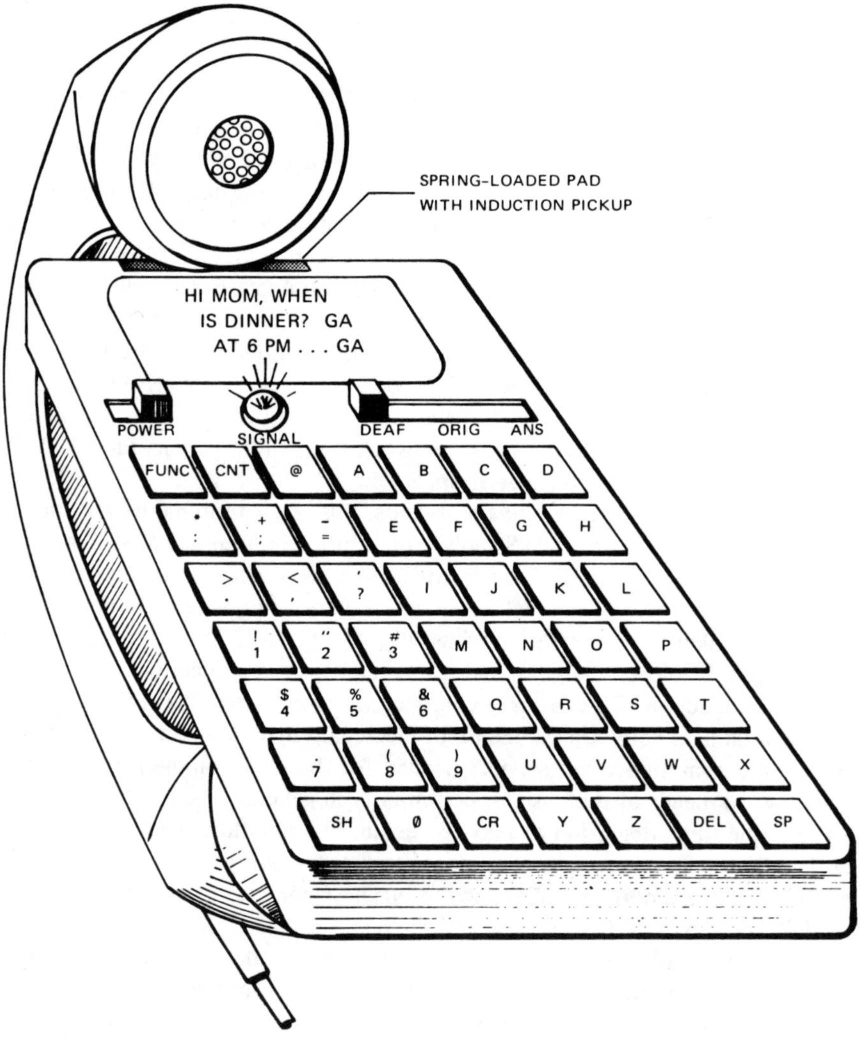

Figure 23-1. Proposed Configuration of Hand-Held Terminal

Table 23-1
Possible Features of a Hand-Held Terminal

Battery operation, rechargeability
High portability
Capability for attachment to standard G handset without intrusion
Full- or half-duplex operation
Printer or LCD display
Scrolling display
Standard character set
Standard signaling frequencies
Bus extender for peripherals
Down-line loadability
Low cost

Table 23-2
Some Uses for the Hand-held Terminal

For the deaf or mute
- Full-duplex alphanumeric dialogue with compatible terminal
- Services
  - Emergency aid
  - Airline information
  - News
  - Weather
  - Catalogue ordering
  - Directory assistance

For general use, remote computer access
- Electronic mail
- Time-share computing
- Stock, insurance quotations
- Inventory taking
- Down-line loading of a wide variety of programs

An important eventual feature of such an intelligent terminal is its ability to be down-line loaded—a term meaning to pass executable code into the terminal's random-access memory over the telephone line. Such a capability means that the terminal may act either as a communication device, using a stored nonvolatile program, or as a data processor—depending on the transferred program.

Such devices give rise to a number of policy issues. Although the hand-held terminal poses no functional uniqueness in this regard, it may come to have substantial policy impact because of its affordability. A few policy factors are listed in table 23-3.

The implications of these factors are extensive for both the deaf and the nondeaf. In the case of the deaf, there has been clamor for some time to give preferential long-distance and MTS pricing to users of the deaf teletype network. This is because connect times are understandably longer than for conventional

Table 23-3
Implications for Policy and Regulations

For the deaf or mute

- Special rates and cross subsidization
- Shared use through a nonprofit resale carrier
- Standards
  - Hindset
  - Coupling techniques
  - Signaling conventions

For general use, remote computer access

- Impact on existing carriers
  - Record carriers
  - Telephone carriers
- Impact on computer inquiry
- Impact on resale issue

voice calls for a given amount of traffic. Connecticut's Public Utility Commission has already invoked a reduction of one-fourth on certain telephone toll charges for deaf users. Impediments to this discrimination are the public acceptance of cross subsidization, the problem of enforcement, and practical billing techniques. Telephone dimension or signaling standards should be strengthened if such terminal equipment is expected to proliferate. It follows that the telephone company must actively avoid any arbitrary or capricious change to the handset and its signaling conventions.

The nondeaf issues are even more difficult to deal with. First, we are talking about almost indistinguishable differences between data processing and communication, and this one small unit contradicts the FCC's present attempt to segregate those functions. Second, the increased use of electronic mail, rather than conventional mail or telephones, may lead to a marketing problem for the Postal Service and the telephone companies, although the extent of the impact has not been gauged. The resale of wide-area telephone service (WATS), currently an open docket, will have a large effect on the number and type of computer-mediated message services that may emerge. Finally, the wide availability of an affordable hand-held terminal may create more pressure from the general public to have such message services than any other single motivation.

In conclusion, it appears that the problems of discriminating between the deaft and nondeaf users of common carrier networks can be largely avoided if the potential services become widely available to the general public. This reasoning presumes that wider use will mean uniformly lower costs and that the time-cost penalty paid by the deaf in their "main" channel of communication will not be unacceptable. That, of course, is pure speculation, but the proliferation of the devices and services considered here, of which the deaf can take advantage, appears to be the most straightforward course for all concerned.

24 Regulatory Barriers to Home Information Services

Joseph Grundfest and
Walter Baer

Many U.S. telephone subscribers are accustomed to dialing the correct time, the local weather report, and other recorded information. Microelectronics now makes it feasible to display such information on a television receiver. The basic technical requirements are a semiconductor chip (containing digital storage, decoder, character generator, and control logic) at the television set and an interconnection device (a modem) between the television and the telephone. The push-button telephone can serve as the system signaling device.

Linking the family television set to computer data banks through existing telephone lines can provide a rich array of information services to the home. Another application is sending and receiving messages. One subscriber can leave a message for another subscriber at the central computer. The second subscriber can then dial the computer for messages at his or her convenience. Alternatively, messages can be stored in digital form at the recipient's television receiver.

Development of these services has progressed most rapidly in Great Britain. The viewdata system developed by the British Post Office underwent field trials during the summer of 1978 with fifteen hundred users. Over one hundred information providers will offer financial data, employment opportunities, sports results, betting odds, and a variety of other information on the viewdata computers. If these field trials are successful, telephone authorities in several other countries plan to offer similar services in the near future.

It may seem curious that the United States has lagged behind Great Britain in developing information services to the home. The U.S. is more advanced in computer technology and data communications, its telephone network is more sophisticated and of higher quality, and it has greater telephone and television penetration and higher per capita income. In the United States, however, policies regarding industry structure, regulatory jurisdiction, and pricing present major barriers to the introduction of information services.

Separation of Communications and Information Processing

The British Post Office (BPO) can offer a complete viewdata service in Britain. It can operate the viewdata computer centers and rent computer storage to information providers, as well as provide the communications links. In contrast,

U.S. policies have sought to separate communications from data processing functions. As a result, no single entity in the United States can assume the strong entrepreneurial role of the BPO in introducing a viewdata service.

The Federal Communications Commission (FCC) has sought to distinguish between communications and data processing in its two computer inquiries. In the first inquiry the commission reasoned that it could exercise jurisdiction over large parts of the data processing industry but that policy considerations called only for regulation of telecommunications and hybrid telecommunications services. Hybrid telecommunications services were defined as services exhibiting both data processing and telecommunications attributes but with the telecommunications element predominating. Hybrid data processing services were defined analogously as services in which the data processing element was predominant. The commission declined to exercise jurisdiction over data processing and hybrid data processing services.

The first inquiry also contains a series of maximum separation rules intended to reduce the potential for cross subsidy from common carrier participation in the unregulated data processing markets. The rules require all common carriers to offer data processing services only through arm's-length subsidiaries that have separate management, directors, facilities, and employees. These rules have been upheld on judicial review.

AT&T is subject to additional antitrust constraints on entering data processing markets. Under the terms of the 1956 Consent Decree, AT&T may not engage "in any business other than the furnishing of common carrier communications services . . . or . . . services incidental to the furnishing . . . of common carrier communications services." The consent decree does not, however, define what is meant by a "common carrier communications service" or what constitutes a service "incidental" to a common carrier communications service. Operationally, the definitions of the first inquiry would likely be used to interpret the terms of the consent decree and thereby exclude AT&T from data processing and hybrid data processing markets—whether or not AT&T establishes an arm's-length subsidiary.

Regulatory Jurisdiction over Viewdata Services

Potential offerors of viewdata services in the United States face uncertainties over the extent to which these services would be regulated—and by whom. Under the terms of the first inquiry, viewdata information services would probably be classified as data processing. They would consequently be unregulated by the FCC and open to unregulated firms and the arm's-length subsidiaries of non-AT&T telecommunications carriers. A viewdata service with a message-forwarding capacity would arguably be classified as hybrid telecommunications service and thus be subject to FCC regulation, even though it does not exhibit economies of scale or other natural monopoly characteristics that presumably provide the rationale for regulation. Although such a classification is hardly certain, it potentially opens the entire viewdata system to FCC oversight. The second inquiry, which is presently under way, may clarify questions of this sort.

Moreover, the FCC has no monopoly over regulation. The Communications Act reserves jurisdiction over intrastate common carrier services to the states. Thus to the extent that viewdata services can be classified as intrastate common carrier services, they will be subject to regulation by state public utility commissions. The system configuration is obviously important. A system based in California with all its customers in California is a strong candidate for intrastate classification. A system with its base in Virginia and customers in Virginia, Maryland, and the District of Columbia may provide interstate services but still be classified as intrastate with respect to its Virginia subscribers. Such jurisdictional questions remain to be resolved.

Communications Pricing

Flat-rate pricing for local residential telephone service in most of the United States also reduces the telephone carriers' incentives to introduce viewdata-like services. In Britain and other countries outside North America, telephone subscribers pay for each call as well as a fixed monthly charge. Home information services thus will generate new communications revenues (to the extent that these services do not substitute for current calling), in addition to revenues from the information providers. A principal motivation behind the BPO's support of viewdata is the opportunity to earn additional revenue from greater use of the existing telephone plant. Without usage-sensitive pricing, U.S. carriers lack similar motivations.

Information Pricing and Billing

In principle, providing information to consumers should be highly competitive, even if the communications facilities needed for delivery are regulated. No natural monopolies exist for gardening advice, restaurant guides, or jokes. (This point was brought home all too clearly to those who called the New York City dial-a-joke in mid April to hear such delights as, "The President's brother, Billy, thinks the CIA is opening his mail. His *Weekly Reader* arrives with the crossword puzzle already filled in.") Viewdata encourages competition by permitting information providers to set their own prices for each frame of information (users see the price on the television screen before purchase). Thus two cookbook publishers can charge different prices for their recipes, while a food company provides free recipes as an advertising promotion and a government agency offers consumers free recipes to promote nutrition. The

BPO charges all information providers at the same rate for access to the system and computer storage.

However, practical problems may arise in billing subscribers for information services. Information providers can set up their own billing procedures for high-value transactions—for example, by requiring each user to log in with an identifying account before the system will respond with information or by providing hardware to automatically identify the caller to the billing system. Computer time sharing and credit card transactions currently operate this way. But such methods do not seem economically feasible for information priced at only a few cents per frame, which may well characterize most information services directed to the consumer mass market. The billing transaction cost could easily exceed the value of the information.

There appear to be several ways to reduce billing costs for information of lower value. Advertiser-supported or government-subsidized information could be offered without charge, of course, as suggested by the recipe examples. A Viewdata service could charge a flat-rate subscription fee for unlimited access to certain items in the data base. But the easiest way to maintain a market in low-cost information is to use the telephone carriers' existing billing systems. Viewdata takes this approach and adds a billing and handling charge of 0.5p (roughly one cent) to each frame purchase (advertisers or other information providers can agree to pay this charge instead of the user). Consequently, information providers can charge less than five cents per frame and still make money.

Could such arrangements be worked out in the United States to permit use of a telephone carrier's billing system by a (presumably separate) viewdata system? The question raises complex issues analogous to those in the recent cable television pole attachment controversy. In that case cable companies desired access to utility poles and conduits so that they could build their distribution plant at lower cost. Telephone and power utilities were not historically in the business of renting pole or conduit space and argued that there was no right of access to their distribution network. In the case of viewdata the information service firms would desire access to the telephone billing system rather than the distribution system, but the carriers might again contend that there is no right of access.

The pole attachment controversy haunted the FCC for years and has only recently been resolved by federal legislation. The FCC repeatedly denied jurisdiction, partially because it believed it had no mandate to set rates for electric utility poles and partially because determining tariffs in each local cable franchise seemed an overwhelming task. Under the new legislation states have the first opportunity to regulate pole attachments. If a state declines to exercise this authority, the FCC can then intervene to set reasonable rates. The statute sets out a range of reasonableness. The minimum reasonable rate is the marginal cost of adding the cable to the pole or conduit, while the maximum rate is the fully allocated cost (the percentage of the usable pole or conduit space occupied by the cable multiplied by the total cost attributable to the pole or conduit). Within this range the FCC appears to have broad discretion.

Some similar resolution may emerge for information service billing. Negotiated contracts between communications carriers and information service firms to use the carriers' billing systems seem the best solution. Failing agreement, firms may petition the states or the FCC to set reasonable terms, conditions, and rates for access to the carrier's billing systems. Although the outcomes of such proceedings are hard to predict, it seems likely that the FCC would claim jurisdiction; and the pole attachment legislation provides a precedent for determining reasonable rates if the commission found in favor of access. In any event, some means to reduce billing costs by piggybacking on the existing carrier billing systems appear important to the development of home information services in the United States.

25 The Demand for Automated Bibliographic Systems: A Diffusion Model

Lester B. Lave,
Samuel Leinhardt, and
Artur Raviv

The rapid increase in scientific literature has necessitated finding more efficient ways to locate the studies of relevance. For example, the number of biomedical journals increased from five in 1799 to more than twenty-three thousand in 1977 (McCarn, forthcoming). For the biomedical literature, the first large-scale attempt to search bibliographic references using a computer was the MEDLARS system of the National Library of Medicine (see McCarn and Leiter, 1973, for a description). As originally set up, the system was designed for batch processing; the user communicated by mail, and a search request typically took several weeks to process.

A more satisfactory successor has been MEDLINE, which provides on-line search and telecommunications. The advantage is not only fast response but an ability to inspect the references immediately to ascertain whether the search produced satisfactory results or must be refined.

Three issues are addressed here. For a new product such as MEDLINE, usage can be expected to increase rapidly as more users learn about the existence of the system, have experience with it, and become proficient in its use. We model this process in terms of a diffusion process. The second section presents some data to support this conceptualization. Initially, at least, there are economies of scale associated with producing a service such as MEDLINE. Furthermore, the social benefit from the system remains low until information about it has disseminated to a large group of users. In the third section we investigate dynamic behavior designed to speed dissemination. In particular, we investigate the desirability of initial subsidy, such as reduced price or some form of advertising. This problem is actually a general one associated with the introduction of any new product or service. The fourth section focuses on some of the operations issues associated with MEDLINE. For example, one question concerns the responsiveness of usage to price. Another concerns the extent to which usage over the day can be smoothed to avoid sharp peaks with their associated

---

The authors' names are in alphabetical order. Support for this work was provided by the National Center for Health Services Research through Grant 5R01HS02122.

frustration at poor service (or the investment in additional equipment that would be used for only a short time each day).

Diffusion of an Innovation

A new product or service cannot be purchased until people know of its existence. If we assume no advertising, few people know about the innovation when it first becomes available. One process for disseminating information is that satisfied users talk with potential users informing them of the innovation. Assuming that satisfied users do not purposely try to contact nonusers to tell them about the innovation, this process is analogous to a contagion model of the kind employed in epidemiology. There, disease carriers are viewed as passing the disease along to randomly contacted noncarriers. The rate at which the number of users (or carriers) increases is the product of the number of users and the number of potential users (who have not yet been contacted). The number of users can be graphed as a logistic function over time, with the initial period of low use representing the small number of users and thus small number of nonusers contacted. The final periods in which few nonusers are converted is due to the small number of remaining nonusers.

This contamination model is too simple for MEDLINE usage, for the following reasons. (1) An initial group of users helped to develop the system, and they were thoroughly familiar with it at the introduction date. (2) The National Library of Medicine explicitly attempted to acquaint potential users with the system. (3) The bottleneck on usage was the rate at which medical librarians could be trained to use the system, since only institutions that had trained librarians could subscribe to the service. (4) The number of entries in the data base increased over time as more journals were added to the base. (5) The number of additional data bases, similar to MEDLINE, provided by NLM, increased over time. All these factors require a more complicated diffusion model, which we are in the process of constructing. However, we discuss some preliminary empirical results.

We have been able to examine utilization data for only a small number of institutions. Shown in figure 25-1 are the utilization rates of five institutions. All have the expected logistic shape over the first months of operation; that is, all experience rapid increases in utilization. However, in all cases utilization overshoots the steady-state rate and subsequently falls. In some cases there is a subsequent increase. Clearly, there is more going on than a simple diffusion model. However, the diffusion framework seems appropriate in the analysis of these data.

A Dynamic Model of Product Introduction

In selecting a product or service, the consumer can choose among only those products that have been used or with which there is some familiarity. Thus

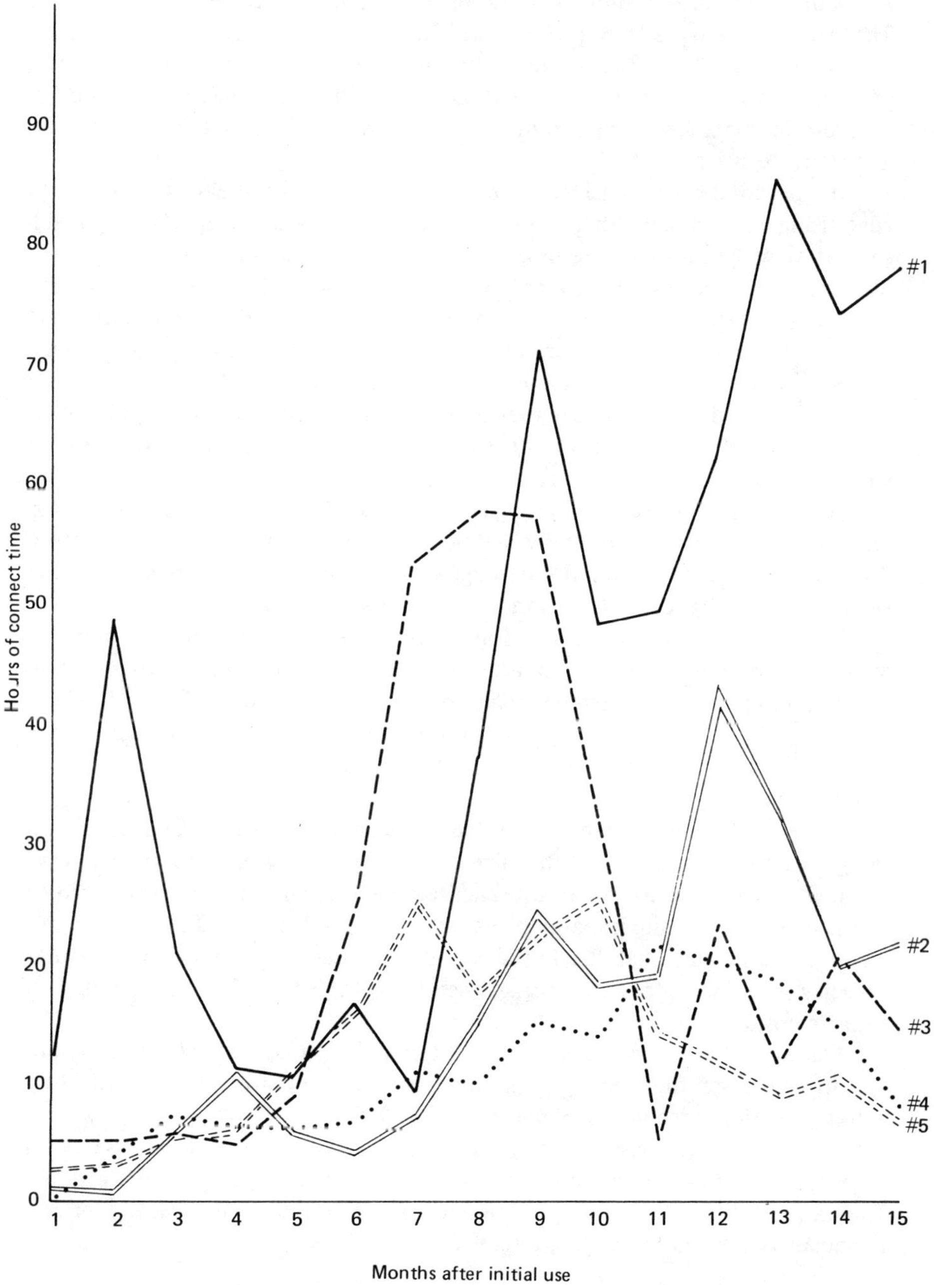

Figure 25-1. MEDLINE Usage During Initial Months

a precondition for a product's adoption is that the consumer know about it. The decision to try a new product will be related to the cost of the product, the expected utility of the product, the quality of information on the product, and the loss if the product is not satisfactory. Probably the best information on a product is a testimonial from a user who has approximately the same needs and characteristics.

In this model we considered both the decision to increase familiarity by advertising and taking other actions to get new users as rapidly as possible so that they will disseminate news of the product to potential users. To speed dissemination, the producer might send out free samples, reduce price (below optimal or even break-even level), or otherwise lower the cost of the user's trying the product. A wide class of possible actions fits this category; all make it easier and cheaper for someone to try the product and thus increase the number of users so that dissemination of the innovation is more rapid. Advertising is conceived to have the effect of getting someone to try the product without being contacted by a user.

If dissemination follows the logistic curve of the contagion model, increasing the number of initial users is especially important because it will speed dissemination. Even within the contagion model other strategies are possible, such as increasing the interaction among users and potential users. However, we exclude these from the discussion. Finally, we note that capacity comes in chunks rather than being continuously variable, and additions to capacity are difficult to make and generally come in even larger chunks. Thus a producer is going to have large excess capacity on introduction, with high average and low marginal cost. This fact provides an additional incentive to be willing to subsidize to speed dissemination.

We want to concentrate on the demand side and so will make many simplifying assumptions. To keep the algebra simple, we assume the time interval extends T periods into the future and set the discount rate equal to zero (in contrast to an infinite-horizon problem with discounting). To simplify production, we assume that the initial capacity chosen is Q_0 and that no further capacity can be added; the initial cost is Q_0C. We further assume that all variable costs are zero.

On the demand side, assume that the population consists of N individuals. Let $n(x)$ denote the density function describing the fraction of individuals with reservation price in the interval $(x, x + dx)$. Clearly $n(x) \geqslant 0$, $\int_0^\infty n(x)dx = 1$. Each individual purchases exactly one unit of the product if his reservation price is above the market price; otherwise he does not purchase the product. Therefore if all individuals are informed of the product, then at any price p the quantity demanded is $Q(p)$, where

$$Q(p) = N \int_p^\infty n(x)dx = NF(p), \quad \text{or} \quad F(p) = \int_o^\infty n(x)dx \tag{25.1}$$

If only $N < \overline{N}$ individuals are informed, we assume that the distribution of reservation prices within this subpopulation is unchanged and thus that the quantity demand is $N\overline{F}(p)$.

Perfect (Instantaneous) Information Dissemination

In this case all N individuals know instantaneously of the product. The demand curve is given by equation 25.1. If the objective is to maximize profits, then we seek to maximize

$$\max_{p(t)Q_0} \int_0^T p(t)Q(p(t))dt - CQ_0$$

$$\text{subject to } Q(p(t)) \leqslant Q_0$$

Rewriting using equation 25.1, we get

$$\max_{p(t),\, Q_0} N \int_0^T p(t)F(p(t))dt - CQ_0 = \max_{p(t)Q_0} N \int_0^T G_1(p(t))dt - CQ_0$$

$$\text{subject to } NF(p(t)) \leqslant Q_0 \tag{25.2}$$

where $G_1(p) = pF(p)$

If we wish to maximize net benefit (the area under the demand curve minus product costs), then the objective function is similar. Define $G_2(p)$ to be the area under the demand curve:

$$G_2(p) = \int_p^{\infty} xn(x)dx \tag{25.3}$$

$Nn(x)$ consumers are willing to pay the price x. If the market price is p, then the area under the demand curve is

$$\int_p^{\infty} Nxn(x)dx = NG_2(p) \tag{25.4}$$

Then net benefit is maximized by

$$\max_{p(t)Q_0} \int_0^T NG_2(p(t))dt - CQ_0$$

$$\text{subject to } NF(p(t)) \leqslant Q_0 \tag{25.5}$$

Notice that equations 25.2 and 25.5 are identical except that G_1 is replaced by G_2.

Both problems are basically static ones; $p(t) = p$ and thus $Q(p(t)) = Q$. Clearly Q_0 is chosen to equal to Q; The Lagrangian is

$$\max_{p, Q_0} \; L = NG_i(p)T - CQ_0 - \lambda(NF(p) - Q_0)$$

The first-order conditions are

$$NTG_i' - \lambda NF' = 0, \qquad -C + \lambda = 0$$

Thus we get $TG_i' - CF' = 0$.

For $i = 1$ (profit maximation), $G_1 = pF$, $G_1' = F + pF'$ and the first-order conditions yield

$$T[F + p_1 F'] = CF' \tag{25.6}$$

Since $F' = -n(p)$

$$T[F - p_1 n(p_1)] = -Cn(p_1)$$

$$F = n(p_1)\,[p_1 - C_1/T]$$

For $i = 2, G_2(p) = \int_p^\infty xn(x)dx, \; G_2' = -pn(p)$

$$-Tp_2 n(p_2) + Cn(p_2) = 0$$

$$-Tp_2 + C = 0 \tag{25.7}$$

$$p_2 = C/T$$

Since equation 25.6 is positive when evaluated at p_2, $p_2 < p_1$. Thus under profit maximization the price is higher and the output lower than under welfare maximization.

A Dynamic Model of Product Dissemination

We now turn to the information dissemination aspect of the problem, assuming that information spreads in two ways: first, by word of mouth from purchasers to those who are unaware of the product, and, second, by advertising, that is, by sending messages that affect the uninformed portion of the population.

If at time t, $N(t)$ individuals are aware of the service and the price charged is $p(t)$, then the quantity demanded is

$$Q(t, p(t)) = N(t)F[p(t)] \tag{25.8}$$

The number of individuals aware of the service grows according to the two diffusion processes. Each individual who at time t was a user (purchased the service and not just knew about the service) contacts γ persons. Thus $\gamma Q[t, p(t)]$ is the number of persons contacted. Of those contacted, $N(t)$ are aware of the service; thus only $1 - N(t)/N$ is the proportion of those contacted that will be newly informed. Thus the number of informed individuals grows by

$$\gamma Q[t, p(t)]\ [1 - N(t)/N] \tag{25.9}$$

Substituting equation 25.8 gives

$$\gamma F[p(t)]\, N(t)\ [1 - N(t)/N] \tag{25.10}$$

Notice that this is not the usual logistic diffusion equation. The "contact coefficient" can be interpreted as $\gamma F[p(t)]$, and thus it depends on the pricing strategy. In particular it need not be constant through time; if it were, the diffusion curve would be the usual logistic curve.

Also because of advertising at rate $A(t)$, the number of individuals who are informed about the product is changing at the rate

$$A(t)[N - N(t)] \tag{25.11}$$

Thus $A(t)$ is the proportional rate at which individuals become aware of the new product. Combining equations 25.10 and 25.11, we get

$$\begin{aligned} N(t) &= \gamma F[p(t)]\, N(t)\ [1 - N(t)/N] + A(t)\ [N - N(t)] \\ &= \left\{\gamma F[p(t)]\, N(t) + NA(t)\right\}\ [1 - N(t)/N] \end{aligned} \tag{25.12}$$

We also assume that $N(0) = N_0$, where N_0 is interpreted as the number of individuals who are familiar with the product at time 0, perhaps those receiving free samples.

Our problem is to find $p(t)$, $A(t)$, and Q_0 to maximize

$$\int_0^T [p(t)Q(t, p(t)) - w(A(t))]\, dt - CQ_0 \tag{25.13}$$

where $w(A)$ is the cost of advertising, subject to the constraint 25.12, the definition 25.8, and $0 \leqslant Q(t, p(t)) \leqslant Q_0$.

Rewriting we obtain the following:

$$\max_{p(t), A(t), Q_0} \int_0^T [p(t)N(t)F[p(t)] - w(A(t)) - CQ_0/T]\,dt \tag{25.14}$$

such that

$$\dot{N}(t) = [\gamma F[p(t)]N(t) + NA(t)]\ [1 - N(t)/N] \tag{25.15}$$

$$N(t)F(p(t)) \leqslant Q_0 \tag{25.16}$$

where $N(t)$ is the state variable, $p(t)$ and $A(t)$ are the controls. If the objective is to maximize net benefit, then

$$\max_{p(t), A(t), Q_0} \int_0^T [N(t)G_2(p(t)) - w(A(t)) - CQ_0/T]\,dt \tag{25.17}$$

where $G_2(p(t))$ is defined by 25.4: $G_2(p(t)) = \int_{p(t)}^{\infty} xn(x)dx$. Assume that $A(t) \equiv 0$, that is, no advertising effects. In this case 25.17 becomes

$$\max_{p(t), Q} \int_0^T [N(t)\,G_2(p(t)) - CQ_0/T]\,dt$$

$$\dot{N}(t) = \gamma F(p(t))N(t)(1 - N(t)/N) \tag{25.18}$$

$$N(t)F(p(t)) \leqslant Q_0$$

The Hamiltonian is

$$H = N(t)G_2(p(t)) - CQ_0/T + \lambda_1(t)\gamma F(p(t))N(t)(1 - N(t)/N)$$
$$- \lambda_2[N(t)F(p(t)) - Q_0] \tag{25.19}$$

Necessary conditions for maximization are

$$\frac{\partial H}{\partial p} = N(t)G_2' + \lambda_1 \gamma F'N(t)(1 - N(t)/N) - \lambda_2 N(t)F' = 0 \tag{25.20}$$

$$\frac{\partial H}{\partial Q_0} = -C/T + \lambda_2 = 0 \tag{25.21}$$

Therefore from 25.20

$$G_2' + \lambda_1 \gamma F'(1 - N(t)/N) - CF'/T = 0$$

Using G_2' and F' from 25.20 gives

$$-p(t)n(p(t)) - \lambda_1 \gamma n(p(t))(1 - N(t)/N) + (C/T)n(p(t)) = 0$$

$$-p(t) - \lambda_1 \gamma(1 - N(t)/N) + C/T = 0$$

We obtain that

$$p(t) = C/T - \lambda_1 \gamma(1 - N(t)/N) \tag{25.22}$$

Comparing 25.22 and 25.7, we see that *p(t) is smaller than the price-solving 25.7 if and only if* $\lambda_1 \geqslant 0$. In what follows we prove this result. (At $T, \lambda_1(T) = 0$; thus $p(T)$ is the price in the static model. Thus we are going to prove that price increases until it reaches the price in the static problem.)

$$\dot{\lambda}_1(t) = -\frac{\partial H}{\partial N} = -G_2 - \lambda_1(t)\gamma F(1 - 2N(t)/N) + CF/T$$

$$= -G_2 - \frac{(C-p)/T}{1 - N(t)/N} F(1 - 2N(t)/N) + CF/T =$$

$$= -G_2 + pF + \frac{N(t)/N}{1 - N(t)/N} F(C/T - p) \tag{25.23}$$

Differentiating 25.22 with respect to time gives

$$\dot{p} = -\dot{\lambda}_1 \gamma(1 - N(t)/N) + \lambda_1 \gamma \dot{N}/N$$

$$= \gamma(1 - N(t)/N)\left[-\dot{\lambda}_1 + F\frac{(C/T - p)}{1 - N(t)/N}\right]$$

$$= \gamma(1 - N(t)/N)\left[G_2 - pF - F\frac{C - p}{1 - N(t)/N}\left(\frac{N(t)}{N} - 1\right)\right]$$

$$= \gamma(1 - N(t)/N)\,[G_2 - pF + F(C/T - p)] \tag{25.24}$$

We want to prove that $C/T - p > 0$ for all $t < T$.

1. Assume that $C/T - p < 0$ for all t. From 25.23 we obtain that $\dot{\lambda}_1(t) < 0$ for all t and from 25.22 that $\lambda_1(t) < 0$ for all t. This contradicts the fact that $\lambda_1(T) = 0$.

2. Assume that $C/T - p$ changes signs. By continuity, there exists t_0 such that $C/T - p(t_0) = 0$. (If more than one exists, choose the largest.) If $C/T - p < 0$ for $t > t_0$ then the argument in part 1 holds. If $C/T - p < 0$ for $t < t_0$ and $C/T - p > 0$ for $t > t_0$, then it must be that $\dot{p}(t_0) < 0$. But from 25.24 if $C/T - p(t_0) = 0$ it follows that $\dot{p}(t_0) > 0$, a contradiction.

To summarize, we have proved that $p(t) < p_2$ for all $t < T$ and $p(T) = p_2$. The price $p(t)$ increases through time. Equation 25.22 is giving also the intuition: $p(t) = p_2 - \lambda_1 \gamma(1 - N(t)/N)$. Since λ_1 measures the marginal contribution of the state variable $N(t)$ to the objective function, $\lambda_1 \gamma(1 - N(t)/N)$ measures the marginal contribution of price reduction toward the objective function.

The Price Elasticity of MEDLINE Demand

MEDLINE was initially put into operation as an experimental system for which no price was charged. When commercial operation was begun, telecommunications costs were paid by the National Library of Medicine. Thus local users or those with WATS lines experienced an increase in price from zero to \$6 per hour, while distant users found their costs falling, since previous telecommunications charges greatly exceeded \$6 per hour. The price was later increased to \$8, and then a category of prime time was defined at a rate of \$15 per hour while nonprime time was still billed at \$8 per hour. Finally, a block of time was shifted from prime to nonprime time. During the period of commercial operation, the National Library of Medicine continued to pay for telecommunications. Since the size of the data base and number of additional data bases available to the user increased during the period, it is difficult to sort out the effects of price increases. The data are summarized in table 25-1.

Any price elasticity estimated from these data is necessarily crude. Over the entire period new institutions were added to the set of those with access to the system. Furthermore, there were exogenous changes in the level of funds available for health research and for patient care and education. However, neglecting these effects, McCarn estimated a crude demand function with elasticities of -0.28, -0.35, and -0.53 at prices of \$6, \$8, and \$12 per hour, respectively. It is surprising that demand should be so price-inelastic.

We have estimated some alternative specifications of the demand function using these aggregate data. The number of additional data bases, the increasing size of the data base, the increase in the number of licensed institutions, and the increase in the number of users cannot be accounted for explicitly. We have attempted to bracket the estimated demand elasticity by ignoring such factors and by representing them by a time trend.

Table 25-1
National Library of Medicine On-Line Service Usage, January 1973 to October 1977

Month	*Total Connect Hours*	*Price ($/hr)*	*Data Bases*	*Searches*	*Pages Printed*	*Connect Hours of Billed Users*	*Total Billing ($)*	*Price Per Hour ($)*
1973								
Jan	2,549	0	2	11,620	69,041	2,266	0	0
Feb	3,443	0	2	15,040	77,578	2,821	0	0
Mar	4,240	0	2	21,402	96,296	3,581	0	0
Apr	3,986	0	2	18,387	63,188	3,130	0	0
May	3,593	0	2	18,469	56,803	3,000	0	0
June	3,312	0	2	17,962	76,856	2,677	0	0
Jul	2,865	0	2	16,124	45,200	2,292	0	0
Aug	3,352	6	2	17,086	56,387	2,646	7,033.91	2.66
Sep	3,087	6	3	13,468	45,492	2,320	14,028.93	6.05
Oct	4,169	6	3	19,835	63,694	3,282	23,184.60	7.06
Nov	3,805	6	3	19,577	60,110	2,974	21,562.75	7.25
Dec	2,664	6	3	15,469	42,866	2,137	14,909.48	6.98
1974								
Jan	4,275	6	3	31,334	66,082	3,240	23,334.03	7.20
Feb	4,251	6	3	23,570	55,181	3,179	22,918.34	7.21
Mar	4,862	6	3	26,562	73,948	3,678	26,149.58	7.11
Apr	4,907	6	4	30,564	79,053	3,778	29,582.26	7.83
May	4,788	6	5	29,960	83,542	3,557	26,824.21	7.54
Jun	4,410	6	5	24,684	77,965	3,180	27,047.14	8.51
Jul	5,232	6	6	29,617	80,741	3,849	32,056.09	8.33
Aug	4,861	6	6	31,092	94,866	3,808	34,939.36	9.18
Sep	5,091	6	6	30,394	89,307	3,831	34,293.37	8.95
Oct	6,525	6	6	37,506	113,750	5,021	42,560.08	8.48
Nov	5,733	6	6	31,677	104,808	4,427	38,547.62	8.71
Dec	4,616	6	7	27,527	91,321	3,534	32,020.68	9.06
1975								
Jan	6,561	6	7	35,946	161,426	5,162	45,822.85	8.88
Feb	5,788	8	7	33,060	140,323	4,483	48,996.13	10.93
Mar	6,086	8	7	34,907	174,008	4,537	50,975.71	11.24
Apr	6,463	8	7	38,079	154,156	4,777	53,478.99	11.20
May	5,903	8	8	35,513	156,746	4,066	49,263.68	12.12
Jun	5,463	8	8	37,338	176,335	4,019	53,140.80	13.22
Jul	6,258	12[a]	8	39,898	182,452	4,525	62,976.94	13.92
Aug	5,251	12[a]	8	34,459	168,138	4,054	58,129.84	14.34
Sep	6,004	12[a]	8	39,610	165,840	4,970	63,556.98	12.79
Oct	6,870	12[a]	8	42,556	187,846	5,182	70,886.89	13.68
Nov	6,291	12[a]	10	37,552	166,445	4,654	63,748.18	13.70
Dec	5,794	12[a]	10	36,805	164,469	4,290	61,952.01	14.44
1976								
Jan	6,914	12[a]	11	44,906	196,934	5,145	73,771.66	14.34
Feb	6,766	12[a]	11	43,640	226,792	5,096	77,897.66	15.29
Mar	7,524	12[a]	11	52,144	250,944	5,758	87,771.04	15.24
Apr	6,923	12[a]	11	44,911	208,627	5,086	76,864.21	15.11
May	5,991	12[a]	11	39,976	205,065	4,361	72,275.13	16.57
Jun	6,751	12[a]	11	41,967	259,664	4,764	78,737.97	16.53

Table 25-1 (Continued)

Month	*Total Connect Hours*	*Price ($/hr)*	*Data Bases*	*Searches*	*Pages Printed*	*Connect Hours of Billed Users*	*Total Billing ($)*	*Price Per Hour ($)*
Jul	7,030	12[a]	11	42,725	242,913	5,151	84,850.12	16.47
Aug	6,930	12[a]	11	45,219	232,941	5,357	87,107.29	16.26
Sep	7,205	12[a]	11	42,926	263,174	5,524	92,576.64	16.76
Oct	7,946	12[a]	11	65,154	246,858	6,090	94,177.06	15.46
Nov	8,326	12[a]	11	68,491	287,602	6,342	103,001.01	16.24
Dec	7,278	12[a]	11	59,609	263,580	5,479	94,542.93	17.26
1977								
Jan	7,918	12[a]	12	72,463	329,968	6,308	123,932.74	19.65
Feb	7,749	12[a]	12	70,794	315,957	5,995	117,228.67	19.55
Mar	9,670	12[a]	12	84,470	339,008	7,413	140,832.07	19.00
Apr	7,680	12[a]	12	71,006	303,644	6,108	119,356.47	19.54
May	7,502	12[a]	11	69,856	300,325	5,678	111,368.11	19.61
Jun	7,413	12[b]	12	71,206	326,401	5,820	116,178.78	19.96
Jul	7,235	11[n]	12	67,674	328,614	5,694	106,911.40	18.78
Aug	8,217	11[b]	13	75,955	341,857	6,769	124,393.78	18.38
Sep	7,790	11[b]	13	55,044	304,654	6,276	113,908.79	18.15
Oct	8,724	11[b]	13	62,901	296,557	6,940	120,822.57	17.41

Source: David B. McCarn, "The Medical Literature Analysis and Retrieval System," in *Development and Utilization*, forthcoming.

[a]Average base on $15/hr in prime time and $8/hr in nonprime time; prime time from 10 a.m. to 5 p.m.

[b]Prime time from 10-11:30 a.m. and 1-5 p.m.

Connect Hours

We begin with an attempt to explain the aggregate number of connect hours. The pattern of data over time seemed to call for a multiplicative, or log-log, specification. A series of models was estimated. The simplest included only the log of price and the log of the number of additional data bases. Successively, we added the log of the number of months that had elapsed since MEDLINE was offered as a commercial service (August 1973) and eleven dummy variables for the months of the year in order to control for seasonal shifts. We discovered that the number of data bases is highly correlated with time ($\rho = 0.98$); thus nothing is gained by putting both variables in the same regression. Since the predominant effect is likely to be time, we chose to keep this variable. Aside from the simplest equation, the estimated coefficient of price hardly varies across the specifications, indicating great stability.

Regressions 2.1 and 2.2 in table 25-2 present the final specifications. All variables are included except the number of data bases. In the estimation of the second regression, the fifth and forty-second months were deleted, since

Table 25-2
Analysis of Medline Use

	Log Connect Hours		*Log Searches*	*Log Pages Printed*	
	2.1	2.2	2.3	2.4	2.5
R^2	0.95	0.96[a]	0.91	0.97	0.98[b]
Constant	7.00	7.11	8.01	7.94	7.77
Log price per hour	−0.14	−0.12	−0.19	0.26	0.12
	(−2.13)	(−2.18)	(−1.39)	(2.28)	(1.20)
Log month	0.54	0.52	0.83	0.96	1.10
	(13.94)	(15.39)	(10.35)	(14.02)	(17.33)
January	0.25	0.18	0.31	0.31	0.31
	(4.98)	(3.72)	(2.95)	(3.53)	(4.27)
February	0.19	0.12	0.18	0.19	0.20
	(3.87)	(2.66)	(1.75)	(2.25)	(2.75)
March	0.30	0.23	0.28	0.33	0.32
	(6.03)	(5.03)	(2.71)	(3.76)	(4.54)
April	0.22	0.15	0.23	0.21	0.20
	(4.41)	(3.30)	(2.19)	(2.37)	(2.75)
May	0.13	0.06	0.14	0.19	0.17
	(2.61)	(1.36)	(1.37	(2.11)	(2.36)
June	0.10	0.03	0.10	0.25	0.22
	(1.98)	(0.70)	(0.92)	(2.79)	(3.12)
July	0.17	0.11	0.14	0.20	0.18
	(3.47)	(2.31)	(1.32)	(2.23)	(2.53)
August	0.15	0.07	0.17	0.26	0.17
	(3.11)	(1.69)	(1.74)	(3.06)	(2.34)
September	0.14	0.07	0.03	0.16	0.16
	(2.93)	(1.52)	(0.34)	(1.85)	(2.31)
October	0.29	0.22	0.24	0.24	0.15
	(6.21)	(5.09)	(2.44)	(2.80)	(2.13)
November	0.22	0.15	0.17	0.19	0.19
	(4.44)	(3.23)	(1.65)	(2.11)	(2.70)

[a]Outliers removed: fifth and forty-second values

[b]Outliers removed: first and third values

they were extreme observations, differing significantly from the bulk of the data, and we did not wish to have this bias the estimated coefficients.

Both models fit the data extremely well since more than 95 percent of the variation in billed connect hours is explained by the variables. The most significant variable is time, indicating a declining growth rate in connect hours. The monthly dummy variables show a strong seasonal effect with December as the lowest month of the year, although in general, September through March are the months of highest usage. Finally, the effect of price on usage is small but statistically significant. A 10 percent decrease in price is estimated to lead to a 1.2 percent to 1.4 percent increase in usage. These values are smaller than those obtained by McCarn.

Searches and Pages Printed

The same models were estimated with two additional dependent variables. A search is defined by the user; it automatically terminates when the user signs off but may be terminated earlier by the user. However, what constitutes a search is somewhat arbitrarily defined, since a user can take advantage of the on-line nature of the system to refine his ideas and get several bibliographies at one sitting. The other dependent variable is the total number of pages printed (with bibliographic entries). Over this entire period, the charge for printing a page was fixed ($0.10). Pages are printed off-line only if there is a greater number of references than can conveniently be printed on a terminal.

One regression for searches (2.3) and two for pages printed (2.4 and 2.5) are shown in table 25-2. The model fits the data well in all cases. When the dependent variable is log of searches, the estimated coefficient of price is extremely stable (after the simplest formulation). Again the coefficient of log time is the most significant, indicating a declining growth over the period. As with connect hours, there is a strong seasonal effect; December is the lowest month and October through March are the highest months. The price variable is the price per connect-hour. The estimated coefficient indicates that a 10 percent increase in price per connect-hour is associated with a 1.9 percent decrease in the number of searches. The coefficient is not statistically significant, since there is an ability to vary the number of searches independently of the number of connect-hours. One would expect the elasticity of connect-hours to be higher than the elasticity of searches, since one could get in the same number of searches in fewer connect-hours by being more organized before sitting down at the terminal. The point estimate is lower, although the point estimates are not significantly different.

The log of pages printed is the dependent variable for the last two regressions. The price per connect-hour is used as the price variable, since the price per page printed was constant ($0.10 per page) over the period. The principal difference with this regression is that an increase in price causes an increase in pages printed. This is plausible for two reasons. First, a higher price for connect-hours should lead the user to print out fewer references on the terminal and thus print out more pages off-line. Second, the higher connect-hour charge might lead the user to be less discriminating and to stop the search sooner, thus necessitating the printing of additional references.

In all these specifications we examined the residuals for an indication of misspecification such as omitted variables or incorrect functional form. For connect-hours and pages printed, a few observations were quite unusual. Since such observations exert a strong influence on the estimated coefficients under least squares, we excluded them and report the results in table 25-2; there was relatively little change. For none of the final regressions was there evidence of autocorrelation in the residuals, nor did a graph reveal any systematic temporal pattern. Finally, a plot of residuals against Guassian quantiles was approximately linear, implying that the residuals are approximately normally distributed and that the F- and t-tests used to evaluate the results are appropriate.

Smoothing Usage

The estimated elasticities indicate that demand is nearly price-inelastic. A price increase will bring forth a small decrease in usage. One might also examine responsiveness to time-of-day pricing. Usage is unlikely to be evenly distributed over a day. Rather it is concentrated during the portion of the day when librarians are normally at work and people come to them with search requests.

Until July 1975 connect time was billed at $8 per hour. During June 1975, 79 percent of billed connect-hours were during the period from 10 a.m. to 5 p.m. Beginning in July 1975 connect time from 10 to 5 was billed at $15 per hour, with nonprime time being billed at the old rate of $8 per hour. During this first month of peak time pricing, prime-time usage fell to 64 percent of total usage with non-prime-time usage almost doubling from 21 percent to 36 percent. The effect of the peak-time pricing was to diminish prime-time usage (from 3,172 hours to 2,899 hours) and increase non-prime-time usage (from 847 hours to 1,625 hours). Total usage rose 12.6 percent, composed of a 9.4 percent drop in prime-time usage and a 91.9 percent increase in non-prime-time usage.

By way of comparison, records exist for National Institute of Health users who receive free service. One should expect that their usage would increase, with no shift toward nonprime time. Total nonbilled usage increased 20.1 percent, with an increase of 30.9 percent in prime time and an increase of 6.4 percent in nonprime time. The abnormally small increase in nonprime usage might be due to the degradation in response time (as many paying users switched) for nonprime time and the associated improvement in response time for prime time.

At 11:30 a.m. usage is now increasing, due to the lower rate, and usage stays high during the lunch period. As before, usage begins a precipitous decline about 3:30, arrested at 5:00 by the off-peak rate.

One way of isolating the effect of making the period 11:30 a.m.-1:00 p.m. nonprime time is to assume that usage grew at the average rate, 19.4 percent, during each hour of the day; we can compare this predicted usage with actual usage. From 7:30 to 9:59 a.m., 183 (the sum of users on the system at each half hour) people used the system, compared with 187 expected to use it. From 10:00 to 11:29, 137 people used the system, compared with 135 expected. Thus prior to 11:30, expected and actual usage were very close. From 11:30 to 1:00, 135 people used the system compared with 101 expected. This increase of 34 percent is the effect of the price decrease. From 1:00 to 2:29, 199 people used the system, with 208 expected; from 2:30 to 4:59, 180 people used the system, compared with 205 expected; from 5:00 to 6:00, 82 people used the system, compared with 79 expected. Thus one effect of the change in prime time was to slow the growth of use in the peak hours by shifting this use to the 11:30 to 1:00 period.

The effect of this 53 percent decrease in price was to increase usage by 34 percent. Thus the price elasticity is roughly -0.6.

Summary

We have modeled the demand for MEDLINE, an automated bibliographic system. The demand for this service is complicated by the fact that use can take place only after an individual knows about the system and its usefulness. Thus we conceive of diffusion of information about the system as playing a crucial role in the level of demand for it. We presented evidence that utilization of MEDLINE by several institutions followed the sort of logistic curve that one would expect from a simple diffusion process. However, the process is inherently more complicated, as shown by subsequent data. We modeled the introduction of a product, focusing on the role of diffusion. A simple dynamic model was presented and revealed that diffusion leads to an initially low price that rises to the profit- or welfare-maximizing level only after many users have been informed about the new product. We estimated the price elasticity of demand for MEDLINE to be roughly -0.1, although other estimates put it somewhat higher. As one should expect, time-of-day pricing leads to shifts in usage over the day. Reaction is swift (within a month) and the data indicate success in smoothing usage over the day.

These results are a rough beginning. We expect to have more detailed models and results shortly.

References

McCarn, Davis B. "The Medical Literature Analysis and Retrieval System," *Development and Utilization*, forthcoming.

McCarn, Davis B. and Leiter, Joseph. "On-Line Services in Medicine and Beyond." *Science* 181 (1973):318-324.

26 Policy Implications of the Use of Telecommunications Technology by the U.S. Postal Service

John F. Lyons, Jr.

The impact of electronic telecommunications on the U.S. Postal Service and the possible future impact of electronic postal services on the private sector are familiar topics to regular attendees of the Airlie Conference. In 1975 Albert Vezza, Cal Vandergeest, and Chuck Jackson discussed the relationship of the Postal Service and the telecommunications industry. They noted that new and useful services could be offered if the respective roles of the traditional common carriers, the value-added carriers, and the Postal Service could be sorted out. In 1976 John McLaughlin, Henry Geller, Chuck Jackson, Gene Hodge, and Ray Bowers presented a very informative session on electronic mail. This session focused more specifically on what the future role of the U.S. Postal Service should be in the electronic message area. Policy, legal, and market considerations were raised.

Since that time some additional information on the research and development efforts of the Postal Service has become available. I would like to summarize briefly this research and development activity. The Postal Service contracted with General Dynamics in 1969 for an assessment of communications technology; with Philco-Ford in 1972 for electronic message service system design; with the Institute for Telecommunications Sciences in 1974 for satellite transmission studies; with Arthur D. Little in 1974 for a technology assessment and market analysis; with Fairchild Camera in 1976 for optical scanner development; with Pitney Bowes in 1976 for high-speed printer development; and with RCA in 1976 for an overall system evaluation. The result of these efforts is expected to be the development of a test bed at the Postal Service research headquarters to test a prototype of an electronic message service system (EMSS). While it is projected that six to ten years will be required to implement a nationwide system, limited commercial service, linking perhaps ten cities, could be in operation as early as 1981. The EMSS proposal would accept and electronically process and transmit conventional letters but would also be capable of accepting messages generated by facsimile machines, computers, magnetic tape, and

This paper does not represent the official policy or position of the Office of Policy Analysis and Development, the National Telecommunications and Information Administration, or the U.S. Department of Commerce.

keyboard terminals. Although the Postal Service stresses that no decision has been made to offer alternatives to conventional physical delivery, the system could be adapted to provide electronic output to the same types of equipment that can generate input.

The House Post Office and Civil Service Committee Report on H.R. 7700, the current postal reform bill in Congress, criticized the Postal Service for its failure to undertake research on electronic communications. This must be read, however, as a criticism of the level of research conducted and not of the lack of research. The Postal Service has been engaged in research and development on electronic communications systems since at least 1969 and has spent more than $10 million on this research over the past five years. While the version of H.R. 7700 approved by the House Committee would have increased the total annual research and development budget of the Postal Service from its current level of $32 million to $360 million by FY 1980, this provision was deleted by an amendment offered on the House floor. Given the resources allocated to planning its EMSS project, the Postal Service appears well on the way to completing the first of four phases leading to implementation.

In addition, since the 1976 session, a number of regulated carriers such as ITT, Tymnet, and Graphnet have offered services that fall within the general electronic mailbox category. These are end-to-end, store-and-forward, electronic communications services that do not ordinarily involve physical delivery. One firm, Scientific Time Sharing Corporation, which has been offering an electronic mailbox service on an unregulated basis as a hybrid data processing activity, has applied to the Federal Communications Commission (FCC) to offer it as a regulated communications service. Although the FCC's Computer Inquiry Rules would require Scientific Time Sharing to set up a separate subsidiary to provide a regulated communications service, Scientific Time Sharing has asked for a waiver of this requirement. Scientific Time Sharing's decision to apply for FCC authority apparently reflects a judgment that the electronic message market is sufficiently feasible that it will soon be unable to characterize its involvement in communications as merely incidental to its data processing activities. I believe that these developments indicate that a market for electronic message services exists and is growing at a significant rate.

Each year new developments shed further light on the implications of the Postal Service's plans for electronic telecommunications. In March 1978 the Postmaster General announced a one-month experiment in the provision of an international facsimile service. The experiment, which will be followed by a one-year field trial, will be conducted under a contract with the Communications Satellite Corporation (COMSAT). The experiment will involve the transmission of facsimile copies between the United States and approximately six foreign countries. It has been reported that West Germany, France, Switzerland, Egypt, Brazil, and Argentina are the likely correspondent nations. The international service is initially conceived as being one with hard copy input and output, but it could be reconfigured to provide electronic input and delivery. This development has implications for the international and domestic communications markets, and it may be an indication of how the Postal Service perceives its role as a telecommunications vendor.

Although the Postal Service has experimented with facsimile services in the past, the COMSAT contract and the follow-up field trial appear to represent a different kind of initiative. Because the Postal Service has been engaged in a rather comprehensive research effort on electronic messages, the international experiment could be viewed primarily as a test of some of the technologies and marketing assumptions that it has developed for domestic application. This could be done, however, on a domestic satellite. On the other hand, the international telecommunications market is certainly a lucrative one, and it would not be surprising if the financially pressed Postal Service were to look to it for revenue relief. In any event this current venture seems to be much better conceived and planned than past experiments. I believe that it makes it quite apparent that the nation must soon make some fundamental policy decisions on the role that the Postal Service is to play in the telecommunications marketplace.

The role of COMSAT under the Postal Service contract appears to be a limited one. COMSAT is essentially providing the Postal Service with technical assistance through its COMSAT laboratories. These consultative services are intended to provide the Postal Service with information and experience relating to the effectiveness of using international commercial satellite circuits for the provision of facsimile service. Questions such as the best types of terminal equipment to use, earth station arrangements here and abroad, and terrestrial circuit arrangements will presumably be addressed. During the one-month experimental operating period, February 1979, no actual mail will be transmitted. Thus no question need immediately arise about whether the service is venturing beyond a postal function and into the electronic telecommunications arena. The follow-up field trial, however, will involve the facsimile transmission of actual mail. According to COMSAT, the satellite circuits used during the one-month demonstration will be leased by the Postal Service from one or more of the international record carriers (IRCs). Thus the authorized user doctrine, which prohibits COMSAT from providing satellite circuits directly to customers, is not at issue. Arrangements for the subsequent one-year field trial are incomplete at this time.

Postal Service representatives have repeatedly stated that the Postal Service should not provide electronic telecommunications services that private industry is willing and able to provide. This sentiment was just recently repeated by Congressman Hanley, chairman of the Postal Operations and Services Subcommittee, during the debate on H.R. 7700. Congressman Van Deerlin, chairman of the House Communications Subcommittee, also shares this view. How would the Postal Service international facsimile service, as it is now proposed, measure up to this standard? The Postal Service's intention to purchase circuits from the IRCs does not answer this question since, as AT&T will certainly affirm, resale carriers are indeed competitors. If the Postal Service simply obtains its circuits from the IRCs, it will be in effect a resale carrier. Recognizing that at a certain level of detail nearly all services become distinguishable, fundamental policy decisions on the nature of the Postal Service's activities must be based on an anlysis of fairly general characteristics that can have some precedent-setting value. Policy decisions of this nature must be fair and reasonable to the Postal Service and to the private sector.

The IRCs essentially provide three basic services: international telegraph message, international telex, and international private line. While these services naturally share some common elements with the proposed Postal facsimile offering, it could be argued that they are essentially different services. The IRCs have shown a desire, however, to provide joint services with the Postal Service. Since the early 1970s the carriers have had applications on file with the FCC to provide an international Mailgram-like service. They have allegedly tried, without much success, to encourage Postal Service support for these applications. While Mailgram-like services are certainly distinguishable from facsimile, the IRCs have clearly indicated their desires for joint ventures with the Postal Service. The Postal Service, on the other hand, apparently did not contact the IRCs prior to its negotiations with COMSAT.

The IRCs are also beginning to develop new services that are nearly identical to the Postal facsimile service. Note, for example, RCA Globcom's Q-Fax service. Q-Fax is a new international digital facsimile service that permits users to send and receive messages, documents, graphics, engineering documents between RCA's Q-Fax center in New York City and the KDD center in Tokyo. It is expected that by the end of 1978 Q-Fax will also be a switched service with dial-up access. In Japan delivery is accomplished by KDD messenger services in Tokyo and by the postal service in areas outside Tokyo.

Regulatory and legal restrictions may, however, prevent Q-Fax services from ever becoming equivalent to a postal facsimile service. Section 222 of the Communications Act which authorized the 1943 Western Union and Postal Telegraph merger also split the telegraph industry into domestic and international sectors. Thus RCA Globcom can offer its Q-Fax service only from certain gateway cities (New York, Washington, San Francisco). Although the parent RCA Corporation has been allowed to establish a subsidiary corporation for domestic data services (RCA Americom), it is not allowed to interconnect its domestic and international subsidiaries. While the FCC presumably could allow them to do this, it has chosen not to do so pending the conclusion of a docket that is reconsidering certain international issues such as the need for more gateway cities. This docket was begun in 1972 and shows no sign of ending soon. The FCC has recommended to Congress that Section 222 be repealed, but this issue has been subsumed under the effort to overhaul the Communications Act.

In summary, at the present time, the IRCs cannot offer a facsimile service that is equivalent to that of the Postal Service if the latter chooses to offer it on a nationwide basis. It would be possible, of course, for a joint service to be configured by the Postal Service and the IRCs. It is not clear, however, that the IRCs would, in that case, be filling a fundamental role that the Postal Service could not fill itself. More important, perhaps, the Postal Service has not shown any interest in initiating such a joint venture. Because of legal and regulatory restrictions in the international record communications industry, it is not difficult for the Postal Service to offer a service that is fairly unique. Should it decide to implement electronic input and electronic delivery, however, its service would be much more comparable to the combined offerings of the domestic and international carriers.

One final aspect of the Postal/COMSAT experiment that deserves note is the ability of the Postal Service to negotiate with the foreign postal, telegraph, and telephone (PTT) administrations without FCC oversight. Regulated communications common carriers, on the other hand, must obtain both FCC authorizations and agreements with the PTT's. In the past the FCC has disapproved agreements that the IRCs have entered into with the PTTs. The PTTs have also refused to enter into agreements with carriers that the FCC has certified for entry into the international market (for example, Telenet and Graphnet). This tension between the FCC and the foreign administrations was evident recently in the TAT-7 cable controversy. The Postal Service's ability to deal directly with the PTTs may give it a degree of leverage that will work to the disadvantage of domestic U.S. carriers that might in the future want to extend their electronic mailbox services to foreign points. The PTTs, in turn, may welcome the opportunity to negotiate arrangements with a quasi-governmental vendor that they might consider more their equal. While this problem may be hypothetical at the moment, I believe it deserves further study on the part of federal agencies concerned with international telecommunications issues.

While the Postal/COMSAT experiment is timely and interesting, the broader question of what the role of the Postal Service should be in telecommunications deserves some comment. In my view this policy issue is no closer to a resolution today than it was two years ago. Some insight can be gained from the debate on H.R. 7700, but the basic questions have not really been resolved by policymakers. Here are some questions that I feel need to be addressed. Should the Postal Service go beyond joint-venture, Mailgram-type services and become a value-added carrier offering telecommunications services in competition with regulated common carriers? Should it offer services that entail electronic input or delivery, or should it employ telecommunications only in the long-haul segment of its network? Should the Postal Service be required to interconnect with carriers other than Western Union so that competition might evolve in joint-venture offerings? How should the Postal Service price its competitive offerings to assure that government subsidies or monopoly revenues do not allow it to engage in predatory pricing? And finally, should a long-standing national policy excluding the Postal Service from the private telecommunications sector be reversed without a full public debate and specific Congressional authority?

These are not easy questions. The Postal Service does, after all, have a unique national delivery network that Congress has indicated an intention to maintain. This network, as Mailgram has demonstrated, can be productively combined with telecommunications technology. The policy question is how to do this in a way that will not stifle innovation and service to the public or disrupt the healthy competitive market that is developing in telecommunications services.

Some insight on how Congress views these issues can be gained by examining the action the House of Representatives has taken on H.R. 7700. The Senate Committee is still considering this legislation. As reported out of the Post Office and Civil Service Committee, the bill contained several significant provisions. One would have required the Postal Service by 1980 to devote an amount equal to at least 2 percent of its previous year's revenues to research and development. It explicitly required the development of systems for the electronic transfer of messages. Because of opposition to the bill, primarily from the administration, H.R. 7700 was blocked in the House Rules Committee until an agreement could be reached on a modified version. One of the modifications reduced the required research and development budget from 2 percent to 1 percent of the previous year's revenues. Another, initiated at the request of Chairman Van Deerlin, entirely deleted the requirement that an electronic message system be developed. Despite the compromise reduction, the entire research and development requirement was deleted by an amendment offered on the House floor. This does not mean that the Postal Service cannot continue to develop its electronic message transfer system, it merely means that it would not be under a congressional order to do so.

Another provision of the committee bill allowed parcel post service to be priced at a rate that only covered its fair share of indirect costs rather than all the indirect costs that can properly be attributed to it. The only limitation on this provision was that public service appropriations could not be used to reduce parcel post rates by more than 7 percent below the rates that would otherwise be applicable. Thus the Postal Service in competing with United Parcel Service, Greyhound, and others could have priced its service below cost. An amendment was offered and adopted on the House floor requiring parcel post to recover all its direct and indirect costs. The argument used to support this amendment was that it would be unfair to use tax revenues to subsidize a government organization in order that it might compete unfairly with a taxpaying private enterprise. This philosophy would presumably be equally applicable should competition develop between the Postal Service and the private telecommunications sector.

The House also amended H.R. 7700 in order to restrain Postal Service attempts to expand its monopoly under the private express statutes. For example, it specifically excluded data processing materials and time-sensitive intracompany business correspondence from the definition of a letter. It also excluded any means of communication not in existence as of January 1, 1977, or not generally considered to be a letter. The exact effect of this last provision is unclear, but it should give some assurance to those in the telecommunications industry who fear a unilateral expansion of the private express statutes by the Postal Service. H.R. 7700 has two major steps yet to complete, however–Senate action and possibly a House-Senate conference–so none of these actions should be considered final.

The next development to look for with regard to postal electronic services, aside from Senate action on H.R. 7700, is a detailed report on the Postal Service EMS System which is expected to be released in the near future. This may answer some of our questions and will no doubt raise others. At the National Telecommunications and Information Administration we will continue our evaluation of the implications of Postal Service involvement in telecommunications. In addition, we shall continue to put the technical expertise of ITS at the disposal of the Postal Service in order that it might have adequate information on the areas where postal resources might be productively directed.

Part III
Spectrum Allocation

Introduction

Herbert S. Dordick

The 1979 meeting of the World Administrative Radio Conference will very likely be one of its most significant meetings since the International Telecommunication Union ITU was founded in 1865. Until relatively recently the allocation of spectrum was a matter for the more developed nations. Developing countries, although quick to join the ITU as well as the United Nations, were slow in recognizing the importance of telecommunications technology in the development process. But once having done so, and at a time when their numbers made them a majority both in the United Nations and the ITU, they quickly developed the necessary political acumen to deal with issues of spectrum allocation.

Perhaps because spectrum allocation and indeed utilization was for so long the prerogative of the developed nations, surprisingly little attention has been paid in the United States to this matter. Very little has been done to structure a theory or model for rationally allocating the use of this growingly scarce resource. The emergence of satellites raised levels of consciousness to spectrum scarcity, and more recently with concern over the flow of data across national boundaries, both by cable and satellite, these levels of consciousness have risen to considerable concern for national sovereignty in the face of the invisible forces of radio and data. Levin's early work was called on by the Communications Act rewrite staffers and has given rise to considerable research on models of spectrum allocation.

George Codding, Jr., reviews the history of the ITU, pointing out the policy consequences of the new world environment occasioned by the emergence and interest of the now majority new nations in the allocation arena. He points out that the developed nations who have ruled so long will have to make way for what might be a new set of ground rules.

Readers are encouraged to read the House Subcommittee paper on Spectrum Options before proceeding to Jackson's review of the reactions to the House's paper. Jackson summarizes the reactions to eight specific options, including those received and proposed by the respondents and commentors on the original paper. His paper offers an excellent overview of the range of interests that must be satisfied or considered when new policy strategies are proposed—proof, perhaps, that the policymaking process in the United States is indeed one of competition, controversy, and confrontation.

It has long been recognized that the comparative hearing proceedings of the Federal Communications Commission to resolve cases of mutually exclusive applications for radio licenses result in a long and expensive process and produce a subjective result. Robinson proposes that a public auction be substituted for the hearing procedure for the selection among competing applicants for multipoint distribution service channels. After discussing the proposed procedure and its pros and cons, Robinson provides additional examples of services to which the proposed concept would be applicable.

In his second paper Jackson argues that the necessity for a market allocation system for international satellite parking spaces is "even stronger than the arguments for the use of market allocation in spectrum uses confined in or to a single country." Recognizing that both equity and technical efficiency considerations are involved in the allocation of orbital-frequency resources, Jackson proposes a formulation that allows for the separation of these two factors such that equity can be taken care of by the initial distribution of spectrum rights traded in the market. Technical efficiency, then, can be achieved after the first step—meeting equity rights—has been taken.

27 The New Nations and the International Telecommunication Union: Some Policy Implications for the Future

George A. Codding, Jr.

In recent years a number of international organizations have come under a great deal of criticism. Criticism of international organizations is, of course, nothing new. The United Nations has been the target of abuse from a number of fringe groups from its very inception. The United Nations Educational, Scientific, and Cultural Organization has always had its critics, as has the International Labor Organization, an agency from which we have just withdrawn. The subject matters with which these agencies deal lend themselves to a wide variety of approaches; thus consensus on the proper approach is often difficult to achieve.

I must admit, however, that I was surprised when the International Telecommunication Union (ITU) was added to that list. Not that the ITU is perfect, of course, but after all the ITU is a technical international organization and it has been operating for over 110 years without any real opposition and without undue criticism. I was especially surprised because these criticisms came from the well informed in addition to the well intentioned. For example, in his report to the secretary of state, the U.S. delegate to the 1973 ITU Plenipotentiary Conference in Torremolinos stated that so much of the time of the conference had been spent on matters having nothing to do with the role of the ITU in promoting international telecommunications that it cast some doubt on "the continuing usefulness of the ITU as an international regulatory agency for telecommunications."[1]

The purpose of this paper is to investigate the traditional role that the ITU has played over the years and the changes that have been demanded over the past few years by the new majority of developing countries and to indicate some possible policy alternatives that could make the ITU more responsive to the needs of the developing nations while at the same time preserving the activities that make the ITU valuable to the United States and the other developed nations of the world.

The ITU's Traditional Role

The role that the ITU was to play in communications for the first one hundred years of its life was established at an early date; to facilitate the interconnection of national telegraph, telephone, and radio systems. This was to be accomplished basically by the calling of international conferences at which delegates from the member states could meet to discuss the problems confronting the international telecommunication network and to draw up the rules and regulations that these individuals in their collective wisdom felt would solve those problems.

All the elements of this role were evident in the first meetings of the International Telegraph Union, the immediate predecessor of the International Telecommunication Union, in Paris in 1865. The electromagnetic telegraph had been demonstrated only a few years earlier but was quickly adapted to national use. Soon it was being extended across frontiers to meet the needs for rapid and efficient long-distance communication by governments and commercial interests. As the European network began to form, it was realized that it would be difficult for it to reach its true potential unless certain problems were resolved, including differences in telegraphic equipment and procedures and a complicated and confusing rate structure.

At the invitation of the French government, a conference was held in Paris in 1865 to deal with the matter. Twenty European countries sent delegations. The result of the conference was a treaty that established common rules for Europe's telegraph network, including a table of rates, which the delegates believed would insure rapid and uninterrupted telegraph traffic between the major European cities. The delegates decided to call themselves the International Telegraph Union and to meet periodically to update the rules and regulations. At the next conference in 1869 it was agreed to establish a small secretariat in Berne, Switzerland, to help prepare for official conferences and meetings and to act as a clearinghouse for information concerning the international telegraph. The expenses of the secretariat were to be met from contributions from the member states in a proportion that would reflect the ability of the member to pay and the benefits received.[2]

The membership of the International Telegraph Union increased slowly but surely throughout Europe and Scandinavia and then to Africa, Asia, and the Americas as telegraph cables were perfected and laid. In 1885 the Telegraph Union also became the regulator of the international telephone.[3]

The first major change in the structure of the Telegraph Union occurred in 1925 when it was decided to add a committee to study the problems concerning the international telephone in the intervals between conferences. Actually such a committee had been organized the previous year by a number of European telephone experts who felt that the European telephone system was at a point in its development that concentrated effort was needed if it were to reach its true potential. A proposal was introduced at the 1925 Paris Telegraph Conference to bring this body into the Telegraph Union to study problems of the telephone and to help prepare a universal set of telephone regulations. The proposal was eagerly received, and a special section of the Telegraph Regulations was added "constituting" an International Telephone Consultative

Committee[4] and charging it with the study of "standards regulating technical and operating questions for international long-distance telephony."[5] The Telephone Consultative Committee was given a free hand with regard to organization, dates of meetings, and questions to be studied. It was agreed, however, that the committee should elect its own director and that all the results of the work of the committee should be transmitted to all the members of the union.

At the same conference Germany presented a proposal to create a similar organ within the Telegraph Union for the international telegraph. A number of problems, both technical and administrative, according to the German delegate, were still hindering the development of a rapid and effective international telegraph service. The proposal was supported by a large majority of the delegates, and the International Telegraph Consultative Committee came into being.

When radio emerged in the latter part of the nineteenth century, it found immediate application in the maritime service. For the first time ships at sea could keep in contact with the land and with each other at long distances. Because of the nature of radio, especially the possibility of interference between stations, common rules governing equipment and procedures were even more necessary than had been the case for either the telephone or the telegraph.

It did not, however, as in the case of the telephone come under the auspices of the International Telegraph Union but under a separate international organization, the International Radiotelegraph Union. This occurred for a number of reasons; the maritime application of radio immediately removed the European emphasis that had dominated the telegraph and the telephone; the technology was such that it did not come under the purview of the Post Office in most countries, as was the case with telegraph and telephone; and the United States could see the need for close regulation of radio, whereas it had left most of the control of telegraph and telephone to private enterprise.[6]

The nations primarily involved in radio began a series of conferences in the early years of the twentieth century which brought into existence the International Radiotelegraph Union and a set of regulations, both technical and procedural for this new technology. While the Radiotelegraph Union convened its own conferences and meetings, it agreed with the International Telegraph Union to have its secretariat perform all the necessary administrative tasks.

In 1927 the International Radiotelegraphic Union created an International Radio Consultative Committee along the lines of the Telegraph Union's telephone and telegraph consultative committees. In contrast to what occurred in the other two cases, however, the proposal to create a radio consultative committee ran into substantial opposition. The United States attacked the idea with the now familiar refrain that since radio was evolving so rapidly such a committee might well hold back progress by establishing too rigid standards; the French because they feared that private enterprise might attempt to gain benefits from committee decisions; and the British simply because they felt it would be useless. The opposition was overcome by a substantial majority and the

International Radio Consultative Committee came into being and rapidly gained the respect of all the members of the International Radiotelegraph Union.

The merger of the International Telegraph Union and the International Radiotelegraph Union into the International Telecommunication Union in 1932 did little to change the nature of the international effort to facilitate the development of the international telephone, telegraph, and radio systems. It was, in effect, an acknowledgment of the growing interdependence of the three types of communication. Few changes occurred in the structure and functions of the International Telecommunication Union from 1932 until well after World War II.

The Atlantic City conferences of 1947 have been hailed for modernizing the ITU; however, while it did fairly extensive work in changing the structure of the union, it did fundamentally little to change its functions. A recognition of the tremendous impact of the developments in communication technology during World War II led the United States to initiate a preparatory conference among the victorious powers to prepare the way for peacetime ITU cooperation. The conference was held in Moscow in 1946 and was attended by representatives of the United States, the Soviet Union, France, China, and the United Kingdom. Most of the proposals considered at Moscow came from the United States and the Soviet Union, but in the end there was common agreement about the future structure and role of the ITU. This done, the United States government issued invitations for interested governments to send delegates to Atlantic City in 1947 to make the final touches to the ITU's basic treaty and to the radio regulations.

The Atlantic City conference made three fundamental changes in the structure of the ITU. First, it created an Administrative Council of eighteen members to meet annually to take charge of the administration of the union between plenipotentiary conferences and to provide liaison with the newly created United Nations Organization. Second, it was decided to internationalize the predominantly Swiss secretariat and to move the headquarters from Berne to Geneva in Switzerland. Third, it created the International Frequency Registration Board (IFRB). The expanded use of radio preceding and during World War II and the anticipation that it would continue to expand led the United States to propose the creation of a master international frequency list in which the actual frequency needs of the various services of the member states of the ITU would be recorded. The United States also proposed the establishment of a highly qualified board of elected members to screen all new frequency requests for conformity with the radio regulations and for usage without causing harmful interference with other radio stations. The board was created and a series of conferences were scheduled to establish the International Frequency List.[7]

Finally, the Atlantic City conference for the first time in the history of the ITU or its two predecessor organizations attempted to describe the aims and purposes of the union for inclusion in the basic treaty. It was agreed that the overall aims and purposes were

> to maintain and extend international cooperation for the improvement and rational use of telecommunication of all kinds;
>
> to promote the development of technical facilities and their most efficient operation with a view to improving the efficiency of telecommunication services, increasing their usefulness and making them, so far as possible, generally available to the public;
>
> to harmonize the actions of nations in the attainment of those common ends.[8]

It was also agreed that to obtain those aims and purposes the ITU and its members would

1. Allocate the radio frequency spectrum and register radio frequency assignments in such a manner as to avoid harmful interference between radio stations in different countries.
2. Establish rates for the use of telecommunication services as low as possible consistent with sound administrative practices.
3. Promote the adoption of measures to insure the safety of life.
4. "Undertake studies, formulate recommendations, and collect and publish information on telecommunication matters for the benefit of all Members and Associate Members."[9]

Although a number of conferences and meetings were held in the interval between Atlantic City and the Geneva plenipotentiary conference of 1965, there were few demands for any substantive change in the structure or function of the ITU. Politics did arise but dealt mainly with nontechnical issues. The conferences that were held to create the master International Frequency List ran into a number of problems; a major one was the exaggerated estimate of future needs, so great that the original concept was abandoned and the IFRB was reduced to a frequency registration agency. This function was not that different from the one that the ITU secretariat had already been performing for a number of years.

One other fact about the first hundred years of the ITU should also be noted. With one exception all the major elected officials of the ITU were citizens of Western Europe or the United States. That one exception was Marco Aurelio Andrada of Argentina, who served as the secretary general from 1954 to 1958.

New Members, New Demands

The developing nations and their demands have brought about the most important change in the function of the ITU since the merger of the International

Telegraph Union and the International Radiotelegraph Union in 1932. For a number of reasons, however, this change was slow in coming. Although the new nations were quick to join the ITU, as soon as they became independent members of the international community it took some time for them to become a majority in the ITU and to realize the fact. It took even more time for the developing nations to recognize the importance of telecommunications technology to the process of development. Finally, the developing nations had to awaken to the need to perfect effective politics of alliance to achieve their desired ends.

Although the postwar process of converting colonies into independent states had made a good start by 1959, the plenipotentiary conference of that year was a normal one in that the developed Western nations tended to manage the course of business without a great deal of opposition. Two decisions were made, however, that can be considered to be of special benefit to the new members but appeared to be more a gift from the Western-dominated majority rather than a victory won by the Third World.

The first was the increase in the number of members of the Administrative Council. Although the original proposal was a Chinese one, once the subject was raised the new members did take an interest and won an argument to create a special African region, with four members to be elected from that region, and to add an additional member from the Asia and Australasia region. The second was an addition to Article 4 of the ITU Convention, Purposes of the Union, to the effect that the ITU would in the future "foster the creation, development and improvement of telecommunication equipment and networks in new or developing countries by every means at its disposal, especially its participation in the appropriate programmes of the United Nations."[10]

The delegates also passed a resolution for an expanded role of the ITU in development assistance. These last two decisions were more a recognition of what the United Nations was already doing, and an acceptance of the responsibility to help, rather than the result of any massive politicking on the part of the new nations, although the American delegate to the conference was forced to admit that "the tremendous interest of the new and developing countries in technical assistance" was "the most significant single trend noted at this Conference."[11]

The major political issue of the 1959 conference was the proposal by the delegate of the Soviet Union to exclude the delegation of the Republic of China and invite instead a delegation from the Chinese People's Republic. After a short discussion an American proposal to make no changes in the status of the Chinese delegation won by a vote of forty-six to twenty with seven abstentions and two invalid ballots.

Although the chief U.S. delegate complained that enlarging the Administrative Council would tend to make it unwieldy and more costly to run, neither that change nor the decision to make cooperation in development assistance one of the major functions of the ITU kept the U.S. delegate from concluding his report on the conference with the reassuring words that the Geneva conference "provided the opportunity to advance with representatives of other countries the position and point of view of the United States on international telecommunication questions generally and to encourage full understanding and friendly cooperation between the participating countries. It is believed that the collaboration between the United States and other countries maintained the best traditions of the ITU in its long record of international cooperation and accomplishment."[12] This turned out to be the last time that an American delegate would be so satisfied with the outcome of an ITU plenipotentiary conference.

By 1965 the environment had changed drastically. Membership in the ITU had grown to 129, and 122 sent delegations to the Montreux plenipotentiary conference of that year (in 1959 total membership was 96 and participation in the conference 88). The new majority was vocal and insistent.

Two important changes were made in the structure of the ITU, and the developing nations were involved in both. One of the first actions of the conference was to enlarge the Administrative Council again. A group of delegates from developing countries proposed that three to six additional seats be created to be given mostly to the African region. After a certain amount of discussion, it was agreed on a secret ballot to increase the council from twenty-five to twenty-nine, with three of the additional seats to go to Africa and the other to the Asia and Australasia Region. The next action was to reduce the IFRB from eleven to five. The United States originally proposed that the IFRB be abolished and that its functions be carried out by the ITU's secretariat. The American proposal was immediately attacked by the delegates from the developing countries, who argued that even if the board was not carrying out the tasks originally envisioned for it, it was carrying out a valuable new function. This new function was aiding the new states in the selection of frequencies, something they found difficult to do in view of their stage of development. In the end it was agreed to reduce the number to five, one from each of the five regions of the world recognized by the ITU, and that agreement was probably reached only because of the desire of a number of the participants to eliminate the seats held by Nationalist China and South Africa.

The important political news at Montreux was no longer the East-West conflict, although there were still some vestiges of it. The main political discussions occurred over the attack by the developing states on South Africa and Portugal. After rather warm debate an Arab-African proposal to exclude South Africa from the conference was passed by a vote of fifty-nine to twenty-seven with seven abstentions and two invalid ballots. Eleven delegations were absent. The United States and fourteen others, including "most Latin American Delegations" refused to participate in the vote because the vote was on "a proposal in direct violation of the Convention which all governments here represented have signed and ratified."[13]

A motion condemning "without appeal the colonial policy of the retrograde Government of Portugal" and demanding that Portugal immediately improve the conditions in the territories under its control, introduced by the Senegalese delegate, was adopted by a vote of sixty-one to thirty-five, with fourteen abstentions. The United States was again in the minority.

If the new politics of the developing nations caused the most excitement at the Montreux conference, their demands for development assistance probably caused the most consternation to the American delegation. The chief American delegate stated: "If there was one dominant note at the Conference it was the oft-stated belief that the ITU had as one of its basic obligations the requirement to assist the new and developing nations with their individual telecommunication problems."[14] This was, emphasized the American delegate, an attempt to make an important and drastic change in the work of the ITU. In the past the main and only proper concern of ITU conferences "was to expand and improve the global telecommunications networks maintained by the leading telecommunication administrations of the world" and not to aid nations "in their regional and even domestic telecommunication problems."[15]

Although the new nations were able to push through a number of proposals increasing ITU activity in development assistance, they failed to achieve their three major goals: the establishment of a special ITU technical assistance fund in addition to that provided by the United Nations Development Program; the establishment of regional ITU offices in the less developed areas of the world to aid the countries in those areas in developing domestic telecommunication systems; and the creation of a new and separate organ in the ITU to deal exclusively with development assistance matters which would be independent of the secretary general and which would have its own elected director as in the case of the CCITT and CCIR.[16] This failure on the part of the new majority can be attributed in about equal parts to the persuasiveness of the developed countries, a fear of an additional financial burden, and the fact that the developing countries were still not well organized.

That the U.S. delegation was not completely aware of what was beginning to happen is evident in the following statement in the conclusions to its report to the U.S. State Department:[16]

> Although not all of the principal objectives of the United States proposals met with approval of the Conference, definite progress towards these objectives was made. There is reason to believe that the next Plenipotentiary Conference, on which it may be expected *more experienced delegations from the developing countries will attend*, will move in the direction of accomplishment of the United States objectives.[17]

The delegates from the developing countries to the Malaga-Torremolinos plenipotentiary conference in 1973 were indeed more experienced but not along the lines hoped for by the head of the U.S. delegation to the Montreux conference, and more numerous. The overall membership of the union had grown from 129 to 146 and attendance was up from 122 to 131. Many of the developing states in attendance had had the occasion in other organizations such as the United Nations to work together and to press their demands on the developed nations. They thus brought to Torremolinos a new sense of power, and they were prepared to use it. In addition.the ITU secretariat had been led for a number of years by a secretary general from a developing state, a person who would at least lend a sympathetic ear to the aspirations of the new majority.[18]

The Torremolinos conference was dominated almost completely by the concerns of the developing nations, both the new politics and the emphasis on development aid. The first issue, as in previous conferences, was to increase the size of the Administrative Council, this time from twenty-nine to thirty-six, to give the developing states more representation. This was followed by a number of political problems that lasted for the duration of the conference. South Africa and Portugal were both expelled; Israel was condemned for a number of reasons, including the alleged destruction of two submarine cables within Lebanese territorial waters; and after extensive debate it was agreed that liberation organizations recognized by the United Nations should be allowed to attend meetings of the ITU as observers.[19]

The expulsion of Portugal from the conference had an interesting spinoff. For a number of years predating World War I, the major Western nations in the ITU were allowed representation and votes for their colonies and territories. By 1973 all that remained were Portuguese, Spanish, British, French, and American territories. During the attack on Portugal the membership in the ITU of the Portuguese Overseas Provinces was also questioned. This was then linked to the general problem of colonialism, and a motion was introduced and approved to eliminate "the last vestiges of colonialism from the Union." As a result Spain, the United Kingdom, France, and the United States, as well as Portugal, lost their additional votes in ITU conferences and meetings.[20]

Since the United States had always paid an additional contribution for the territories, the deletion of all territorial membership also reduced the U.S. contribution from 11½ percent of the budget of the ITU to 6½ percent.

This unforeseen result prompted the delegates at Torremolinos to consider the ITU's voluntary contribution scheme of financing the work of the union. A proposal introduced by Argentina and sponsored by twenty-six other delegations to adopt the UN scale of contributions, which would have raised the U.S. contribution to 25 percent, was turned down by a vote of forty-six in favor, sixty-seven opposed (with three abstentions) after strong opposition was expressed by delegates from a number of states including both the United States and the Soviet Union. The United States in fact threatened not to sign the final acts if they were passed.[21] The conference did, however, request that the Administrative Council continue to study the matter of union finances for discussion at the next plenipotentiary conference.[22]

The main issue at Torremolinos, as it was in Montreux in 1965, was how the ITU could be used to upgrade the telecommunications networks of the developing nations. Indeed, in his opening address to the conference, which immediately followed the welcoming speech of the director general of the Spanish Posts and Telecommunications Administration, the secretary general of the ITU made a point of praising the ITU for what it had done in the field of development assistance since the last plenipotentiary and to thank his Technical Cooperation Department for its hard work.[23]

Fifteen major proposals concerning development assistance were discussed in committee and plenary, thirteen of which were approved. The two proposals that were defeated included one similar to that defeated at Montreux to establish regional ITU offices in the developing countries and one that would have created a new consultative committee to deal strictly with development matters. Of those receiving approval, probably the most important was the one creating a special ITU technical cooperation fund to be met by voluntary contributions from member states to meet urgent needs for assistance by developing countries. As at Montreux, the U.S. delegation argued in opposition, suggesting that all technical assistance should be coordinated through United Nations Development Program and that if the proposal were to pass the United States would not contribute to it.[24]

Since it dealt primarily with the ITU convention, the Torremolinos conference did not spend too much time on technical matters. However, during a discussion of the structure and functions of the IFRB a proposal was made to give the IFRB the authority to allocate orbital slots for communication satellites to member states. The developing countries that supported the proposal expressed the fear that the developed countries might use so many of the slots that none would be available to the developing countries when the time came that they could use them. After "strenuous opposition" on the part of the United States it was agreed not to give the IFRB such powers but instead to charge it only with recording positions of satellites notified to it in a manner not unlike that it used to record radio frequency assignments.[25]

Although the U.S. delegation to Torremolinos recommended that the United States approve the new convention and continue its participation in the work of the ITU, the criticisms of the conference were harsh. Rather than looking at the effectiveness of the ITU's structure and the performance of its organs, the U.S. delegate charged that the conference "spent an inordinate time on issues that had little to do with the role of the ITU in furthering international telecommunications." Unfortunately the emphasis of the conference was on political matters and technical assistance. "The obtaining of more funds from the major contributors to the ITU," continues the report, "seemed more important to many of the developing countries than furtherance of the basic purposes of the Union." According to the U.S. delegate, this tended to place in doubt the continuing usefulness of the ITU as an international regulatory agency for telecommunications and therefore it might be useful to look elsewhere for an alternative. While these attitudes of the developed countries probably represented more "frustration than a serious desire to withdraw from the ITU," warned the U.S. delegate, nevertheless, "it is a danger to which many of the developing countries ought to give serious thought if they are interested in continuing to have an effective worldwide organization in the international telecommunications field."[26]

Two events have occurred since Torremolinos that have a bearing on this discussion, one external to the ITU and one internal. The external development was, of course, the 1974 Sixth Special Session of the United Nations General Assembly and the demand for a massive reallocation of the world's resources that came from that meeting in the form of the Program of Action on the Establishment of a New International Economic Order. Although telecommunication was not made a specific part of these demands, the assembly did set a priority for all types of assistance from international institutions. That the ITU should make a contribution in this effort was immediately accepted by the ITU's secretary general. In an editorial in the *Telecommunication Journal* of August 1974, after hailing the Sixth Special Session of the General Assembly as "a decisive turning point in the history of mankind," he stated that the International Telecommunication Union "welcomes enthusiastically the opportunity offered to cooperate, within the sphere of its competence, in implementing the Program of Action on the establishment of a new international economic order."[27]

Although it is still not completely clear what the specific applications of telecommunication technology may be to the needs of the developing countries, it is obvious that the decisions of the Sixth Special Assembly provide additional legitimation to the demands of the developing nations for an increased effort on the part of the ITU for developmental assistance and that they have a willing advocate in the secretary general.

The second event has been the focusing of the attention of the developing countries on the radio frequency spectrum as an important world resource to which they have a claim. This was seen probably for the first time at the World Maritime Radio Conference in Geneva in 1974. In essence a large bloc of countries led by the Algerian delegation and including the Arab States as well as African and South American states demanded that new frequency notifications be given the same priority as older ones. "The procedure adopted by the Conference," states Commissioner Robert Lee in his report, "under which future allotments for coast stations are to enjoy the same status as existing ones, abolishes the existing system which affords seniority to allotments/assignments on the basis of a date associated with initial frequency use"[28]

As we are aware, the most recent manifestation of this policy of obtaining an equal share for possible future use occurred at the 1977 World Administrative Radio Conference for the Planning of the Broadcasting-Satellite Service in the 12-GHz band. At this conference the developing countries, with much assistance from a number of delegations from developed countries, insisted on and obtained a fixed frequency and orbital slot allocation for ITU members in regions I and III, despite U.S. arguments that the satellite technology was new and that any fixed plan would tend to freeze the technology too soon.[29] As pointed out by Commissioner Lee, who also headed the U.S. delegation to this conference, the developing countries "wanted to go home with their names on a slot," and that is what they did.[30]

Just how is the ITU changing under pressure of demands of the new majority, and how can these demands be met without destroying the value of the ITU to its traditional clients?

Problems and Prospects

The ITU was born to meet the needs of twenty European industrial nations. A rapid and reliable means of linking the major cities of Europe was important to governments and individuals, but it was essential to the commercial and trading interests of the time. The ITU was the instrument that was created to achieve this goal.

The major function of the ITU for many years was to provide a meeting ground where representatives of member states could meet to draw up the rules and regulations that were considered vital to the extension and maintenance of the international communication network. A small secretariat, located in Berne, Switzerland, helped prepare for the conferences and meetings of delegates and acted as a central clearinghouse for information concerning the network. Research and the development of standards was later added to the functions of the ITU in the form of the international consultative committess.

Slowly but surely the international network of telegraph lines was extended to cover Europe. With the advent of the underseas cables, it was extended to Africa, Asia, and the Americas. For a number of reasons the telephone remained a secondary means of communication in the growing international communication network. The development of radio brought new problems and concerns. In addition to the issue of interconnection there was the problem of preventing harmful interference between stations of different countries. Even later, it also became clear that although new technologies could push back the limits of the radio frequency spectrum, portions of it were becoming more and more crowded. Thus the emphasis after World War II was on the drafting of an "engineered" list of frequencies to make certain that the needs of the member states would be met.

It was this system that the new states encountered when they became independent in the 1950s and 1960s. It had its primary focus on the maintenance and improvement of an international telecommunications network. The system, and the ITU, were dominated by the major Western trading nations, and it functioned mainly to meet their requirements and perceived needs.

On gaining independence the new states joined the ITU almost as quickly as they joined the United Nations. They found that they were expected to provide a connection to the international system and in general to conform to its rules. In return they were permitted to send delegates to conferences and meetings, to receive information concerning the international communication system, and to have their frequency requirements registered by the IFRB in the international frequency register.

While much of the work of the ITU was of great value to the new nations, it was soon obvious that they had needs in the field of telecommunications that were not being met. In many cases the newly independent states found themselves with little in the way of internal communications and the only foreign links they had were with the cities of the former colonial power. A great deal of effort and expense would be necessary to provide even the most basic national telecommunication systems and to communicate with neighbors without using the old colonial link. In its traditional role, the ITU could do little to help.

Three events took place in the latter part of the 1960s and early 1970s that led the new nations to turn to the ITU to solve their communication problems. The first was the linking of telecommunications to the process of development which was promoted by a number of agencies including the telecommunication industry and the ITU itself. The second was the awakening of the developing countries to the concept that in unity there is strength, thus the creation of voting alliances in international organizations among the developing states of Africa, Asia, and Latin America. The third was the strong articulation by the developing states in the United Nations of a demand for an immediate and large-scale redistribution of the global economic wealth.[31]

The impact of these developments in the ITU was the overwhelming demand on the part of the developing countries that development assistance be added to the ITU's list of primary functions. This was to consist of supplying information on technologies appropriate to the needs of the new states, the use of organs of the ITU to teach representatives of the developing states how to use these new technologies, and above all the establishment of a special ITU developmental fund, which would be controlled by the new majority, to aid the developing states in their quest for rapid and efficient domestic telecommunication systems and links with neighboring states.

Another demand is for a share of the radio frequency spectrum. Although the developing countries have used the terms *equity* and *equal shares* in a number of other contexts, these terms have taken on a special meaning with regard to the radio frequency spectrum. If the new states are going to create efficient internal telecommunication networks, there will be a growing need for radio frequencies. If one thing has been made clear in ITU circles, it is that the radio frequency spectrum is both finite and, in many areas, subject to overcrowding. It is not difficult to sympathize with the developing countries in their demand for a share of international resources as different portions of the spectrum come under consideration in ITU conferences—thus the demand for frequencies and orbital slots for communication satellites from nations that may never have the capability of launching a satellite or even paying for such a launching.

Another ramification of the changing demands on the ITU is the interjection by the developing countries of their own brand of North-South politics in the ITU's conferences and meetings. Some of the new politics is simply a reflex action to any circumstance that can be labeled colonial or neocolonial and some is a part of the payoff that is necessary in any system of alliances.

Now with regard to the future. If the developed countries of the world wish to preserve the traditional functions of the ITU—and the ITU appears to be the only place where these functions can be effectively carried out—they must be prepared to accept an additional burden. They must be prepared to increase their contribution to the union budget to pay for the additional costs that stem from the mere presence of so many additional members, such as those attributable to the enlarged Administrative Council and conferences and meetings, and to pay for the additional development-related tasks that have been given to permanent organs of the ITU. The cost probably also includes a contribution to the ITU's own development fund.

The developed countries must pay the price of seeing many additional instances of the new politics being interjected into ITU meetings. The developed nations do not have to automatically agree with the objectives of the new politics, but they must take these objectives seriously. And they must on occasion be prepared to lose.

The developed nations must also pay the price of learning how to work with the new nations in order that both may achieve their legitimate goals. No longer can the United States or other developed countries expect to have their policies accepted only on the basis of power, extent of telecommunication systems, or even logic. The developed nations must approach the new majority not as adversaries but as partners in a global enterprise that is in the interest of both.

Notes

1. Department of State, Office of Telecommunications, *Report of the United States Delegation to the Plenipotentiary Conference of the International Telecommunication Union, Malaga-Torremolinos, Spain, September 14-October 25, 1973* (TD Serial No. 43), Washington D.C., 1973, (mimeographed) p. 8.

2. The early history of the ITU, unless otherwise referenced, is from George A. Codding, Jr., *The International Telecommunication Union: An Experiment in International Cooperation*. Leiden: E.J. Brill, 1952. Republished in 1972 by Arno Press.

3. For a number of reasons, including linguistic diversity and the fact that for a number of years in Europe the telephone was considered only a supplement to the telegraph (and both were nationalized), not much effort was made to create a truly international telephone network even in Europe. As a result, the early work of the Telegraph Union on the international telephone was minimal. The conference of 1885, for instance, confined its work to the addition of a five-paragraph chapter on telephone to the Telegraph Regulations. It was not until 1932 that separate Telephone Regulations came into being.

4. The committee was originally named the Comité consultatif international des communications téléphoniques à grande distance. Later it was changed to Comité consultatif international téléphonique. To differentiate its initials from that of the Comité consultatif international télégraphique (originally the Comité consultatif internationales des communications télégraphiques) which was also added to the union in 1925, it was designated the CCIF and the telegraph committee the CCIT.

5. See Codding, *International Telecommunication Union*, p. 36.

6. The United States never became a member of the International Telegraph Union, nor did it subscribe to the Telegraph and Telephone Regulations until after World War II. Its reason was that the telegraph and the telephone in the United States were primarily the concerns of private enterprise, and thus the government did not have the right to regulate them as would be necessary if the United States became a party to the regulations.

7. For a comprehensive account of the IFRB and its work to 1969 see David M. Leive, *International Telecommunications and International Law: The Regulation of the Radio Spectrum*. Leyden: A.W. Sijthoff, 1970.

8. ITU, *International Telecommunication Convention, Atlantic City, 1947*. Atlantic City, 1947, Art. 3.

9. Ibid.

10. ITU, *International Telecommunication Convention, Geneva, 1959*, Geneva, 1960, Art. 4, para. 2, d).

11. U.S. Department of State, Telecommunications Division, *Report of the Chairman of the United States Delegation to the Plenipotentiary Conference of the International Telecommunication Union, Geneva, Switzerland, October-December, 1959* (TD Serial No. 905), January 15, 1960 (mimeographed), pp. 5 and 16.

12. Ibid., p. 29.

13. U.S. Department of State, Office of Telecommunications, *Report of the United States Delegation to the Plenipotentiary Conference of the International Telecommunication Union, Montreux, Switzerland, September 14 to November 12, 1965* (TD Serial No. 973), Washington, D.C., December 15, 1965 (mimeographed), p. 10.

14. Ibid., p. 8.

15. Ibid. This unfortunate tendency in the conference, according to the U.S. delegate, left it "to the members which are dominant in global

telecommunication the development and integration of the global communication network consisting of long distance radio circuits, high capacity cables, and the new global satellite system" (ibid., p. 9).

16. The International Telegraph Consultative Committee (CCIT) and the International Telephone Consultative Committee (CCIF) were merged by the Geneva Plenipotentiary Conference in 1959 to form the International Telegraph and Telephone Consultative Committee (CCITT).

17. Ibid., p. 46 (emphasis added). In another part of the report these objectives were referred to by the U.S. delegate as "forward looking proposals to improve the operations of the Union" (ibid., p. 8).

18. The American secretary general, Gerald C. Gross, had retired in 1965 to be replaced in 1966 by an Indian, Manohar Sarwate. Sarwate died in 1966 and was replaced by Mohamed Mili from Tunisia.

19. See ITU, *International Telecommunication Convention, Malaga-Torremolinos, 1973*, Resolutions 29, 30, 31, and 48.

20. See the debates in ITU, *Minutes of the Plenipotentiary Conference of the International Telecommunication Union, Malaga-Torremolinos, 1973*, Geneva, 1974, pp. 428-441. At present the only nonsovereign members of the ITU are Bielorussia and the Ukraine.

21. Ibid., pp. 410-412, 415-419, and 469-472.

22. *Torremolinos Convention*, Resolution No. 8.

23. The secretary general also stated: "I am glad to take advantage of this solemn occasion to express from this rostrum on behalf of the I.T.U. our sincere and heartfelt thanks to all those who are assisting us and lending their support to help the new or developing countries. This homage is addressed in particular to countries which are highly developed in the field of technology." See ITU, General Secretariat, *Minutes of the Plenipotentiary Conference of the International Telecommunication Union, Malaga-Torremolinos, 1973 (Addendum)*, Geneva, 1974, pp. 11-12.

24. *Report of the United States Delegation to the Plenipotentiary Conference of Torremolinos*, p. 43.

25. Ibid., p. 55.

26. Ibid., p. 8.

27. M. Mili, "Editorial," *Telecommunications Journal* 41, no. 8 (1974): 468.

28. U.S. Department of State, Office of Telecommunications, *Report of the United States Delegation to the World Maritime Administrative Radio Conference of the International Telecommunication Union, Geneva, Switzerland, April 22-June 7, 1974* (TD Serial No. 50), Washington, D.C., 1974, p. 28. See also the treatment of the 1974 Maritime WARC in Martha Jane Wheaton, *The Preparation for and the Implications of the General World Administrative Radio Conference of 1979* (AD/A-017 531), National Technical Information Service, U.S. Department of Commerce, Springfield, Virginia, September 1975, pp. 48-53.

29. The U.S. was able to postpone the creation of a plan for Region II, the Americas, until 1982.

30. See Wilson Dizard (Rapporteur), *WARC: 1977: Implications for the Future*. CRIS Notes, Center for Strategic and International Studies, Washington, D.C., March 1977, p. 2.

31. C. Clyde Ferguson likens the anger and rage among the developed nations that this demand first engendered with the reaction of many Americans to the civil rights movements in the United States in the late 1960s. See his "The Politics of the New International Economic Order," in *The Changing United Nations*, ed. David A. Kay, New York: American Academy of Political Science, 1977), p. 152.

28 Reactions to the Spectrum Options Paper

Charles L. Jackson

About a year ago the staff of the House Communications Subcommittee prepared a series of options papers for the subcommittee discussing policy alternatives for communications legislation. These papers were made public and comments received.

This paper is a subjective report of my reactions to the spectrum options paper. I say subjective because the report is subjective in two different ways. First, it is based on the reactions to the spectrum options paper as those reactions filtered through to me. Since I was the author of that paper, I may have received a censored version of the reactions. Second, I am reporting those reactions I found most interesting or most important.

The options paper put forward eight basic options:

1. No change; maintain current policies.
2. Provide more explicit standards for spectrum resource allocation than the current "public interest, convenience and necessity."
3. Provide more resources to the spectrum resource management task, including resources for systems design.
4. Create a "transition trust fund," financed by radio license fees, to aid individuals whose investment in radio equipment is destroyed by changes in spectrum management policy.
5. Allow the Federal Communications Commission to use auction techniques to choose among alternative claimants for the spectrum resource in non-broadcast services.
6. Allow the Federal Communications Commission to use auction techniques to choose among licensees in all services, including broadcasting.
7. Allow the Federal Communications Commission to use a lottery mechanism to choose among competing applicants for a license. (This reduces administrative costs in the same fashion as auction-based alternatives, but it has quite different distributional and dynamic impacts.)
8. Regulate some characteristics of receivers.

The options paper also included some background discussion of the spectrum resource and of spectrum regulation and spectrum regulatory institutions. This background discussion was brief and only sketched broad outlines.

There were several kinds of visible responses to the spectrum options paper, including written comments, conversations, positions taken by industry associations, and testimony before the subcommittee.

Responses can also be characterized as surprising or as expected. First, I will highlight the responses that I found most interesting. Then I will go through each option and summarize reactions.

The third option, calling for more resources for system designs (especially for distributed systems) was based on my own observations of a major failure in communications policy. But it was not an issue that had received much discussion in the literature. Nor did I see that any natural constituency existed to support the idea. Hence I expected that this option would be ignored. Perhaps it might get a few negative comments, but it got some positive comments as well. In particular the American Waterways Operators, an industry association representing a group of users of a distributed system, sent in extensive comments detailing some of their headaches and lamenting the lack of systems design. In contrast, the Bell System supported changes that would create improved incentives to create new designs in the private sector as opposed to resorting to public funding to augment systems design. I think the Bell System is somewhat myopic here. Their own achievements in improving the efficiency of microwave and mobile communications depend at least in part on the fact that those systems are not distributed systems with divided ownership.

The fourth option, creation of a transition trust fund, was based on my observations while at the FCC of the great reluctance of regulators to make equipment obsolete or unusable when reallocating spectrum.

Nobody who responded tried to discuss this issue quantatively. Most people who responded, including all spectrum users, felt that the option had more minuses than pluses. I was surprised at the lack of positive response to this proposal. Everybody seemed to think the only way it would affect them was through fees or taxes to support the fund.

The options paper varied from much of the traditional literature on the spectrum resource by referring to the myth of spectrum scarcity (with appropriate qualifications) rather than to spectrum scarcity.

I expected that the word *myth* would be the focus of much adverse comment. In fact it did not receive such criticism. One witness before the subcommittee went so far as to say, "From a purely engineering point of view, there is no scarcity of spectrum in the United States today . . . unmet needs are traceable to a regulatory structure."[1]

One last interesting story. The options paper stated that the current regulatory system delays and denies the development of new systems and new technology. The Bell System strongly supported this statement giving several examples. To quote them, "In one case, an entire development became obsolete during the decisionmaking process."[2]

Now to consider each specific option.

1. *No change*. Many groups liked this option. Especially if it could be linked with a mechanism that would force the FCC to act faster on their own requests for spectrum.

2. *More explicit standards.* No one suggested specific, explicit standards.

3. *More resources for spectrum management.* This was supported, in low-key fashion, by a number of groups.

4. *Transition trust fund.* This was discussed before.

5, 6. *Use of economic incentives in spectrum management.* This concept was supported by a number of witnesses at hearings before the subcommittee including Richard Wiley, Glenn Robinson, and Raymond Spence. It was opposed by two common carriers and by some users of private systems. (Ironically, one of the private user's main objections was that the major carriers were strongly favored by such a system.) In addition the National Association of Broadcasters has indicated its opposition to such a concept.

7. *Lottery techniques for choosing licensees.* No one seemed to like this option. Those who felt it might make some sense felt that market mechanisms were better.

8. *Regulate receiver characteristics.* This option has received relatively little discussion so far. I expect that consideration of the UHF taboos issue, the UHF noise figure issue, and the Senate hearings on S.864, Senator Barry M. Goldwater's bill on interference reduction, will generate further discussion of this issue.

Perhaps the best way to end this paper is to quote AT&T's summary of their comments on the options paper.

> The Bell System recognizes that the present process used in the U.S., although imperfect, has not led to chaos, but, in fact, has resulted in the most advanced and extensive use of the radio spectrum. Consequently, changes on purely hypothetical grounds or solely for the sake of change should be avoided. It may be better to merely strengthen the present process where needed than to attempt wholesale changes.

That represents the most common reaction to the spectrum options paper.

Notes

1. Hartman and List, prepared statement, 4 August 1977, before the House Communications Subcommittee, p.1.

2. Comments of American Telephone and Telegraph Company on behalf of the Bell System "Policy Options for the Spectrum Resource," p. 1.

29 Assignment of Radio Channels in the Multipoint Distribution Service by Auction

John O. Robinson

In his policy options paper to the House Subcommittee on Communications, Charles Jackson noted that "spectrum policy consists of two basic choices: What kinds of services shall the spectrum provide? Who gets to use the spectrum?"[1] The Federal Communications Commission (FCC) devotes a considerable portion of its resources to both questions. This paper is concerned primarily with the second one.

After the FCC has decided the first question, which in spectrum management parlance is known as allocation, it addresses the second question through a process termed assignment. This process consists of issuing licenses to eligible users to operate specific radio stations. The frequency on which a particular station will be authorized to operate is selected from within the spectrum block allocated to the type of radio service that the applicant proposes to initiate.

The nature of the communication provided by certain of the radio services administered by the FCC requires exclusive use of a frequency channel over some specified geographic area. In other words, it is not operationally feasible for licensees to time-share the channel. If there is more than one applicant for one of these radio channels, then the assignment of the channel becomes administratively complicated. The reason for this is embedded in the Communications Act of 1934. Section 309(e) of this act states that in any case where the Commission cannot find it in the public interest to grant a license for a radio station, the applicant is entitled to a hearing. In addition, a court decision, known as the Ashbacker doctrine, holds that in cases where the grant of a license to one party precludes the grant to another, all applicants are entitled to a joint hearing.

Clearly, if more than one party is applying for a license to operate a radio station on a channel that cannot be shared, the applications are mutually exclusive. Unless the matter is resolved by private negotiation among the parties involved, the commission must designate the applications for a comparative hearing. Cases of mutually exclusive applications occur frequently in the broadcast service and certain of the common carrier and safety and special radio services.

This paper outlines an alternative to the comparative hearing for mutually exclusive applications in one of the common carrier radio services, the multipoint distribution service (MDS). Specifically, it is proposed that an auction

---

The opinions expressed in this paper are solely those of the author and should not be interpreted as representing the views of the Federal Communications Commission or its staff.

be substituted for the hearing to make a selection from among the competing applicants for a MDS license.

The paper is not intended as a specification around which an auction can be designed. In some respects it raises more questions than it answers. Certainly there is a need for better information pertaining to the cost of the comparative hearing as well as that of an auction. The discussion of legislation required to implement the auction does not contain any of the legal analysis and detail that will be needed. Finally, vital issues that must be considered may have been missed altogether. Nonetheless, I believe that this paper provides a framework for discussion and critique of the concept of substituting an auction for the comparative hearing to select operators of MDS stations. There are other radio services to which the concept can be extended.

The Multipoint Distribution Service

The MDS is a common carrier service employing omnidirectional transmission to distribute multiple-address broadband communications for simultaneous reception at several fixed receiver points. The message content is the responsibility of the customer using the service. It may consist of various forms of private television, computer data, facsimile, or other communication adaptable for broadband radio transmission. One of the most popular uses has been transmission of specialized entertainment programming for reception on standard television receivers by subscribers in multiple-dwelling buildings.

At the present, two 6-MHz channels are allocated to the MDS at 2,150-MHz-2,162-MHz. One channel is available anywhere in the country. The other channel is available only in fifty designated metropolitan areas. In addition, a 4-MHz channel is available without geographic restriction. Because a bandwidth of 6 MHz is required for television transmission, the two 6-MHz channels are in greatest demand.

The MDS in its present form was inaugurated in 1970 with the FCC's ruling that a channel bandwidth sufficient to carry a television signal could be used in the 2,150-MHz-2,160-MHz band. This ruling authorized the first 6-MHz channel, as well as the 4-MHz channel. In 1975 the 4-MHz channel was expanded to 6 MHz in fifty metropolitan areas. Mutually exclusive applications occurred almost immediately and increased to the extent that the FCC faced 147 possible comparative hearings—97 for channel 1 and 50 for channel 2. Processing these mutually exclusive applications began in mid 1975 with the adoption of the first hearing designation order, and a year later the commission adopted new procedural rules to encourage private settlements.

At this time nearly half (seventy) of the possible hearings have been settled; of these twenty-five occurred prior to designation for a hearing. Of forty-nine cases designated for a comparative hearing, forty-five have been or are in the process of being settled. To date, only four cases have actually resulted in hearings. Currently seventy-three mutually exclusive cases remain to be settled.

Although many of the cases of mutually exclusive applications in the MDS have been settled, this service still provides a good model around which to discuss the auction. First, there are seventy-three cases yet to be resolved. Also the long period of time required to gain the present number of settlements will have significant cost implications. Further, as the MDS stations become more profitable, mutually exclusive applications may result at the time of license renewal. Finally, there remains the question whether settlement negotiated under the duress of an impending comparative hearing is the most economically efficient solution.

The Comparative Hearing

The comparative hearing is a costly and lengthy process that produces at best a subjective result. Unfortunately, it is not possible to describe a typical hearing in terms of its cost or time required because of the nature of the hearing. Since it is a legal process, it provides many avenues for change in the proceedings and allows several levels of appeal. The unpredictability of the length of the hearing makes any estimate of its cost equally uncertain.

The major elements of a comparative hearing are easily identified. The initial step is formal designation of the mutually exclusive case for a hearing. In preparation, the issues on which the competing applicants are to be compared have been defined by the staff. Although the applicants have individually met the qualifications contained in the FCC's Rules, additional criteria must be specified to provide a basis for comparison among the applicants for the specific license in question.

Next is a prehearing conference that may lead to disagreement with the issues as prescribed. This can result in petitions for change, including extension of the issues. After these matters are resolved, the hearing is scheduled. The hearing itself is not usually a lengthy process and, barring interlocutory arguments, is concluded in a matter of days. The next event is the law judge's decision. As with the hearing itself, the time required for preparation of this decision is relatively short. However, the appellate route subsequent to the law judge's decision can consume considerable time. The appeal steps are the FCC Review Board, the commission itself, and finally the courts. Most appeals are terminated with a Review Board decision. In its entirety, it is not uncommon for the complete hearing process to extend over a period of years.

The expense of a hearing includes both direct and indirect costs. The direct costs are those incurred by the applicants and the FCC in the conduct of the hearing. The FCC has only recently begun to segregate its hearing costs. In general, several high-level professional staff members will be involved for some indefinite period of time. Applicant's costs are available only as estimates. For a MDS hearing these estimates range from a few thousand to tens of thousands of dollars. As would be expected, the funds that an applicant will commit appear to depend on the anticipated profitability of the intended operation and thus the value of the license. For some television broadcast cases the applicant's costs have been alleged to be over a million dollars.[2]

A large number of MDS mutually exclusive cases have been settled by private negotiation after designation for a hearing. It can be reasonably assumed that the expected cost of the hearing provided the major incentive for these settlements.

The indirect costs are in the form of the value of communication services foregone during the time occupied by the hearing. These costs are borne primarily by the public, although the foregone profit of the ultimate licensee must also be included.

In a recently concluded case of mutually exclusive applications for a license to operate a MDS station in San Diego, the applications had been on file for more than six years. Less than a year of this time was occupied by the hearing. The remainder was consumed by regulatory matters associated with the establishment of the multipoint distribution service. Nevertheless, following adoption of the commission's first designation order in 1975, service to the public was delayed for at least three years because of mutually exclusive applications and the consequent requirement for a hearing. As an incidental note, the law judge's decision in this case has been appealed to the Review Board.

Merit of an Auction

The comparative hearing is by nature a subjective process. The resulting decision must be based largely on promises made by the competing applicants. The only function of the hearing is to acquire sufficient information, relative to the issues prescribed, to allow the commission (law judge) to identify the applicant whose proposed services will be most in accord with the public interest.

Because a common carrier cannot influence message content, the public interest in an assignment is determined when an applicant is found qualified according to the FCC Rules. If any one of a group of mutually exclusive applicants had been the only applicant, a license would have been granted on the basis that it would have been in the public interest. The operation of a radio station in a common carrier service is a regulated commercial venture, whose communication product is to be determined by the marketplace even though rates are regulated. A most important measure of service in the public interest is the level of economic efficiency at which the service is provided. In general, the common carrier operation that results in the highest-valued communication service for any given amount of resources used (including spectrum) will, in the economic sense, be acting most in the public interest.

While the applicant who will actually operate with the greatest economic efficiency cannot be predicted with absolute certainty, an auction would determine the most likely prospect. It must be assumed that the candidate that offers the highest bid expects to put the spectrum to its highest-valued use. If this were not true, there would be no rational basis for the bid.

An auction, then, would provide an objective comparison among the applicants, with the sincerity of the selected applicant evidenced by the willingness to commit financial resources in the form of a bid. In addition, the auction would be a much less expensive process. The open-ended nature of a hearing allows the proceeding to extend over long periods of time, resulting in very high costs to the applicants, the government, and the public. The auction, on the other hand, would be a straightforward process requiring relatively little time.

It cannot be claimed, of course, that an auction-and-lease process can be introduced and maintained at no cost. The important difference is that the major part of such cost is a one-time cost to introduce the process. The cost of administration would be minimal.

Legislation will be required to authorize the FCC to establish an auction-and-lease procedure. Given this enabling legislation, there will be a one-time cost to design the procedure, conduct a rule-making proceeding, and incorporate the procedure into the FCC Rules and Regulations.

Conceivably, the initial cases under the new procedure could result in hearings and appeals to test the concept. However, these early cases would soon establish precedent, and future cases would involve no more than advertising the channel for which bids were being solicited, verifying the eligibility of bidders (which must be done under the present system), and reviewing the bids. Once established, the time required to execute an auction-and-lease procedure would be a matter of a few months, at most, compared with perhaps years for the hearing process.

Rights To Be Auctioned

The proposal discussed in this chapter does not imply free-market conditions. Instead, it suggests the application of a market technique within the constraints of administrative control of spectrum allocation and use. However, in determining the rights to be auctioned, a general guide has been to seek answers to the question, How would the market resolve the matter?

Insofar as operating, or technical, rights are concerned, there appears to be no alternative to those rights now specified in the FCC Rules. That is, such matters as frequency bandwidth, types of emission, power allowed, and antenna specifications would be prescribed by the FCC as they are now. In the economist's terms these are input rights. They may be changed, but only by administrative action, not as market forces might dictate.

The reason that no alternative is apparent is because the spectrum is administratively allocated by blocks to radio services. Provision for interservice bidding across blocks would entail far more market involvement than envisioned here. If complete freedom to choose operating characteristics was to be permitted within the spectrum block allocated to the MDS, some mechanism would be required to resolve any instances of radio interference that might occur both inside and outside the MDS blocks of spectrum. In other words, output rights are not feasible even within the MDS frequency bands under the present system of administrative allocation.

Given the operating rights as just defined, there is no obvious reason, except for possible economic rents, that these rights should not be freely transferable to any qualified party. Qualification would be as specified in the present FCC Rules; that is, the buyer must satisfy certain legal requirements and be financially and technically capable of installing and operating the radio station. The only requirement would be notification of the impending transfer to permit the FCC to certify that prospective buyer's qualifications. It might be argued that in the case of such a sale all original bidders should be given the opportunity to buy the rights. From a market point of view this would seem unduly restrictive. However, it would allay to some extent the often expressed concern for equity among bidders.

One of the most difficult elements of the rights package to be determined is the period of the lease. It would seem that if a market prevailed, something longer than the present five years would emerge. Since a longer period would increase the certainty of the right, it should make the right more valuable. On the other hand, the FCC must set a termination date consistent with the necessity to periodically review whether the spectrum allocated to the MDS continues to be used for a purpose that is most in the public interest. A period of fifteen to twenty years seems reasonable to satisfy both considerations. Yet it must be recognized that this range of numbers is an arbitrary selection.

Another point relative to the period of the lease that must be addressed is whether there should be provision for renegotiation or cancellation of the lease before its expiration by either the FCC or the operator. The only reasons apparent at this time for the FCC's desire for such provisions are inadequate performance by the operator or a rate of return high enough to imply that the spectrum was considerably more valuable than had been bid. As a common carrier regulator, which is independent of its role as a spectrum manager, the FCC is in a position to control both the performance and the profit of the operator. However, if the spectrum had in fact increased in value, the subscribers to the MDS would be subsidized by the public at large. Further, if resale were permitted, economic rents could accrue to the seller. In the longer term the FCC could control these rents by allocating more spectrum to the MDS.

Since as a common carrier the operator can recover all costs, it would seem that the operator would desire renegotiation or cancellation privileges only in the event that business was so unprofitable as to result in its cessation. Clearly, the entire matter or provision for renegotiation or cancellation requires more detailed study.

A final item associated with the rights to be conveyed is the matter of additional spectrum allocation to the MDS. Since the FCC is substituting for a market in allocating the spectrum, it can have considerable influence on the value of any block of spectrum. It can use this influence to counter the existence of economic rents. It can also, however, depress the value below an original purchase price. It seems important, therefore, to consider the inclusion of a provision to compensate successful bidders in any case where allocation of additional spectrum reduces the value of the leased spectrum below the price that was bid.

Auction Procedure

Unlike the definition of rights to be auctioned, there is considerable experience with assignment of other public resources on which to base an auction procedure. The Agriculture and Interior departments have been involved for many years in the auction and lease of access to publicly held grazing land, timber, oil, coal, and minerals. If it can be accepted that concern for the public interest is essentially no different in the case of the spectrum than it is for these other resources, then their experience can provide useful guidance.

The following procedure is outlined in terms of the major items that must be considered. It is not intended as a finished operational procedure.

1. *Identification of channel to be auctioned.* The discussion thus far has proposed the auction as a substitute for comparative hearings. However, regardless of the reason for introducing the auction, there is no obvious reason for limiting its application to a particular circumstance. If it is decided that MDS stations should be assigned by auction for any purpose, it is not at all obvious that all MDS stations should not be so assigned. It is true, of course, that where there was only one interested party and this fact became known, the amount bid for the spectrum would be expected to be very low or even zero.

The FCC Rules specify permissible locations for MDS stations, and FCC assignment records indicate occupied locations. Station locations available for auction could be made known by publication of this information at frequent intervals. However, such a procedure would be expensive and seems unnecessary. A more practical means would be to announce the auction of a station location when the first application for the station was received. In other words, the first application for a MDS station would be used to initiate the auction of that station.

2. *Notification of auction.* To ensure equal opportunity for all interested and qualified parties to bid and to attract as many bidders as possible, notification

of an impending auction should be distributed nationally and within the local area of the station location. National distribution could be accomplished by publication in the Federal Register. Local notification would be best performed through announcement in local newspapers.

3. *Scheduling of the auction.* The time between notification of the auction and opening of bids must be a compromise between expediting the process and permitting the less sophisticated bidders to prepare their bids and arrange financing. FCC review of bidders' qualifications must also take place during this period. Ideally, bidders would be notified that they were qualified to bid before expending effort to prepare their bids. Yet specifically stating and publishing the qualification requirements may pose some risk to the bidders. This approach would have FCC review of qualifications in parallel with bid preparation.

4. *Qualification of bidders.* At the present time, eligibility requirements for MDS license applicants are stated only in general terms. In effect, the FCC Rules indicate that an applicant must show legal, financial, and technical qualification to provide MDS service. In the interest of expediting the auction process and minimizing any basis for appeal, the terms *legal, financial,* and *technical* will need to be more specifically defined.

The qualifications for eligible bidders should not be enlarged beyond what is now required for an applicant. If these qualifications are deemed adequate to satisfy public interest requirements under the present system where only one applicant is involved, there is no need for additional qualifying criteria when an auction is used. Such additional criteria are necessary in the case of a hearing only to provide a basis for comparison. The auction provides its own basis.

5. *Lease charge.* The lease charge to be bid could be a periodic payment (such as an annual rent), a single charge for the period of the lease, or some combination of both. Provision for renegotiation of the charge might be included. No reason to prefer one option over another is apparent at this time. Fixing the total cost would increase the certainty of the transaction, but the question remains whether this is the most desirable procedure.

Whether a minimum acceptable bid should be specified must also be considered. Since the value of the spectrum to be auctioned is currently unknown, there would seem to be no basis on which to establish a minimum bid initially. However, it may be desirable to include provisions for introducing such a minimum after some experience with the process.

6. *Frivolous bidding.* It is possible that a party might submit a winning bid without intending to erect a MDS station. One reason for doing so would be to prevent or delay the operation of the stations by others. A second reason would be to acquire rights to the MDS channel as a speculative venture, anticipating a capital gain. Bidding for such purposes could be controlled and its effect minimized by requiring a sizable deposit from the winning bidders at the time of selection and establishing deadlines for beginning station construction and operation.

Identifying speculation as frivolous may appear somewhat harsh when viewed in terms of normally accepted market activity. However, because in this instance a scarce public resource is involved, and this resource is *not* allocated by the market, the inference seems appropriate.

7. *Bidding collusion.* The possibility that bidders may conspire to artificially establish a bid price is, like any tendency toward anticompetitive practice, one of the risks that accompanies a market. There can be no guarantee that bidding collusion will not occur. However, discussion with personnel in the Agriculture and Interior departments suggests that large numbers of bidders and the use of sealed bids tend to discourage collusion. It makes sense that collusion is practical only when the potential bidders can be identified and the number who must cooperate is small. This would argue for wide geographic dispersion of the bid request and the use of sealed bids rather than oral auction. In any case where collusion is suspected, prompt investigation and prosecution under the antitrust laws, if the facts so indicate, would also be a deterrent in future auctions.

Legal Concerns

The Communications Act is silent on the matter of a market approach to spectrum management. It neither authorizes nor prohibits the implementation of market mechanisms such as auctions or user charges. However, discussion with lawyers indicates a general consensus that legislation would be required to permit the FCC to proceed with confidence in this area. There are several reasons for this opinion.

First, there is the fact that the Communications Act explicitly precludes private ownership of the spectrum.[3] While the approach suggested in this paper does not propose private ownership as such, in the legal sense an auction or rental normally implies an acquisition of some form of right in the property involved, even if only temporarily.

The history of congressional concern that the spectrum might be monopolized by private interest, and its determination that such an eventuality not be allowed to occur, predates the Radio Act of 1927. It would seem then that legislation would be required to clarify this point by specifying that an auction of a radio channel does not convey any form of ownership.

Another reason that legislation would be required is to permit the auction to replace the hearing. Even if it were decided that the FCC has the authority to institute an auction process, Section 309(e) of the Communications Act as now written would entitle any losing contender to a hearing. Finally, the application of economic factors, in any form, to spectrum management would be an extreme departure from the fifty-year tradition of allocating and assigning the spectrum as a free good. Thus the prudent course for the FCC seems to be to request legislation providing explicit authority to institute such changes, rather than to effect the changes and confront legal challenges that would have to be resolved in the courts.

While legislation giving the FCC explicit authority to substitute an auction for the hearing process does seem to be desirable, if not actually necessary, no legal obstacles of a more fundamental nature have become apparent. Further, the present Congress may be more favorably disposed to enact such enabling legislation than was the case even two or three years ago.

Other Services

The discussion in this paper is directly applicable to an auction as an assignment mechanism for any radio service in which the public interest determination does not involve message content. This includes all common carrier services. In this respect, there are frequent instances of mutually exclusive applications among the radio common carriers. These are nonwireline common carriers who provide mobile radiotelephone and paging services.

The landmobile cellular radio system now undergoing development at 800 MHz may also become a candidate. Although the large capital investment required for these systems will tend to limit the number of license applicants, there may well be cases of more than one. Since only one system per geographic area is to be permitted, the applications would be mutually exclusive. Once again, although the strongest immediate argument favoring the auction is that it would eliminate the costly comparative hearing, there is no reason to restrict its application to mutually exclusive cases.

Examples of non-common-carrier services for which the auction might be considered are the aeronautical advisory stations and subscription television (pay TV). The aeronautical advisory stations provide private aircraft with information such as ground services available, status of runways, wind and weather conditions, and dispatch service. Only one aeronautical advisory station is permitted to operate at a landing area. As a consequence, there is competition for these stations, and multiple (mutually exclusive) license applications are not uncommon.

Subscription television stations (STV) broadcast programs intended for members of the public in return for a fee. Authorization to operate a subscription television station is granted only to a licensee of a commercial television station or to the holder of a construction permit for such a station. An applicant for a construction permit may also apply for STV authorization, but such authorization will not be issued before the construction permit is granted.

Two quite different forms of mutually exclusive cases may arise in the STV service. In one instance all parties are applying for STV authorization. In the other, contenders for the same license or construction permit may include parties proposing to provide STV and others who wish to provide conventional television service. Where all competing applications are for STV authorization, selection by auction would clearly be appropriate. The public interest aspects are satisfied in that the applicants have met basic qualifications and message content is intended to be determined by the market.

The desirability of an auction becomes less clear when one or more of the applications for a construction permit is to operate a STV station while the same channel is being sought by at least one applicant to provide conventional broadcast service. Here the public interest matters involve the relative social value of conventional television versus STV, as well as the attributes of the applicants. This case, in fact, introduces the question whether the auction can be applied in the broadcast services in a manner that would be deemed to be in the public interest.

The public interest in broadcasting involves matters that are not encountered in other radio services. Some of these are the concept of localism, the diversification of media control, the fairness doctrine, news and public affairs programming, and children's programs. The validity of some of these items as measures of public interest has been the subject of some controversy. There is also the question of the extent to which the FCC can influence programming without violating First Amendment constraints. Of greater significance here, however, is whether the public interest would be better served by the present comparative hearing process using the items noted as indexes for comparison or by an auction among otherwise qualified contenders.

A thorough discussion of this subject is beyond the scope of this paper. A good analysis of the auction as a replacement for the comparative hearing in broadcast cases is that by former Commissioner G.O. Robinson in his dissent in *Cowles Florida Broadcasting, Inc. et al.*[4] While the case in point was concerned with a license renewal, the commissioner's analysis was applied to the general problem of broadcast licensing. His conclusion is that the auction would be preferable to mutually exclusive applications in the television broadcast service.

Additional Benefits

Although the auction has been proposed primarily for the purpose of eliminating costly comparative hearings, other benefits would accrue. One important additional benefit is that the process would provide a measure of the value of the spectrum involved. Obviously, this value information would be of limited form. Specifically, it is the value to a potential operator in the MDS service of a radio channel at 2,150 MHz-2,162 MHz in some particular geographic area.

Despite the limited nature of the information, the almost complete lack of knowledge of the value of any part of the spectrum makes this information important. In this respect the auction procedure, as proposed, includes provisions that correspond somewhat to Webbink's level-3 constraint on spectrum transfer: "the FCC would begin to obtain more information on the value of spectrum to users."[5]

Another benefit would be the return of at least a part of the spectrum's value to the public. This appears to be consistent with the long-standing congressional judgment that the spectrum is a public resource. In this regard, a staff member for the House Subcommittee on Communications, in a response to an expression of concern by a subcommittee member that deregulation would obviate the notion that radio is a public resource, remarked that a spectrum-use fee would answer the member's concern about returning some of the value to the public.[6]

In the longer term, the greatest dividend that might result from assignment of MDS stations by auction could be its role as an introductory experiment in the application of market characteristics to spectrum management. Many still view such measures as incompatible with allocation and use of the spectrum in the public interest. If an auction could be used successfully as an assignment mechanism for the MDS, its extension to other services, as well as the introduction of user charges, may be considered in a more favorable light.

Summary

A public auction has been proposed as a replacement for the comparative hearing to resolve cases of mutually exclusive applications for a radio license in the multipoint distribution service. The open-ended nature of the comparative hearing leads to a lengthy and expensive process and produces a subjective result. The auction, on the other hand, would be relatively straightforward and objective.

The rights to be auctioned and the major elements of an auction procedure were described in a general fashion. A need for legislation to authorize the auction was indicated. Examples were given of other services to which the proposed concept would be applicable. Also benefits that would be in addition to the avoidance of the comparative hearing were discussed.

Douglas Webbink has correctly noted that many of the questions raised are really matters of equity rather than efficiency.[7] This includes reference to such items as economic rents, frivolous bidding, collusion, and lessee losses due to allocation of additional spectrum for the MDS. Classifying these items in this manner serves to place them in proper perspective and emphasizes the necessity to clearly distinguish between issues of equity as opposed to efficiency in the determination of the important concerns in establishing a spectrum auction.

Notes

1. *Options Paper* prepared for the Subcommittee on Communications, House of Representatives, 95th Congress, 1st session, Committee Print 95-13, May 1977, p. 1.

2. *Report of the Federal Communications Commission to the Congress of the United States Re: The Comparative Renewal Process*, September 1976, p. 47.

3. Communications Act of 1934 (47 U.S.C.) Sections 301, 304, and 309(h).

4. G.O. Robinson, "Dissenting Statement in Dockets Nos. 19168, 19169, 19170," *FCC 60 2nd (1976)*, p. 435.

5. D.W. Webbink, "The Value of the Frequency Spectrum Allocated to Specific Uses," *IEEE Transaction on Electromagnetic Compatibility*, 19, no. 3 part 2 (1977): 351.

6. *Broadcasting* 94, no. 11 (1978): 24.

7. D.W. Webbink, "Issues in Spectrum Allocation: Comments," paper presented at Sixth Annual Telecommunications Policy Research Conference, May 12, 1978.

30 A Market Alternative for the Orbit-Spectrum Resource

Charles L. Jackson

This paper presents a formulation of a market system for allocating orbital space to geosynchronous communications satellites. Although based on a simplified model of satellite communications systems, the market system retains enough flexibility to be applicable to the allocation problems of the next decade or two. Ultimately, of course, the definitions of usage right that are put forward here might have to be modified to explicitly include additional aspects of the orbital-spectrum resource. This formulation, however, includes in the definition of the usage right most of the information actually used today in picking an orbital location for a satellite.

The purpose of this paper is to argue, by example, that technically efficient market systems can be designed for satellite systems. *Technically efficient* refers to the fact that the proposed market system does not rule out important technical alternatives, thereby creating potential inefficiency.

The arguments for the necessity and possibility of a market allocation system for international satellite parking spaces are even stronger than the arguments for the use of market allocation in spectrum uses confined to a single country. Both equity and efficiency considerations are involved in the allocation of the orbital-frequency resource. A well-designed market system should be able to separate these two problems. The efficiency equation—what services should be provided by satellite and how should these systems be built—is essentially a question of engineering and economics. The equity question—which nations should get the benefits from use of a limited international resource—is essentially a political problem.

The virtue of separating these two problems should be obvious. The efficiency argument is the usual one for market allocation of a resource. Explicit resource prices simplify the task of the system designer. For example, consider the problem of justifying a communications system that uses small-diameter ground station antennas requiring greater-than-normal satellite separation and using more of the orbital resource. A market would make it possible for the system designer to measure the social costs of increased separation and to compare these costs with the benefits created by using cheaper equipment. If the costs were less than the benefits, then the small-diameter earth station antennas would be justified. Without a market, the system designer proposing small-diameter earth station antennas would face potential criticism for wasting the spectrum. And, without a market, it would be very hard to defend a system design against such criticism. Essentially the defense would consist of identifying any services made more expensive or impossible because of this wider spacing, identifying the costs so imposed on others, and comparing these costs with projected benefits to the proposed system. This process, of course, is just a simulation of the information-transfer and distributed decision-making functions of the market. Clearly, if there were a shadow price for spectrum, as suggested by Levin, the process of calculating net social costs would be simple.[1] Net social costs would equal the shadow price times the quantity of the resource consumed by the wider spacing. Nonmarket solutions also create distributional problems. Even if a proposed system has net costs far below net benefits, the costs and benefits can flow to different individuals or groups. There is no formal mechanism in current spectrum regulation for transferring the benefits of low-cost ground stations to those other system operators who face increased costs. And, of course, under the current regulatory scheme, users get a gift of the frequency they use. A market mechanism solves part of the problem. No one gets the resource free; hence they lose less if their frequency is transferred to another user.

The equity question underlies the need for a spectrum market in this service. How do we determine which nations benefit from access to the orbital resource? One way to insure that all nations get equal shares of the benefits from the potentially scarce orbital resource is to preallocate the orbit equally to all nations. Such a preallocation might be accomplished by dividing the orbit into 180 parking slots and assinging one slot to each nation. The ITU has more than 150 member-nations. Some nations wanted to do this or something similar at the 1971 World Administrative Radio Conference, and a step toward this occurred at the 1977 WARC when the plan was developed for broadcasting satellites.

Clearly such a narrow preallocation would greatly limit technical flexibility and would lead to inefficiencies. Either useful systems could not be built or the technical constraints of preallocation would require more expensive system designs. Perhaps nations desiring to operate many satellites could arrange to use other nation's parking slots. This would be cumbersome and inefficient. But preallocation would assure all nations equal access.

This issue has not been taken lightly by Third World nations which draw an analogy between the orbital spectrum and the high-frequency region of the spectrum. The high-frequency spectrum was licensed on a first-come, first-served basis, and it has ended up predominantly in the service of the developed countries. Third World nations want to prevent a similar capture of the orbital resource by developed nations. A spectrum market would, of course, be able to do just that without requiring narrow technical constraints. Any scarcity rents from the orbital spectrum would flow to the rights holders. Hence the equity problem can be taken care of by the initial distribution of the spectrum rights to be traded in the market Clearly technical efficiency is served if this first step, rights distribution, can be divorced from system design. This paper tries to describe a market system that would enable just that separation.

Overview

The proposed market system uses a simplified model of the orbital resource. The orbit space is considered to be a one-dimensional ring about the earth. Only the spatial aspect of the resource, the separation between satellites, is considered here, for two major reasons. First, with current technology and frequency allocations, for example using the 500-MHz wide bands at 4 GHz and 6 GHz, building and launching a satellite that can work efficiently using the whole band is only slightly more expensive than building and launching a satellite that uses a small part of the allocated band. Second, the provisions that allow orbit users to combine systems into compatible, closely packed arrangements in order to reduce their usage fees give the proper incentives for efficient use of the bandwidth component of the resource. Therefore system rights are defined only by satellite location and by the extent of a band of angular separation inside which other satellites cannot be located and outside of which other satellites are protected against interference.

The orbital ring is divided into segments; for the sake of discussion assume ten 36 segments, as in figure 30-1. This division is necessary to allow the market to reflect variations in the value of different sections of the orbital arc. Each segment is divided into segment shares (SSs), which are the actual rights to be marketed. For the sake of discussion assume that a SS is equal to one degree of arc. Thus there are thirty-six SSs per segment.

A system operator who wanted to operate a satellite in segment 3 requiring two degrees of protection on either side, for a total of four degrees of occupied arc, would go into the market and buy four SSs for segment 3. (A system operator might be a nation, an international organization, or a national corporation. There are international political problems involved in defining the allowable system operators. Some countries might object to allowing private corporations or individuals to be actors in such international activities.) The system operator would then go through a simple registration process for his system similar to today's frequency registration process. He would announce the location of his proposed system, which would necessarily be in an unoccupied portion of segment 3. But it would be a precondition to registration that the operator have sufficient SSs to operate the system.

At this point, I imagine that an economist reading this will say, "Seems nice. So what?" And an engineer will say, "But, but, but . . . what about everything you have left out? What about bandwidth, power limits, modulation format, the difference between uplink and downlink operation, interference to and from terrestrial facilities, polarization discrimination, spot beams on

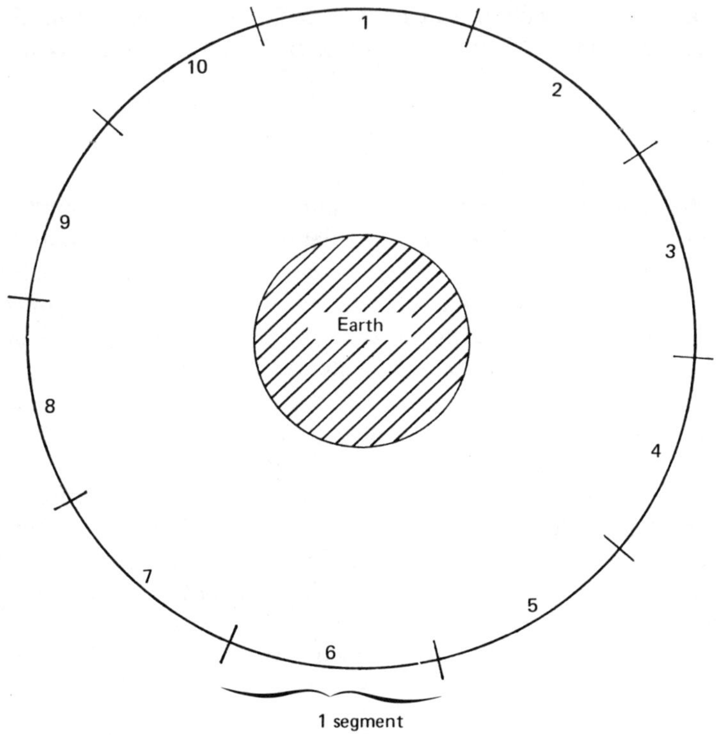

Figure 30-1. An Example of Orbit Segments

downlink? It won't work!" The following sections consider the definition of market rights, user rights, and market operation in greater detail. They try to establish that the market system so defined would be good enough for the essential decisions to be made in assigning current-generation satellites, such as Intelsate IV, to orbit locations.

Market Rights

The market rights, or segment shares, attempt to represent what is scarce and unique in the orbital resource. At the same time they have to be uniform enough to let a market develop easily, even though different portions of the orbital arc have different utility. For example, only a narrow portion of the arc (from 119° to 132° west longitude) is best suited for satellites serving all fifty U.S. states and Puerto Rico. The same general region is also best suited for service to other North American countries. Clearly any system of property rights must be able to reflect scarcity in one portion of the arc but not another.

In this example the arc is arbitrarily divided into ten equal segments. The segments need not be of equal size, nor must there be ten. Segment size is an important system variable. As segments become bigger or the separation requirements of satellites smaller, a market can be expected to work better. Each system would occupy a smaller portion of a segment. Thus the purchase and sale of segment shares would more nearly approximate a competitive market. As segments become bigger, any end effects caused by the transition between segments would be reduced. Unfortunately, as segments become bigger, their subdivisions become less and less uniform commodities. Varying value within a segment is undesirable, since the market described here depends on substantial uniformity within a segment. Seats in a theater or sports arena provide a good analogy. A seat on the fifty-yard line at a football game is usually preferred to one in the end zone, and ticket prices reflect this. But even though a seat on the fifty-yard line might be preferred to one on the forty, both seats are priced the same. Some regions of the orbital arc are like the fifty-yard seats and some are like the end-zone seats. Segments need to be small enough to detect such major differences.

With all this background, let us describe the resource units, SSs, more explicitly and talk about a market.

The orbital arc would be divided into ten segments of 36° each. Each segment would be a stand-alone market region. Usage rights to a segment would be divided into segment shares, or SSs, and ownership of the SSs would be distributed among the various nations. For example, if there were twelve nations below a segment, they might each receive segment shares for one-twelfth of the segment, or three degrees' worth. These usage rights apply to the segment generally, *not* to any specific subsection or subdivision.

The initial allocation of SSs is the fundamental political problem. The only economic constraint is that ownership of SSs be distributed among enough nations to preclude monopoly-created market failure. Clearly if all SSs in a segment were given to a single nation, that nation could charge monopoly prices. Such monopoly prices would probably result in inefficient resource allocation and should be avoided.

Subject to this constraint, SSs could be distributed arbitrarily. If the system works properly, they represent proportionate shares in the scarcity rents associated with the segment. If innovative and intelligent system design keeps this scarcity value low, then the rights are almost valueless and no one would care strongly how they are allocated. If some segments become full and access must be rationed, then the SSs would become valuable. Another approach would be to give ownership of all SSs to the IFRB and require the IFRB to rent out SSs on a nondiscriminatory basis. The political problem would then be one of dividing up the IFRB's surplus income (should any such surplus exist).

Spacecraft Rights

A satellite communications system needs protection against interference from other communication systems, both spaceborne and terrestrial. This proposal concentrates on protection against other spaceborne systems and assumes that nonmarket regulations provide protection against terrestrial systems. The two dominant forms of interference from other satellite systems that must be considered are uplink and downlink interference. Consider first interference on the uplink. Figure 30-2 shows this possible interference between two satellite communication systems. Although a highly directional antenna is used on the uplink from ground station A to the desired satellite destination, some energy travels off the desired path and may interfere with satellite B's reception of uplink signals. Neither the design of satellite B, nor the design of ground station A, nor the separation of the two satellites alone determines the potential for harmful interference. Rather it is a complicated function of all three factors.

This interference is a problem for the operators of both system A and system B. Before designing and building system B the operator must be assured that such interference is unlikely or impossible. Otherwise the investment may become useless. Similarly the operators of system A would like to understand

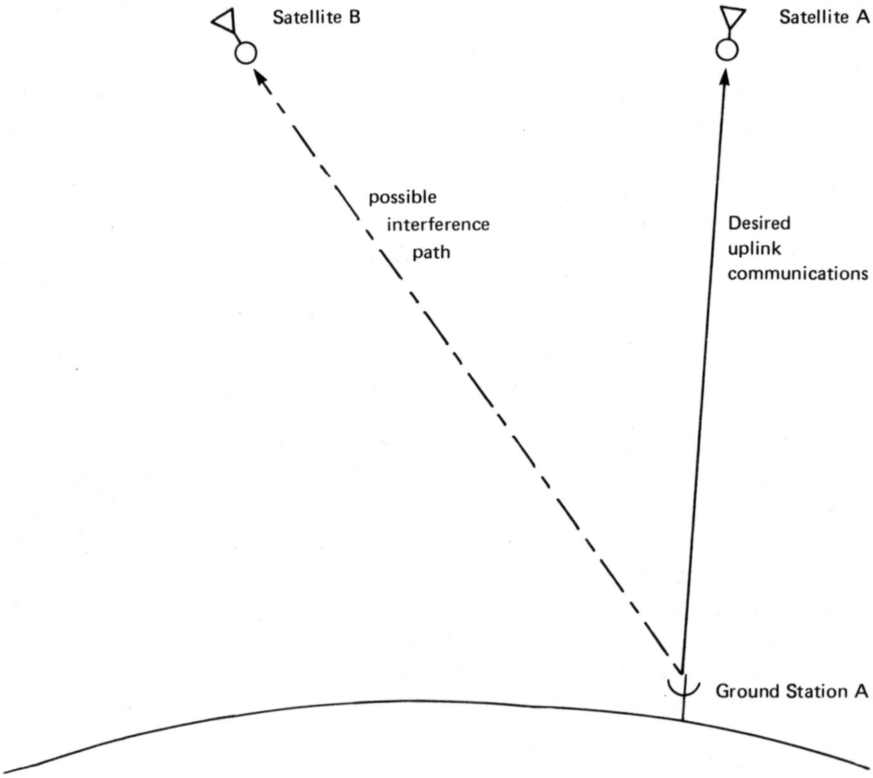

Figure 30-2. Interference between Two Satellite Communication Systems

clearly their obligations to avoid interfering with satellite B. One purpose of a spectrum market, really of any system of spectrum regulation, is to provide this information and such assurances. If the designers of system A knew in advance all the properties of satellite B, they could calculate the separation required between satellites A and B for any specific design choice. Or a standard satellite could be defined, and the system designer for A could be required to avoid using system designs that would interfere with the standard system given the separation.

The spacecraft rights (SCRs) proposed here attempt to clarify such obligations. A system operator would be required to buy rights to the region (bounded by an east and west longitude on the arc) where the satellite would cause harmful interference to a standard satellite. Harmful interference could be defined as (1) E.I.R.P. $> X$ dBW/4KHz in any 4-KHz band, (2) a signal that would cause more than X pW noise power in a reference receiver, or (3) a signal that degrades the error probability of a reference digital system by more than 10^{-8} in absence of other interferers.

These are all measures, in some sense, of equivalent noise power of the system into neighboring systems.

This concept is illustrated in figure 30-3, which shows a satellite centered in a section of arc labeled SCR. It would be the responsibility of the system designer and operator to insure that the SCR was chosen wide enough that a "standard" satellite could be located right at the edge of the SCR and not receive excessive interference. The detailed composition of the SCR is shown in figure 30-4. Using terminology analogous to parking slot terminology, one can say that a spacecraft right is broken down into three regions: a garage, where the satellite would be parked, and front and back lawns that isolate a satellite from its neighbors. One of the major determinants of the width of the lawns is the directivity of the antennas on earth. A weak analogy can be drawn here to residential buildings. Someone who throws a lot of noisy evening parties should buy a house on a big lot or soundproof the house so that party noise does not wake up the neighbors. Similarly someone who is much more sensitive to noise than the average person should buy an isolated home. Noise-sensitive homeowners parallel another aspect of SCRs. Satellite systems need protection on the downlink, from satellite to earth, against interfering signals from nearby satellites. This difference is illustrated in figure 30-5 which shows that, in general, quite different protection may be required on uplink and downlink. Thus the width of the SCR might be different if defined on only an uplink or downlink basis. For now assume that whichever of the two is larger determines the SCR's size. Later we will consider how the market system defined here can avoid this and other seeming deficiencies in the property right.

Once a system user has defined an SCR and found an unused spot in a segment, the user goes to the market and buys sufficient SSs to cover the SCR. At this point the general SSs are in essence converted from floating status, where

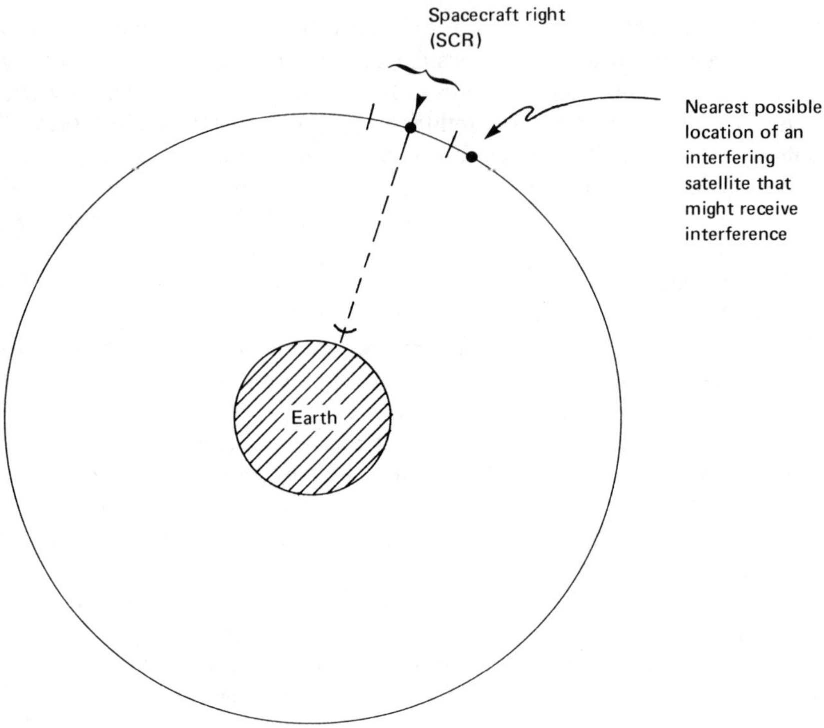

Figure 30-3. Satellite Centered in Spacecraft Right (SCR)

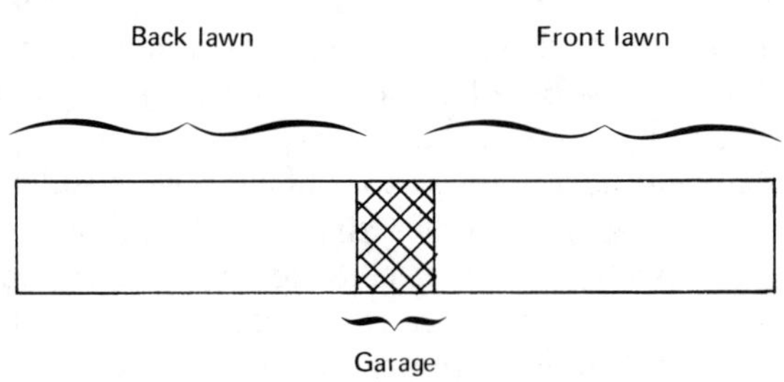

Note: Lawns are wide, perhaps 1°-3°, depending on system design. The garage is relatively narrow, perhaps 0.1° or less. Width is primarily a function of station keeping accuracy of spacecraft and pointing accuracy of ground atennas.

Figure 30-4. Detail of Spacecraft Right (SCR)

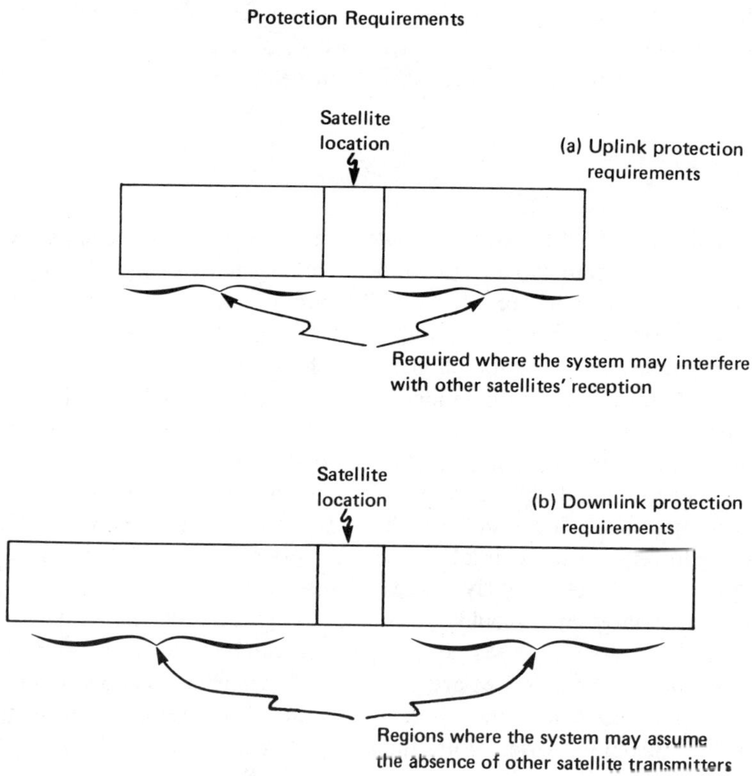

Figure 30-5. Protection Requirements

they represent no specific part of the segment, to an assigned status, where they are associated with a subsection of the segment. The user then registers the SCR with the IFRB. Several possible problems that might lead to market failure occur here. Although access to a segment is rationed by a willingness to pay usage fees for SCRs, there is first-come, first-served allocation within a segment. Since some parts of a segment may be more useful than other parts, this variation might lead to either an inefficient or an inequitable situation. An analogy can be made with a municipal parking lot near a shopping center on a rainy day. Parking places near the stores are more valuable than those farther away, but all parking meters charge the same.

Another problem, far more serious, is the possibility that one user might register SCRs in a speculative attempt to capture the market. The system of floating rents destroys some incentives for such speculation. Another way to destroy such harmful speculation is to have a policy that systems can operate on a secondary basis, without SCRs, as long as they do not interfere with other operating systems and do not claim protection from interference by other systems. Speculation, however, has its positive side for it would also bring benefits to users by protecting them against uncertain prices.

The Market

Careful definition of the rights associated with the purchase or leasing of SSs and the market forum where transactions take place can significantly affect the efficiency of a spectrum market. This section describes a market system that tries to serve efficiency by reducing the power of any single buyer or seller of rights. Reasons for the choice of each alternative aspect of the market are discussed and the choice defended. These are not the only possible choices. Other alternatives could be chosen to tailor the system to meet special needs.

The market operates using only leases of SSs; no outright sale is allowed. There is a central marketplace where all transactions occur and must be registered. The most likely location for such a market would be at the IFRB. Owners of SSs would send to the market their asking price (in units of \$/year/SS) and asking prices would be publicly posted. Those who lease SSs could bring SSs back to the market and would be allowed to re-lease the units. This allows users to sell back surplus SSs. Posting prices publicly solves some problems and creates others. Since prices are public, every seller knows how much prices must be cut to go below the competition. So a traditional open auction is created. The problem arises from the easy identification of the low offerer and the impact the low offer can have, through the system of floating rates, on current lessors and lessees of SSs. The low offerer could become the target of pressure to raise or lower its offer. If a system operator then put in an order for four SSs, the four SSs with the lowest asking price in the desired segment would be assigned to the system operator. The transaction price or usage fee would be the asking price of the fourth or highest-price SS to be transferred. The usage fee of this transaction would then become the usage fee for all SSs in that segment currently assigned to users; SS usage fees would continue to fluctuate with the market.

This point must be emphasized. The lease fee for all SSs is the price determined in the most recent transaction. It is the best measure of the marginal value of the resource. After leasing the SS, the user would then register a SCR that required the four SSs. Different users would be allowed to combine adjacent SCRs into a single SCR. That is, users of two adjacent satellite systems, each requiring six SSs for a total of twelve SSs, might be able to share adjacent lawns and require a total of only nine SSs for the combined SCR. The users could register this joint SCR and re-lease the three surplus SSs. This would allow users to share front lawns and back lawns and to work toward a minimum practical satellite spacing.

Four major points are touched on in this description: lease-only rights, a central market, floating usage fees, and combined SCRs. The discussion that follows looks at each of these points individually and identifies the strengths and weaknesses it adds.

Lease rather than sale of rights appears essential. Sale of rights has two flaws, one economic and one political. If SSs are rented rather than purchased, their current cost (scarcity value) will show up in the annual budgets of the users. Many of the potential users are highly bureaucratic government or quasi-governmental organizations that lack internal mechanisms for enforcing efficient use of capital assets. Leasing rights will force every user to review SS consumption and justify it frequently should rights ever become significantly expensive. Thus leasing, especially as described here with floating prices, is an attempt to make prices less "sticky" and better signals of the scarcity of the resource. Furthermore, future technology and needs are unknown. Irrevocable sale of rights might create obstacles to future efficient use that cancellable or expiring leases would never do. A central market is made necessary by the need for registering SCRs at a central location. In addition, a central market serves other purposes. If necessary, it could help enforce anonymity in the sale and purchase of rights and thus ensure that rights were efficiently leased, independent of the political relations between lessor and lessee.

The price-setting mechanism has flaws, but I think the flaws are minimal and more than offset by the benefit of implicity. Recall the mechanism. Rights holders send their asking prices to the central market where they are publicly posted. Users send in requests for SSs and the lowest-priced SSs are assigned to satisfy the requests. If there were many buyers and sellers, this procedure would yield an efficient market. Unfortunately, in the next few decades there will probably be only ten to twenty buyers in any one segment; hence the many-buyers assumption is not quite correct. Second, and more important, the use of floating rates creates special pressures on the low-priced seller. The higher that price, the better off all the current lessors would be and the worse off all the current lessees would be. Hence both groups would have an incentive to pressure the low bidder to change its bid. However, as long as the market is working, such pressures would tend to be ineffective. For example, a user could buy rights to an SS and immediately turn around and put the SS back on the market. Since the price follows the latest transaction, the user cannot lose more in the transaction than the lease payments while holding the right. But this provides an easy mechanism for users to reduce prices.

If all SSs lease at the same price, then there is no need for explicit identification of who leases to whom. Uniform, floating rates seem particularly useful although they have some drawbacks. They destroy any economic incentive that a rights holder might have to keep SSs off the market, waiting for a better price. The price received two years from now for an SS will be the same whether or not the SS is leased out during the intervening two years. (Actually that is not quite true. Holding an SS off the market could increase only the current price of SSs. This would stimulate design of more efficient technologies and reduce demand for SSs in other ways. Hence two years later the SS would produce less revenue than if it had been on the market during the two years. But such second-order effects would probably be small if rights were widely distributed.)

Uniform rates for all SSs in a segment also destroy any incentive that rights owners might have to associate their rights with more desirable sections of a segment. This is both good and bad. Reservation of SCRs in a segment is on a first-come, first-served basis paralleling current frequency management practice. So to the extent that a segment is a nonuniform commodity, this rights definition does not properly capture all aspects of the "good." But uniform rents will assure all rights holders that there is no need for their rights to be associated with any portion of the segment.

Finally, allowing shared SCRs or the use of combined SCRs is essential. Such sharing will allow the market system to include technical trade-offs not explicitly part of a SCR as defined here. The basic is simple. If two specific satellites can operate closer together than the standard SCR assumes, the system operators should be encouraged to do this and their joint rent reduced to reflect the reduced resource consumption. Such cooperation is important and must be allowed and facilitated by any market system. Consider some examples. SCRs were defined with "lawns" that established a zone of protection from a satellite all the way to the next satellite. Clearly the front lawn of one satellite can be shared with the back lawn of the adjacent satellite, as shown in figure 30-6. (If all satellites had identical separation requirements, we could define a "half lawn" and use nonoverlapping SCRs. But this would sacrifice much technical flexibility.)

One might ask whether such pairwise packing improvements are sure to get to the most efficient or tightest overall packing within a segment. The answer is no. Finding the most efficient packing of satellites requires, at least in principle, simultaneous consideration of the properties of all satellites in a segment. The market system described here does not preclude and, in fact, encourages such segmentwide planning. If all users in a segment get together and plan for the tightest possible packing, the packing requiring the fewest SSs, they all receive a usage fee reduction from the fact that they, as a group, require fewer SSs. Given the conservative, overprotective nature of the design of the SCRs, there should be a way to distribute this overall usage-fee reduction so that every user is better off than in any other configuration.

Not only do usage fees fall as fewer SSs are required, but pressure is put on the SS market to drive the price down. Hence tighter packing brings a twofold benefit reducing the SS requirement and reducing the price of the remaining SSs. The problem of the proper division of the rent reduction made possible by such sharing is in general intractable, being a problem of bilateral monopoly

(a) Joint SCRs: A shared configuration giving needed protection, 25 percent smaller than unshared SCR

(b) Separate SCRs: An unshared configuration giving overprotection

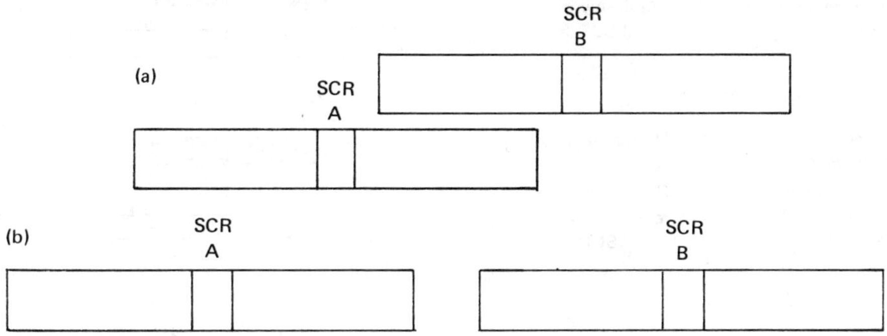

Figure 30-6. Sharing of Spacecraft Rights (SCRs) by Standard Satellites

in the care of two sharing organizations. However, I suspect that automatic procedures at the central market for dividing equally the usage fee reduction made possible when two users combine SCRs would serve efficiency in most cases by speeding up decision making and removing considerations of pride and strength from bargaining. In essence this is a political rather than an economic design choice. The political purpose is to remove an element of economic undeterminism that would lead to bargaining and conflict. In the economists' terminology, this procedure transforms the bargaining set from a continuum into a two-point set containing the Nash solution to the original bargaining problem and no agreement on sharing. Clearly this new bargaining problem has a single dominant solution—the efficient sharing arrangement. (There is a slight flaw in this argument. If a system can be modified, at some expense, to allow closer orbital spacing, then the market should encourage it, even if the cost of system improvement eats up more than half of the joint-usage fee reduction. That alternative is not encouraged by the automatic splitting method.)

Specialized satellite systems give more interesting examples of sharing. Consider two one-way satellite systems, system A, a data collection system and the other, system B, a data dissemination system, whose uses of the orbital resource are complementary. A data dissemination satellite serving many distribution points from a single central source would probably use a high-gain, narrow-beamwidth uplink transmitting antenna but smaller, less directional downlink receiving antennas. This system is quite asymmetric in its consumption of the orbital resource. System B, serving the dual function, data collection from many small-diameter antennas to a single point, would have a complementary pattern of use. Thus, as illustrated in figure 30-7, the satellites of system A and B could fit even more closely together than the direct overlapping of their

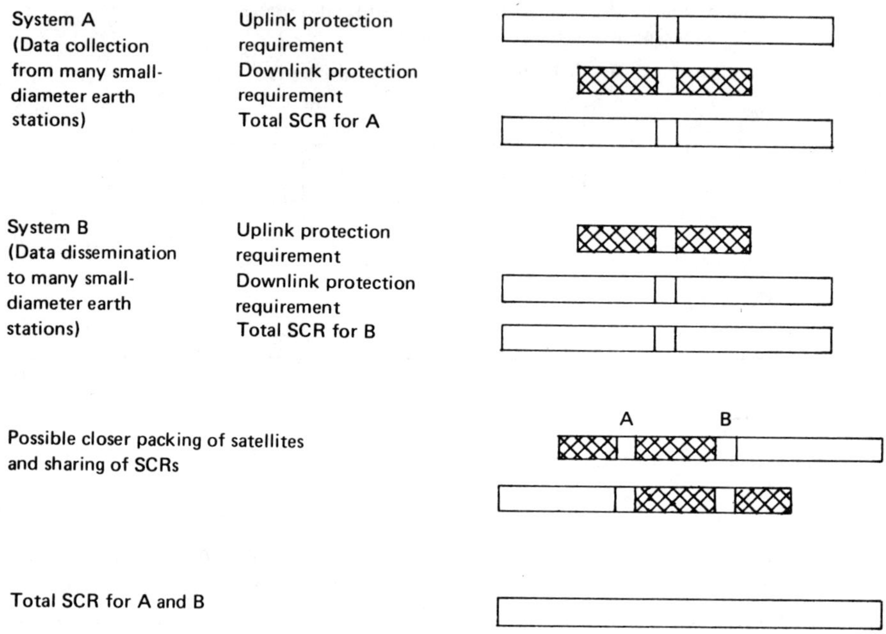

Figure 30-7. Sharing of Spacecraft Rights for Atypical Satellites

front and back lawns would indicate. The closer they fit together, the more they reduce their joint-usage fee. Both system operators have every incentive to find such opportunities for tighter packing and to exploit them.

This kind of analysis can be extended to almost any technological alternative. Use of spot beams on downlink or polarization discrimination will also create opportunities for closer packing of satellites and hence usage fee reduction.

This discussion tries to make two points clear. First, sharing SCRs creates an opportunity for the careful use of technology to pack satellites closely together. Sharing makes this possible even if the specific technological alternatives are not reflected in the SCR definition. Second, these examples should make clear to an engineer how a market in orbital rights can aid the system designer by making explicit the value of extra engineering or components that let satellites pack more closely together in orbit.

A mechanism is needed for protection of satellite systems from terrestrial systems and vice versa. The best mechanism appears to be administrative rules just as we have today. In the market context these rules may be compared to zoning regulations in land use.

Discussion: Defects, Advantages, Open Questions, and Research Problems

A market system might cost more to run than the present administrative system and might provide less technical flexibility. But the market described here is simple, so operating the market would add very little to the cost of running the IFRB and dealing with the market would add little to the costs of satellite system operators. Clearly this calculation excludes the "cost" of any use charges, which would be a transfer payment and not a true social cost.

Diminished technical flexibility seems to be the greatest possible harm this market can bring. I have tried to establish that joint-use agreements can lead to more efficient outcomes than the narrow definitions of SCRs would suggest. I would thus argue that technical flexibility is not especially limited by this proposal.

Of course, one must ask, Diminished technical flexibility with respect to what alternative? If a segment of the orbital arc becomes full, then there will necessarily be some allocation of scarcity rents. In fact, the international political process may try to anticipate such scarcity rents and impose some form of rigid preallocation on the orbital resource before scarcity is present. The proposed market system looks quite flexible when compared with preallocation or any system of preengineering and assignment of the orbital resource in advance of actual use.

But there are also advantages. This proposed market separates the benefits of the ownership of potentially scarce resources from the use of the resources. This cleaves technical and economic problems of system design from the political problem of resource distribution and thus leads to the major benefit of this market proposal. It serves dynamic efficiency by removing any need for preallocation, preserving flexibility, and allowing for the easy introduction of new designs and new technology. This proposal would also serve static efficiency by providing system designers with explicit measures of the cost of additional orbital resource or of the value of using it more efficiently. The proposed market might actually simplify some of the operations in international spectrum managment but these benefits, if any, would be small relative to the other benefits.

A basic question left undiscussed is the assignment of rights (SSs, or shares in IFRB net income) to countries. Many ways of doing this exist, and all could be called fair. Just consider: equal shares to each nation; equal per capita shares; shares proportional to a nation's current telecommunications usage; shares representing a vesting of current orbit use; shares in proportion to the length of orbital arc visible from any point in the nation; shares in proportion to the fraction of the equator running through a country. Compensatory policies are also possible. Shares could be given out in inverse proportion to high-frequency circuit holdings or in inverse proportion to per capita gross national product. Some weighted combination of all these criteria could be chosen. The choice is arbitrary as long as shares in any segment are widely held and monopoly or oligopoly in the sale of SSs cannot exist.

Floating rents or lease prices create another problem. Some users may be unwilling to invest in satellite systems because of uncertainty about costs. Perhaps rights vendors, as a group, could agree on a contract for long-time sales. Or perhaps someone would offer insurance against price fluctuation. Note that such insurance is really a form of futures market. Futures markets bring to mind a whole host of questions about the allowable resale of rights, reservation of rights for future use, and in general the nature of the resource market.

I would argue that while such extensions of the market would serve economic purposes, they would work against the political viability of the market. Consider one example. Suppose an entrepreneur were to buy all the SSs in a segment early, before scarcity became a problem, at a fixed, not a floating, price. He would then be a monopolist and would be able to extract monopoly rents from users. An event such as this would probably damage the image of the market system among rights holders and users alike. With floating rates the speculator's problem is quite different. Most, if not all, of the monopoly rents would be taken away by a rise in the floating price. This price rise would also apply to any user who claimed an extra-large SCR in order to reserve orbital space for future bargaining. Similarly, the policy of allowing systems to operate without SCRs as long as they caused no interference to existing systems would weaken or destroy any speculator's hold on the market.

The use of segments may be unnecessary. It may be possible to design a pricing mechanism that avoids the sharp segment boundaries and yet reflects variation of value with position on the orbital arc.

Conclusions

A market system in the geostationary orbit is technically feasible, at least in the short run for the kinds of satellites now being built. This market would serve two primary needs. It would separate the political problems of allocating the right to use a scarce resource from the engineering and economic problems of using the resource. And it would give system designers an explicit yardstick, the price of the orbital resource, for trading off the use of this resource with system costs elsewhere.

Notes

1. H. Levin, *The Invisible Resource*, (Washington, D.C.: Resource for the Future, 1971.), pp. 132, 133; for a criticism of the use of shadow prices in spectrum management see J.H. Alleman, "The Shadow Price of Electromagnetic Spectrum: A Theoretical Analysis," Office of Telecommunications OTR 74-42, July 1974.

Part IV
The Economics of Regulation and the Economics of Pricing Telecommunications Services

Introduction

Herbert S. Dordick

In recent years two policy issues have emerged that may radically alter the context in which all regulated activities, telecommunications included, will be carried out in the United States. One focuses on the cost of regulation. While regulation is often considered necessary for the public interest, there is an associated cost, not simply in terms of increased development times and costs and possibly longer periods for market introduction, but in the actual price of the service to the subscriber. The other issue is equally significant to the subscriber's pocketbook and deals with the pricing of publicly owned or regulated multicommodity services such as those delivered by the telephone companies. This is not a new problem; regulators have been concerned with this since the introduction of railway rate regulation in the 1880s. Both are issues that concern other nations as well as the United States.

Using data from 1971, 1974, and 1976, Braunstein finds that cable television systems under state regulations charged subscription rates greater than those of comparable states without state regulation. But growth rates and rates of penetration appeared not to be affected by regulation at the state level.

Panzar compares flat-rate tariffs with usage-sensitive pricing and finds that in the absence of metering costs, it is always possible to design a tariff structure that makes all consumers better off than they were under the original flat-rate tariff and increases the profits to the firm. While the paper presents a possibility theorem, the author strongly recommends that barriers to implementation such as metering costs be examined. The critical point is that these costs are incurred only by those customers who elect to avail themselves of the new tariff offering. Panzar suggests several extensions to the research with emphasis on empirical observations of practices and preferences.

Mitchell surveys pricing policies in five European telephone systems—Norway, Sweden, Switzerland, the United Kingdom, and West Germany. After reviewing the rate structures in these countries along four dimensions, Mitchell discusses the development of an applied model of peak-load pricing for telephone calls.

31 Economic Effects of State Regulation of Cable Television

Yale M. Braunstein

Cable television systems often competition from over-the-air television broadcasters while being regulated by the federal, state, and local governmental jurisdictions in which they operate. The Federal Communications Commission regulates content, technical standards, the importation of distant signals, and some of the terms of the local franchise agreements. In most situations the locality negotiates a franchise agreement with the operator that may delineate specific terms covering services to be offered, prices, complaint procedures, and the fee to be paid to the local government. In eleven states, public utilities commissions or separate agencies have authority over the prices that can be charged, the size of franchise areas, and other terms of operation. This paper looks at the economic effects of regulation by the states on the prices charged by the cable systems for their services, on the "penetration" of the systems, and on the growth of the systems. Although the presence of state regulation does not have a statistically significant effect on penetration or on growth, it does have a positive and significant effect on the monthly subscription fees charged by the cable systems.

The Model

Because we were interested in the effects of state regulation on the monthly subscription fees charged by cable systems and on the penetration of the systems and because it is more likely that regulation affects costs and supply rather than demand, we estimated reduced-form equations rather than the demand equations used in previous studies. The reduced-form equations indicate the overall effects of the exogenous variables (including a dummy variable representing the presence of absence of state regulation) on the endogenous variables (monthly fee and penetration). This approach has two major benefits: the

The research reported in this paper was completed for the project on state regulation of cable television conducted by the Harvard University Program on Information Resources Policy and by Kalba Bowen Associates, under a grant from the National Science Foundation. Konrad K. Kalba and Larry S. Levine participated in the research design and analysis, and Anne E. Birinyi and Kenneth Griggs served as research assistants. The author is solely responsible for any remaining errors.

estimated equations are free from simultaneous-equations bias, and the failure to completely specify the structural model does not affect the estimates of the individual reduced-form equations.

The conceptual structure model has a demand and a supply equation. In each the number of subscribers S is a function of the monthly subscription fee P, the number of homes passed H, and several exogenous variables. Therefore we have

$$\text{(demand)} \qquad S_i = f(P_i, H_i, NS_i, Y_i, A_i, B_i, M_i) \qquad (31.1)$$

$$\text{(supply)} \qquad S_i = g(H_i, P_i, A_i, B_i, R_i, M_i') \qquad (31.2)$$

where

S_i = number of subscribers of system i

P_i = monthly subscriber fee (price)

H_i = number of homes passed by the cable

A_i = age of system i

B_i = number of channels provided on the cable

NS_i = number of television broadcast signals that can be received over the air in the franchise area of system i

R_i = the presence of state regulation

M_i, M_i' = variables reflecting management ability, promotional activity

Several assumptions are implicit in this model. First, we assume (realistically) that the franchise area is fixed and determine outside the model. Second, we have not specified how cable extension decisions are made. Although H_i is clearly under management control, we will leave the exact relationship between H_i and the other variables unspecified. What we will do is to transform the equations by dividing both sides by H_i, thereby converting the left-hand sides to S_i/H_i, the "penetration" of each system.[1] Finally, we shall assume a linear, additive relationship between penetration and the other variables and that the variables reflecting management differences (M_i and M_i') are unimportant.[2]

As a result of these assumptions, the demand and supply equations become

$$Q_i = S_i/H_i = a_{11}P_i + a_{12}NS_i + a_{13}Y_i + a_{14}A_i + a_{15}B_i + a_{16} \qquad (31.3)$$

$$Q_i = S_i/H_i = a_{21}P_i + a_{22} + a_{23}R_i + a_{24}A_i + a_{25}B_i \qquad (31.4)$$

The reduced-form equations for this specification are (we drop the subscript i)

$$Q = \frac{a_{21}a_{12}}{(a_{21} - a_{11})} NS + \frac{a_{21}a_{14} - a_{11}a_{24}}{(a_{21} - a_{11})} A + \frac{a_{21}a_{15} - a_{11}a_{25}}{(a_{21} - a_{11})} B - \frac{a_{11}a_{23}}{(a_{21} - a_{11})} R + \frac{a_{21}a_{13}}{(a_{21} - a_{11})} Y + \frac{a_{21}a_{16} - a_{11}a_{22}}{(a_{21} - a_{11})} \tag{31.5}$$

$$P = \frac{a_{12}}{(a_{21} - a_{11})} NS + \frac{a_{13}}{(a_{21} - a_{11})} Y + \frac{a_{14} - a_{24}}{(a_{21} - a_{11})} A + \frac{a_{15} - a_{25}}{(a_{21} - a_{11})} B - \frac{a_{23}}{(a_{21} - a_{11})} R + \frac{a_{16} - a_{22}}{(a_{21} - a_{11})} \tag{31.6}$$

(The measures used for each of the variables and the data sources are described in the appendix to this chapter.)

Estimation Procedures and Results

The reduced form equations were estimated by an ordinary least-squares technique using data from 1971, 1974, and 1976. In addition, because of the use of a number of dummy variables, an analysis of variance and covariance was used so that interaction effects of age, number of channels, and regulation could be detected if they were present. Furthermore, the reduced-form quantity equation was also estimated using a logit specification; namely log $[Q/(1 - Q)]$ was tried as the dependent variable.

In both 1974 and 1976 there were eleven states in which cable television was regulated at the state level (see the list in appendix 31A). The formulation using a dummy variable to indicate the presence of this regulation caused concern about whether this variable could be acting as a proxy for other variables or whether there were differences between the regulated and unregulated systems that predated state regulatory control. To see whether these concerns could be reduced (if not eliminated), a pre/post design was employed. We estimated the same equations for a pre-state regulation year and again examined the coefficient of the state regulation dummy variable. Now its value would not indicate the effects of regulation but would reflect any differences between the systems prior to regulation. The earliest year for which we had comparable data was 1971. (In that year four states had regulation, but this affected only twenty-four of the systems in our sample.)

Effects on the Monthly Subscription Fee

The regression results for 1971, 1974, and 1976 are presented in tables 31-1 and 31-2. Although the overall explanatory power is low, in several instances we have been able to isolate the effects of the explanatory variables. Looking at the price equations (table 31-1) first, one finds that in 1974 the presence of state regulation (d_4) had a positive and significant effect on the monthly subscriber fees. (This amounts to approximately 20 to 22 cents per month.) In 1976 this effect is still positive but is no longer statistically significant. However, the interaction between system size (d_3) and the presence of regulation does have a significant effect. These differences are illustrated in figure 31-1. By comparison, in the control year 1971 neither the regulation variable itself nor the interaction term is significant.

In all three years both per capita income Y and the number of "significantly viewed" over-the-air signals NS have statistically significant coefficients. As one might expect, the effect of income on price is positive, and the effect of the number of over-the-air signals is negative. The regression results also indicate

Table 31-1
Regression Results, Price Equations

	1971	*1971*	*1974*	*1974*	*1976*	*1976*
Dependent variable	*P*	*P*	*P*	*P*	*P*	*P*
Constant	5.076	5.052	5.470	4.985	6.528	5.415
Y (per capita)	–	0.129 (2.85)**	–	0.218 (5.31)**	–	0.228 (7.13)**
N. Sig. viewed	–	−0.076 (3.17)**	–	−0.086 (3.91)**	–	−0.062 (2.82)**
Age (d_1)	0.204	0.217	−0.0458	−0.0186	−0.259	−0.244
Age (d_2)	0.260	0.244	−0.0377	−0.0331	−0.301	−0.264
B-#ch. (d_3)	−0.006 (0.379)	0.0049 (0.276)	0.165 (1.46)	0.157 (1.39)	0.302 (2.68)**	0.257 (2.41)*
Reg'l (d_4)	−0.032 (0.744)	0.0037 (1.25)	0.202 (2.12)*	0.221 (2.61)**	0.053 (0.814)	0.0255 (0.833)
($F\, d_1\, d_2$)	(1.779)	(1.593)	(0.825)	(1.25)	(1.151)	(1.132)
Interactions	–	–	–	–	B-Reg'1 (2.21)*	B-Reg'1 (2.63)**
R^2	.019	.058	023	.084	.029	.107
N	428	428	653	653	653	653

Note: t-statistics are shown below each coefficient in parentheses except for d_1 and d_2; the F-statistic for the test of both coefficients = 0 is reported for d_1 and d_2.

$*p < 0.05$
$**p < 0.01$

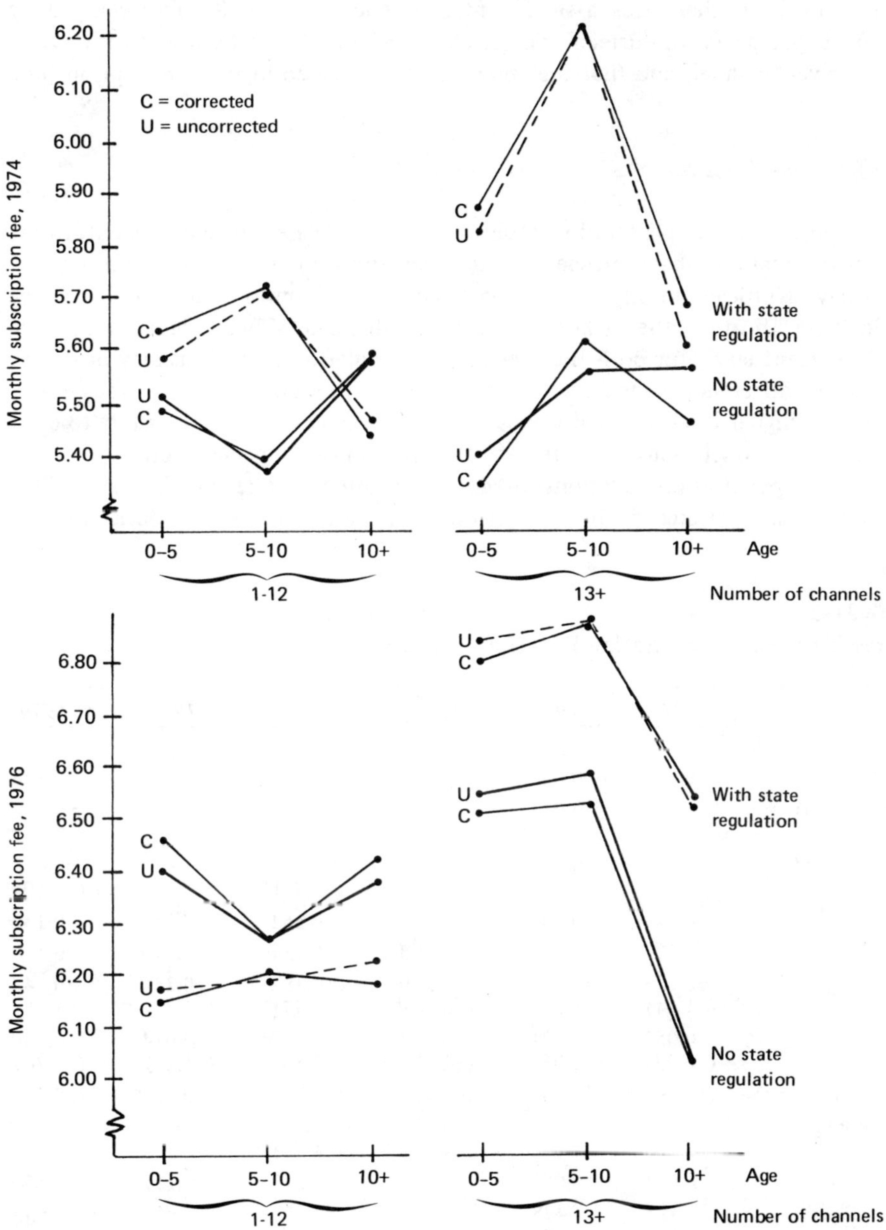

Note: Corrected values are residuals of regression on per capital income and number of significantly viewed over-the-air television signals (plus the overall mean).

Figure 31-1. Impact of Regulation on System Size

that in 1976 there was a significant difference between the fees charged by larger systems (with thirteen or more channels) and those of systems with twelve or fewer channels and that the larger systems charged higher subscription fees.

Effects on Penetration

The results of the penetration equations (table 3-2) are not so consistent. In the linear equations the coefficient of the regulation dummy variable is not statistically significant in any of the years with one exception. However, with the logit specification, the *t*-statistic is just on either side of being significant (at the 95 percent level) for both the pre- and postregulation years. This may be due to existing differences or to other factors. Both the borderline nature of the *t*-statistic and higher levels of significance of other variables make us hestitate to draw any strong conclusions about the effect of regulation on penetration.

The penetration equations provide other interesting results. The age of the system has a strong, positive effect on the penetration. This is consistent with

Table 31-2
Regression Results, Penetration Equations (Linear)

	1971	*1971*	*1974*	*1974*	*1976*	*1976*
Dependent variable	Q	Q	Q	Q	Q	Q
Constant	0.406	0.681	0.535	0.826	0.521	0.828
Y (per capita)	–	−0.013 (.999)	–	−0.022 (2.25)**	–	−0.019 (2.03)*
N. Sig. viewed	–	−0.008 (1.11)	–	−0.014 (2.75)**	–	−0.017 (3.19)**
Age (d_1)	0.101	0.108	0.164	0.161	0.133	0.130
Age (d_2)	0.287	0.280	0.298	0.281	0.290	0.198
B-#ch. (d_3)	−0.123 (1.64)	−0.060 (1.34)	−0.210 (6.34)**	−0.193 (5.77)**	−0.220 (6.74)**	−0.146 (6.16)**
Reg'l (d_4)	0.0525 (0.937)	0.071 (1.06)	0.0228 (1.87)	0.0405 (2.54)*	0.0144 (1.11)	0.099 (1.76)
($F\,d_1\,d_2$)	(17.96)**	(17.24)**	(46.4)**	(42.9)**	(30.1)**	(28.5)**
Interactions	–	–	A-B (4.44)**	A-B (4.71)**		
R^2	0.091	0.093	0.189	0.212	0.152	0.159
N	428	428	653	653	653	653

Note: *t*-statistics are shown below each coefficient in parentheses except for d_1 and d_2; *F*-statistic for the test of both coefficients = 0 is reported for d_1 and d_2.

$^{*}p < 0.05$
$^{**}p < 0.01$

the theoretical work on the diffusion of innovation over time. There is an inverse relationship between penetration and the number of significantly viewed over-the-air signals. In the later years there is also an inverse relationship between the number of channels and penetration. This may be due to the positive relationship between the number of channels and the monthly fee or to the fact that the larger systems are generally the newer ones. The interaction between system age and number of channels is statistically significant in two of the cases.

One of the more intriguing findings is that the level of per capita income has a negative coefficient in the penetration equations, even though this is not statistically significant in most cases. Previous studies that directly estimated the demand for cable television generally found the income elasticity to be positive and significant [1, pp. 57-63]. An exception is the Commanor-Mitchell study [2] which finds an income elasticity not significantly different from zero [2, p. 160]. There is similarly a possible contradiction with the results of our price equation. From equations 31.5 and 31.6 respectively we see that the coefficients of the per capita income term in the penetration and price reduced-form equations are $a_{21}a_{13}/(a_{21} - a_{11})$ and $a_{13}/(a_{21} - a_{11})$. If the second of these is positive, then the first can be negative or zero only if a_{21} is negative or zero. Since $a_{21}P/Q$ is the price elasticity of supply (for the linear specification), it is unlikely that a_{21} is negative. It may, however, be very close to zero, causing our estimate to be insignificant.

Effects on Growth of Penetration

These results are of the comparative statics and cross-sectional variety. We also examined the cable systems to see whether state regulation had any effect on the growth of penetration.[3] We examined changes between 1971 and 1974 and used two different measures of growth. These were the change in penetration between 1971 and 1974 (G_1) and the logarithm of the 1974 penetration divided by the 1971 penetration (G_2). (See table 31-3.) Thus the two dependent variables were

$$G_1 = Q_{74} - Q_{71}$$

$$G_2 = \ln (Q_{74}/Q_{71}) = \ln Q_{74} - \ln Q_{71}$$

The regression results are shown in table 31-4. The presence of state regulation has no significant effect on the growth measures, and the overall explanatory power of the equations is very low. The independent variables that do have significant effects are the number of channels (in two cases) and the age of the

Table 31-3
Growth Equations

	1971-1974	*1971-1974*	*1971-1974*	*1971-1974*
Dependent variable	$(Q_{74} - Q_{71})$	$(Q_{74} - Q_{71})$	$\ln(Q_{74}/Q_{71})$	$\ln(Q_{74}/Q_{71})$
Constant	0.165	0.153	0.578	0.273
Y (per capita)	–	−0.001 (0.089)	–	0.001 (0.032)
N. Sig. viewed	–	−0.006 (1.08)	–	−0.005 (0.352)
Age (d_1)	0.0044	0.0016	−0.264	−0.262
Age (d_2)	−0.057	−0.0596	−0.415	−0.418
B-#ch. (d_3)	−0.089 (3.43)**	−0.083 (3.34)**	0.0392 (0.00)	0.042 (0.03)
Reg'l (d_4)	−0.012 (0.212)	−0.011 (0.367)	−0.034 (0.122)	−0.031 (0.071)
($F\ d_1, d_2$)	(3.428)*	(3.658)*	(13.52)**	(13.63)**
Significant Interactions (*F*)	–	–	–	–
R^2	0.060	0.063	0.078	0.079
N	428	428	428	428

Note: t-statistics are shown below each coefficient in parentheses except for d_1 and d_2; *F*-statistics for the test of both coefficients = 0 is reported for d_1 and d_2.
*$p < 0.05$
**$p < 0.01$

system. The signs indicate that in general the smaller, younger systems tend to grow faster than those with more channels or those that are older.

Conclusion

The reduced-form estimation procedure enables us to draw conclusions both about the effects of state regulation of cable television systems and about the supply and demand for cable television in general. We have shown that state regulation has a positive, statistically significant effect on the monthly subscription fee. The effect on penetration, on the other hand, is not statistically significant. In fact, there is little change in the coefficient of the regulation variable in the penetration equations from the preregulation year to the years with regulation. This leads to the conclusion that regulation causes a price increase that represents a transfer from consumers to the system operators; there is no social welfare loss because there is no reduction in quantity.

Table 31-4
Regression Results, Penetration Equations (Logit)

	1971	*1971*	*1974*	*1976*	*1976*	*1976*
Dependent variable	$\ln\left(\frac{Q}{1-Q}\right)$	$\ln\left(\frac{Q}{1-Q}\right)$	$\ln\left(\frac{Q}{1-Q}\right)$	$\ln\left(\frac{Q}{1-Q}\right)$	$\ln\left(\frac{Q}{1-Q}\right)$	$\ln\left(\frac{Q}{1-Q}\right)$
Constant	−0.514	1.072	0.256	2.049	0.168	2.136
Y (per capita)	–	−0.09 (1.03)	–	−0.086 (1.15)	–	−0.116 (1.171)
N. Sig. viewed	–	−0.018 (.389)	–	−0.128 (3.13)**	–	−0.128 (3.42)**
Age (d_1)	0.616	0.556	1.018	1.004	0.746	0.720
Age (d_2)	1.760	1.731	1.878	1.768	1.733	1.622
B-#ch. (d_3)	−0.547 (1.37)	−0.569 (1.14)	−1.30 (4.85)*	−1.188 (4.41)**	−1.278 (4.89)**	−1.156 (4.38)**
Reg'l (d_4)	0.280 (1.98)*	0.287 (1.99)*	0.198 (1.55)	0.326 (2.20)*	0.156 (1.18)	0.292 (1.82)
($F\ d_1\ d_2$)	(17.4)**	(16.9)**	(25.1)**	(23.0)**	(18.8)**	(17.9)**
Interactions	–	–	A-B (1.72)	A-B (1.75)	–	–
R^2	0.108	0.107	0.115	0.097	0.088	0.1086
N	428	428	653	653	653	653

Note: *t*-statistics are shown below each coefficient in parentheses except for d_1 and d_2; *F*-statistic for the test of both coefficients = 0 is reported for d_1 and d_2.
$^*p < 0.05$
$^{**}p < 0.01$

These findings imply an inelastic demand for cable television. From the linear demand equation (31.3), we see that the elasticity of demand is

$$\epsilon_d = -\left(\frac{\partial Q}{\partial P}\right)\left(\frac{P}{Q}\right) = -a_{11}P/Q$$

Although a_{11} is not estimated directly, it can in principle be calculated from the ratio of the coefficients of the regulation variable in the reduced-form quantity and price equations (31.5 and 31.6 respectively). Since the estimate in the quantity equation is not significantly different from zero, the estimate of the demand elasticity is approximately zero. As with the income elasticity, this finding is consistent with that of Commanor-Mitchell [2] but not with the other studies reported in Besen et al. [1].

Still to be resolved is the explanation for the differential in the monthly subscription fee between the systems under state regulation and those in states without regulation. Among the possible hypotheses the following seem to be the most plausible and promising for further research.

1. In many states the regulatory commissions are financed by fees levied on the systems they regulate. If the fees are passed on to the subscribers, higher prices may result.
2. Cable systems in states with regulation incur additional legal and administrative costs, and these costs are borne by the subscribers.
3. State regulation is unnecessary or ineffective and possibly reduces the bargaining power of the localities granting the franchises. The higher monthly subscription fees result in higher profits for the cable systems.

Detailed cost and profit data are needed if one is to discriminate among these explanations of the observed price differential.

Notes

1. Penetration has been used as the dependent variable in many of the previous studies; see Beser et al. [1]. Unfortunately actual data on the size of the franchise areas are not available. If they were, one could also test to see whether regulation influences cable extension policies.

2. If M and M' are predetermined (either because they are exogenous or lagged endogenous variables) and uncorrelated with the included exogenous variables, the estimated coefficients of the included variables will be unbiased [3]. If M and M' are endogenous, their omission may cause the estimated coefficients to be biased and inconsistent.

3. Here the problem of differences in the number of homes passed may be important, especially since we do not control for changes in the physical plant of each system.

References

[1] Besen, S.M., et al. "Economic Policy Research on Cable Television: Assessing the Costs and Benefits of Cable Deregulation." In *Deregulation of Cable Television*, ed. P.W. MacAvoy (Washington, D.C.: American Enterprise Institute for Public Policy Research, 1977).

[2] Comanor, W.S., and Mitchell, B.M. "Cable Television and the Impact of Regulation." *Bell Journal of Economics* 2 (1971):154-212.

[3] Theil, H. "Specification Errors and the Estimation of Economic Relationships." *Review of the International Statistical Institute* 25 (1957):41-51.

Appendix 31A Definition of the Variables and Sources of the Data

P = monthly fee for first set subscription [4]

S = number of first set subscribers computed as the number of individual-dwelling subscribers plus the total revenues from bulk-billed subscribers divided by the monthly fee for individual-dwelling subscribers [4]

H = number of housing units passed by the plant [4]

Q = penetration = S/H

Y = 1974 per capita income [5]

NS = number of "significantly viewed" signals in the county in 1972 [6] [1]

d_1 = 1 if the age of the cable system was greater than five and less than ten years on June 1974 (otherwise $d_1 = 0$); calculated from start date in [4] [2]

d_2 = 1 if the age is greater than ten years (otherwise $d_2 - 0$)

d_3 = 1 if the system's channel capacity was thirteen or more channels (otherwise $d_3 = 0$); from [4]

d_4 = 1 if system operated in a state which regulated cable systems; namely, Alaska, Connecticut, Delaware, Hawaii, Massachusetts, Minnesota, New York, New Jersey, Nevada, Rhode Island, and Vermont (otherwise $d_4 = 0$)

G_1 = change in penetration = $Q_{74} - Q_{71}$

G_2 = logarithm of growth rate of penetration = $\ln(Q_{74}/Q_{71}) = \ln - Q_{74} - \ln Q_{71}$

The 1974 sample consisted of (1) all 231 cable systems with 500 or more subscribers in 1974 in those states with state regulation and (2) a random sample of 422 chosen from those systems reporting the necessary data for 1974 in the states without state regulation. The 428 systems that also provided complete data for 1971 composed the 1971 sample. The 1976 sample was the same as the 1974 sample (although some of the data changed over time).

Notes

1. See Appendix B of *Reconsideration of Cable Television Report and Order in Docket 18397 et al.*, 36 FCC 2d 326 at 378-463 (1972). A significantly viewed signal is defined by the FCC on countywide basis in the following way:

A significantly viewed signal is one that is viewed in other than cable television households as follows: (1) for a full or partial network station, a share of viewing hours of at least 3 percent (total week hours) and a net weekly circulation of at least 25 percent; (2) for an independent station, a share of viewing hours of at least 2 percent (total week hours) and a net weekly circulation of at least 5 percent. See § 76.54.

As used in this paragraph, "share of viewing hours" means the total hours that noncable television households viewed the subject station during the week, expressed as a percentage of the total hours these households viewed all stations during the period. "Net weekly circulation" means the number of noncable television households that viewed the station for five minutes or more during the entire week, expressed as a percentage of the total noncable television households in the survey area.

2. For 1971 sample, 1974 age was used; for 1976 sample, 1976 age was used and d_1 and d_2 were recomputed.

Data Sources

[4] Television Digest. *Television Factbook* Services Volume, Edition #44 (Washington, D.C.: Television Digest, 1974-1975).

[5] U.S. Department of Commerce, Bureau of Census, *Current Population Reports, Series P-25*, May 1977.

[6] U.S. Federal Communications Commission, *Reconsideration of Cable Television Report and Order in Docket 18397 et al.*, 31 FCC 2d 325 (1972).

32 The Pareto Domination of Usage-Insensitive Pricing

John C. Panzar

Usage-sensitive pricing (USP) has become increasingly important in the regulatory arena because of the current reevaluation of how best to obtain the revenues needed to provide local telephone services. Economists of (nearly) all persuasions have argued that it is inefficient to charge a zero price for a service that has a positive cost of provision. Clearly a movement to a positive per call charge would increase aggregate economic efficiency. Yet nearly all proposals for a move to USP have met stiff consumer resistance. The reluctance seems to persist even when customers face the prospect of a USP plan that would, on average, result in a lower monthly bill. There are two obvious sources for this phenomenon. First, customers may be uncertain about the number of calls they will wish to make and, being risk averse, prefer the flexibility to costlessly respond to random events. Second (and this will be the focus of this paper), while aggregate economic efficiency or welfare may improve as a result of a move of USP, large subsets of consumers may find themselves damaged by such a change.

In this paper I propose a simple, intuitive pricing structure that is a form of USP which simultaneously makes *all* consumers better off and raises the profits of the firm providing the service. In other words, for any flat-rate tariff there exists (in the absence of metering costs) a Pareto superior nonlinear price schedule. While the assumptions underlying the analysis limit its immediate potential for practical implementation, it does suggest a methodology that, when combined with the relevant data, should prove useful in developing politically more acceptable usage-sensitive tariffs.

The second section provides a brief discussion of the relevant results of recent work on nonlinear pricing. In the third section I develop a simple model of a local-service telephone market and derive the basic proposition of this paper. The fourth section concludes the paper with some suggestions for extensions and possible empirical implementation.

Flat-Rate Tariffs and Nonlinear Pricing

Despite the technical sounding name, nonlinear pricing merely denotes any pricing scheme having the property that a consumer's total expenditure on

---

The views expressed are my own and do not necessarily reflect those of Bell Laboratories or the Bell System. I would like to thank R.D. Willig and E.E. Zajac for helpful comments and suggestions.

a good or service is not simply equal to price times quantity. The simplest form of nonlinear price schedule is the two-part tariff, which requires the customer to pay a fixed entry charge if he consumes any positive quantity (regardless how small) of the commodity and an additional charge per unit. Thus a flat-rate tariff is merely a two-part tariff with a zero per unit price. Figure 32-1 depicts the outlay functions (consumer expenditures as a function of quantity) for uniform, two-part, and flat-rate tariffs.

Because two-part and other nonlinear tariffs typically involve more than one parameter (price), efficient implementation requires more detailed information about consumer preferences than that provided by the usual price-quantity market data. Although the demand for connection and local calling can be viewed as merely depending on the level of the monthly charge, a move to an optimal USP tariff would require much more information. What would be the number of new subscribers? How many calls would they make? What would happen to the calling patterns of existing subscribers after a positive per call charge was introduced? To answer these questions requires a knowledge of individual consumer preferences, or at least the distribution of such preferences over the population. In other words, analysis of two-part pricing hinges on differences in tastes among the consuming public.

While the data necessary to perform such an analysis are, in general, unavailable, Mitchell (1976) developed and calibrated a simple model of consumer preferences for local telephone service, obtaining suggestive values for the unit price and monthly charge of an optimal two-part tariff, that is, an optimal USP scheme. Although the detailed model and analysis are somewhat complex,

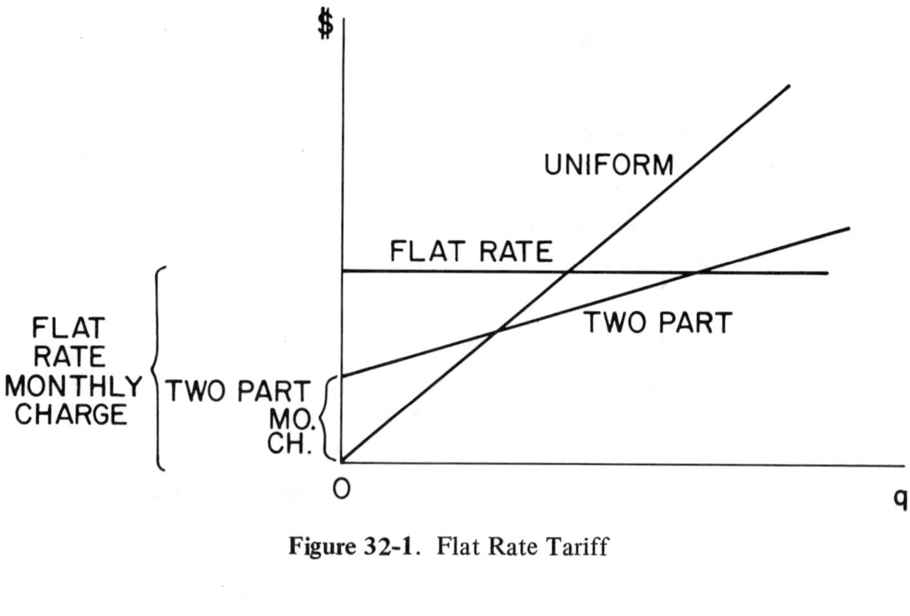

Figure 32-1. Flat Rate Tariff

the underlying intuition is quite simple. When selecting a two-part tariff for a firm with increasing returns to scale, one is, perforce, trading off two distortions. The first involves pricing usage greater (less) than marginal cost, causing users to under- (over-) call. Similarly, when the entry fee (monthly charge) is greater than per subscriber traffic-insensitive cost, subscribers willing to bear such costs may nevertheless be excluded.

The result of optimally balancing such trade-offs is typically (as Mitchell found) a price greater than marginal cost combined with a monthly charge in excess of per subscriber fixed costs.

The problem with such optimal (usage-sensitive) two-part tariffs is that while total consumer surplus is thereby maximized, individuals who were making a large number of calls under a flat-rate tariff may find their welfare seriously reduced. This situation is illustrated in figure 32-2. Consider two types of individuals: "large" consumers with demand curves given by D_L and "small" consumers whose preferences are represented by the demand curve D_S. Suppose that the flat-rate is initially equal to some E^0 which is less than the surplus of small consumers when there is a zero usage charge (E^0 is less than the triangular area A plus the trapezoidal area B), but large enough for the firm to break even. Now assume that the optimal two-part tariff (which also allows the firm to break even) has a usage price of p' and a lower fixed charge E' such that $B + C > E^0 - E' > B$. In words, the reduction in the fixed charge may be large enough to more than offset the surplus lost by small consumers as a result of the move to a positive price (area B) yet still be less than the amount (areas B and C) necessary to similarly compensate large consumers. Presumably, it is

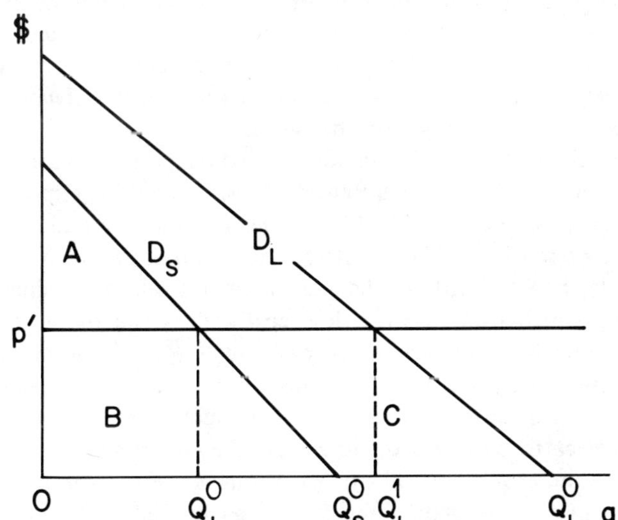

Figure 32-2. Two-Part Tariff

these large consumers (who may constitute a sizable proportion of the population) that are responsible for much of the opposition to USP.

This heuristic discussion serves to illustrate the distributional consequences of a move to USP. If it can be plausibly argued (see Mitchell, 1976, p. 13) that low-income consumers who choose to subscribe to telephone service make, on average, more calls than high-income subscribers, then a move to USP might be opposed on equity grounds. In any event, although a change of pricing policy may be efficient in the sense that those who gain from the change could, in principle, compensate those who lose, it may be effectively blocked in the regulatory arena if, as is almost always the case, no such compensation actually takes place. Any attempt to implement a usage-sensitive two-part tariff will face this hurdle whenever prospective losers can bring significant influence to bear on regulatory decision making.

Fortunately, this policy problem is *not* necessarily associated with *all* usage-sensitive nonlinear pricing schemes. In particular, a simple, intuitive generalization of the two-part tariff can be used to overcome any adverse distributional effects. This extension involves nothing more than allowing consumers to select their own rate schedule from one of two or more possible two-part tariffs. Although the basic principles of such a nonlinear price schedule have been employed for some time,[1] its properties have begun to be studied only recently; and even then only for the purpose of characterizing total surplus-maximizing schedules (Murphy, 1977; Faulhaber and Panzar, 1977). However, a recent path-breaking paper by Willig (1978) has demonstrated that a simple declining block tariff can be used to Pareto dominate any uniform price greater than marginal cost. That is, such a tariff can make all consumers better off *and* increase the profits of the firm. In the next section I demonstrate that a similar scheme can be used to Pareto dominate any flat-rate tariff. Here I shall intuitively explain why a self-selection scheme can always be used to make all consumers (at least weakly) better off.

In figure 32-3 the outlay function $R^0(q)$ represents customer expense as a function of quantity under a flat-rate tariff while $\overline{R}(q)$ expresses customer expense under a two-part tariff with a lower fixed charge and a positive price (equal to the slope of $\overline{R}$). The quantity $\overline{q}$ is the point at which tariff $\overline{R}$ is just as costly as tariff R^0. Neither schedule dominates the other, since there exist quantities for which each tariff is less costly than the other. However, the tariff represented by a choice between R^0 and $\overline{R}$ does dominate R^0. This tariff is represented by the solid portions of R^0 and $\overline{R}$ and is thus at least as low as R^0 for all quantities. When faced with such a composite tariff, all consumers who elect to choose a quantity less than $\overline{q}$ are strictly better off than they would be under R^0, and all other consumers are no worse off. Reasoning of this sort seems to lie behind the suggestion by Kahn and Zielinski (1976) that usage-sensitive pricing could be implemented without coercion, by merely offering the consumers the option of a usage-sensitive rate without forcing

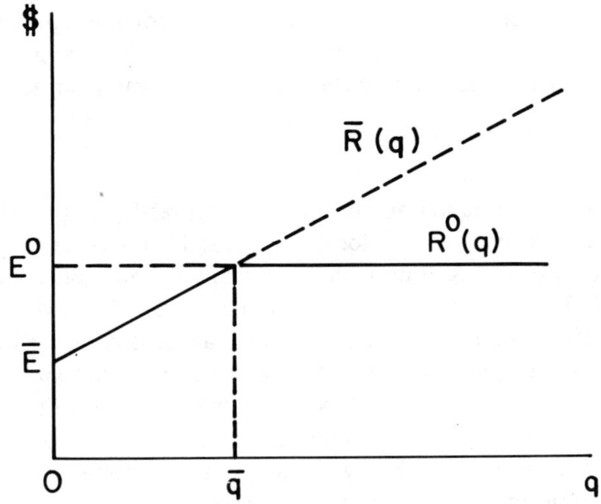

Figure 32-3. Customer Expense as a Function of Quantity under Flat Rate Tariff

it on them. The only difficulty with this argument is that there is no direct indication of what would happen to the firm's profits under such a scheme. At first glance it would seem that since the firm was presumably making zero profits initially, it would tend to lose money under this self-selection scheme, because high-usage consumers would choose the flat rate while low-usage consumers would reduce the firm's revenues by selecting the usage-sensitive tariff. The surprising result of the following analysis is that under plausible assumptions it is always possible to artfully construct a simple self-selection scheme that increases the firm's profits and makes all consumers better off.

Constructing a Pareto-Superior Tariff

I have chosen to work with a very simple model of cost and demand conditions in a local service telephone market, making as few restriction assumptions as possible:

A1: Local calls are produced at a constant marginal cost c.

A2: Either demand for calls or the demand for subscription (or both) are responsive to prices.

A3: Access to the network is priced separately at a price greater than or equal to its marginal cost.

A4: All consumers willing to pay the fixed fee for the right to make local calls choose to make a strictly positive number of calls at a zero price per call; that is, their initial consumption is bounded away from zero.

A5: There are no metering costs.

Assumption 1 is certainly an abstraction from reality in that it ignores the capacity constraints of providing local service. It is best viewed as an approximation to shadow costs resulting from a process-oriented model such as that of Littlechild (1970). At a more abstract level, assumption 1 could be replaced by an arbitrary differentiable cost function, and a construction along the lines of Willig (1978) could be performed. I have not chosen that approach because Willig's Pareto superiority results derive from making only infinitesimal moves from the existing tariff. Since it is more relevant for policy purposes to admit at least the possibility of achieving finite gains in surplus or profits, I have adopted the assumption of constant marginal cost.[2]

Assumption 2 is self-explanatory, requiring nothing more restrictive about consumer preferences than either a downward sloping demand curve for each consumer or that some consumers who are not willing to pay the intial subscription fee can be induced to subscribe by a more attractive tariff offering.

Assumptions 3 and 4 are closely related. The former is made to focus attention on the market for making local calls. Since one effect of moving to a Pareto-superior tariff may be to induce new customers to subscribe, it is important to insure that this does not adversely affect the firm's profits. If, as is likely, access and installation are typically priced below marginal cost, Pareto superiority can be preserved by not offering the new tariff to new subscribers. If one takes a consumers' surplus view of the consumer's decision process (although the analysis does not require it), assumption 4 merely requires that it is impossible to achieve the surplus necessary to justify paying the local calling entry fee without making a strictly positive number of calls. Here the interaction with assumption 3 is crucial; without this interaction, it is quite possible that a customer might be willing to pay an access charge merely for the purpose of receiving calls. The present somewhat artificial distinction between access and entering the local calling market makes assumption 4 eminently reasonable and helps guarantee that new local callers add to the profits of the firm.

While assumption 5 is clearly unrealistic, it is made to separate the Pareto superiority result, which is a nonlinear pricing phenomenon, from real-world complications. Clearly, if metering costs are sufficiently high, any movement to USP, Pareto superior or otherwise, would be futile. The empirical feasibility question raised by the presence of metering costs is addressed later.

Before constructing the Pareto-superior tariff, it is necessary to develop some notation. Since the proposed tariff consists of a choice between the existing flat-rate tariff and an alternative two-part tariff with a lower entry fee and a positive usage charge, it is a simple matter to divide consumers into three categories: (1) original consumers, (2) original consumers who select the usage-sensitive tariff, and (3) new subscribers who are attracted by the new tariff offering. The numbers of people in each category are designated by N_o, N_s, and N_n, respectively. The total number of calls made by original consumers under the flat-rate tariff is given by Q_o, some portion Q_s^o of which is attributed to those consumers who will switch to the new tariff. The quantities of calls made under the new usage-sensitive tariff by new and "switching" consumers are given by Q_n and Q_s', respectively. I can now state the following result.

Proposition. Given assumptions 1-5, for any feasible flat-rate tariff, (one that allows the firm to at least break even) there always exists a nonlinear outlay schedule that (at least weakly) increases the profits of the firm, leaves all consumers at least as well off, and at least one consumer strictly better off.

Proof. (i) *Construction*: Define q_m^0 as the smallest number of calls made by an original subscriber under the initial flat-rate tariff. Then the Pareto-superior tariff is denoted by the outlay schedule resulting from offering consumers a choice (figure 34-2) between the original flat-rate tariff and a second two-part tariff with a usage charge equal to c and an entry fee given by

$$E' = E^0 - cq_m^0 \tag{32.1}$$

In terms of figure 31-3, q_m^0 becomes the break-even quantity $\overline{q}$. Since the outlay schedule under this new tariff is everywhere at least as low as that of the original flat-rate tariff, all consumers are at least as well off as before.

(ii) *Profits*: With the notation as defined, profits before (π^0) and after (π') introduction of the new tariff option are, by definition, given by

$$\pi^0 = E^0 N_o - cQ_o - F > 0 \tag{32.2}$$

and

$$\begin{aligned}\pi' = {} & E'(N_n + N_s) + E^0(N_o - N_s) + p' \cdot (Q_n + Q_s') \\ & - c(Q_n + Q_s' + Q_o - Q_s^0) - F \end{aligned} \tag{32.3}$$

where F is the level of overhead costs and the usage price is momentarily given its own symbol (p') so that the ex post revenues and costs may be more clearly distinguished. After subtracting 32.2 from 32.3 and noting that $p' = c$, we obtain

$$\Delta\pi \equiv \pi' - \pi^0 = E'N_n - (E^0 - E')N_s + cQ_s^0 \tag{32.4}$$

Substituting 34.1 into 34.4 yields

$$\Delta\pi = E'N_n - N_s c q_m^0 + cQ_s^0 = E'N_n + cN_s(\overline{q}_s^0 - q_m) \geqslant 0 \qquad (32.5)$$

where $\overline{q}_s^0$ is simply Q_s^0/N_s, the average number of initial calls made by customers who choose to switch. Since this must, by definition, be at least as great as the smallest number of calls made by any customer, the second term on the right-hand side of 32.5 is nonnegative. Assumption 3 guarantees that the local calling entry fee paid by new subscribers represents additional revenue without offsetting costs. To see that E' is positive, substitute 32.2 into 34.1, obtaining

$$E' \geqslant c(Q_o/N_o - q_m^0) + F/N_o > 0 \qquad (34.6)$$

Thus $\Delta\pi \geqslant 0$.

(iii) *Strict Improvement*: If any new customers subscribe the condition is, by definition, satisfied. If not, then by assumption 2 consumers are responsive to usage prices. Consider the consumer who initially made q_m^0 calls. His demand curve D_m is depicted in figure 34-4. He will find it strictly in his interest to select the new tariff option because the reduction in his fixed charge (area $Ocdq_m^0$) is greater than the consumer's surplus that he loses by facing a positive usage price (area $Ocaq_m^0$). Q.E.D.[3]

The formal proposition has, in fact, been somewhat understated. This was done in order to demonstrate the strength of the Pareto superiority result

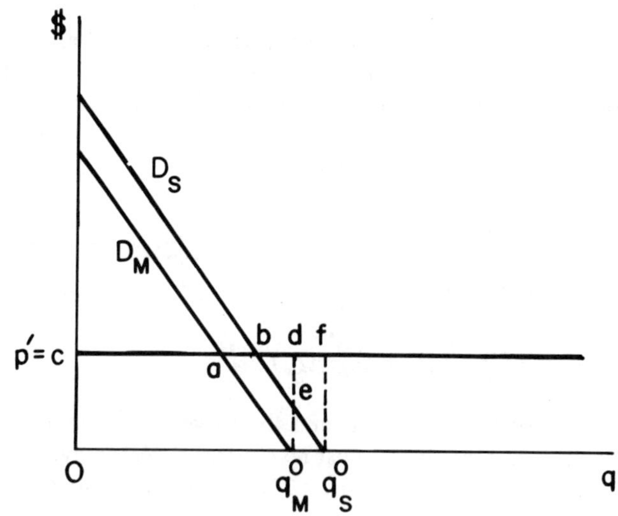

Figure 32-4. Customer Demand Curve—Responsive to Usage Prices

without resorting to special assumptions about the continuity of the distribution of consumer preferences. In the case of a very large number of consumer types whose preferences can be continuously related, the result can be strengthened in such a way that all participating consumers are strictly better off and the firm makes strictly greater profits. To see this, note from equation 34.5 that $\Delta\pi$ is strictly positive if either any new customers subscribe or more than one type of consumer switches (because then $\bar{q}_s^0 > q_m^0$). If there is a continuum of consumer types, one would in fact expect the finite reduction in the effective entry fee to attract new consumers.[4]

Figure 32-4 can be used to illustrate why more than one consumer type is likely to switch and, at the same time, provide some intuition about why the firm makes a profit on those who do. Again D_m represents the demand curve of the consumer who initially made the smallest number of calls. As argued earlier, he has a strictly positive incentive to switch. But so do consumers with slightly larger initial demands. In fact, if for simplicitly one assumes that there is some parameter that indexes consumer types and causes the demand curves to shift to the right, then all consumers who are between D_m and D_s will have a positive incentive to switch. Demand curve D_s is drawn so that the reduction in entry fee just equals the loss in surplus from facing a positive price; that is, "triangles" *bde* and $q_m^0 e q_s^0$ have the same area. A consumer with a slightly larger demand curve would find switching unprofitable.

Consider now the profit effect of an individual consumer switching. The firm breaks even on individuals with demand curve D_m, since the reduction in the entry fee it collects is exactly equal to the costs incurred by the consumer's unpriced initial consumption. (Ex post usage costs and revenues net out, since calls are then priced at marginal cost.) Now consider a switching consumer with a larger demand curve. The loss from the reduced entry fee remains the same, but the cost savings from the elimination of unpriced calling are greater, since the initial number of calls was greater. Thus the firm makes a profit from the switch of each consumer with a demand curve larger than D_m. In the case of consumers with demand curve D_s the per person profit increase is given by area $q_m^0 df q_s^0$.

Thus with "continuity" it is clear that the firm and all new and switching consumers are strictly better off than before. All that remains is to find a way to increase the welfare of consumers who choose not to switch. However, since $\Delta\pi > 0$, one can, by continuity, find a reduction in E^0 sufficiently small that $\Delta\pi$ remains positive. In that case all consumers (who participate either ex ante or ex post) are strictly better off, since the new effective outlay schedule lies everywhere strictly below the old.

Qualifications, Extensions, and Conclusions

The result of this paper is best interpreted as a possibility theorem. It suggests a way to improve economic efficiency without damaging any of the interested parties. However, the resulting tariff structure is by no means the Ramsey optimal such self-selection tariff offering.[5] While the analysis is geared toward making a finite change in the tariff structure, and thus a finite improvement in profits and welfare as well, the magnitude of the changes and the feasibility of implementation are unavoidably empirical questions.

The first barrier to practical implementation is metering costs. As Mitchell points out, any welfare gain resulting from a movement to usage-sensitive pricing must be reduced by the metering costs necessary to implement USP. In making such a determination with respect to a proposed Pareto-superior tariff, such costs must be incurred only for customers who choose the new tariff offering; hence the costs associated with the change are typically much less than if *all* customers had to be metered. In fact, the algorithm for constructing a Pareto-superior tariff can be extended to the case of positive metering costs by altering equation 34.1 to

$$E' = E^0 - cq_m^0 + M \tag{32.7}$$

where M is the (monthly payment equivalent of the) one-time costs of installing customer-specific metering equipment. The usage charge, instead of being set equal to per call marginal cost, is given by

$$p' = c + m \tag{32.8}$$

where m is the level of per call metering costs. This modification of the new tariff offering clearly protects the firm from any adverse profit effects, since each consumer who selects the new tariff is, by construction, covering any metering costs resulting from the conversion. The issue that is raised, however, is whether such a new tariff offering, when combined with the initial flat-rate tariff, results in an effective outlay schedule that lies anywhere below the initial one. If not, no one is damaged, since the new tariff is optional; but the whole exercise would be vacuous, since no one would choose it.

Therefore the first step toward practical implementation of this idea must be an empirical determination of the profit- or surplus-increasing potential of the scheme; at a minimum it must be sufficient to cover any required metering costs. This question can, in principle, be answered from empirical observations of costs and consumer preferences. In attempting such a task, it is important to explore possible Pareto-superior tariffs other than the one formulated in this paper. Here I have merely constructed a tariff option that guarantees that the firm's profits do not decrease. It may be possible to devise offerings that provide larger reductions in the monthly charge and nevertheless protect the profits of the firms. Indeed, if one has the appropriate data, this paper can be viewed as the first step toward developing a methodology that allows one to determine how much it is possible to increase any aggregate welfare measure without damaging any individual economic agent.

I conclude with a brief discussion of possible extensions of this line of research. In principle, it seems that these ideas can be used to achieve a Pareto improvement in situations in which a (nonresalable) good or service possesses multiple costly attributes, only some of which are effectively priced. The most obvious such examples occur when some services are priced without regard to intensity of use, as is the case when message unit calls are priced independently of their duration. If one allows each customer to choose between a fixed and measured rate each time he makes a call, the results of this paper go through essentially unchanged. The more practical and interesting case in which the consumer must make an either-or choice that applies to all his calls is the subject of further research.

Notes

1. They are analytically quite similar to the familiar declining block tariffs long employed by electric utilities. See Leland (1977) and Faulhaber and Panzar (1977) for further discussion.

2. When a new tariff offering admits the possibility of a finite increase in usage, the firm's profits may be adversely affected by the possibility that total costs may be dramatically increased unless some restrictions are placed on the rate of change of marginal costs.

3. If one objects to the consumers' surplus format, a simple revealed preference argument can be used to establish the result as long as consumer preferences are convex.

4. See Faulhaber and Panzar (1977) for a formal discussion of consumer behavior with respect to participation and tariff choice in a model with self-selection.

5. For example, Faulhaber and Panzar have shown that when consumer demand curves do not cross, the Ramsey optimal (surplus-maximizing) pair of tariff options must have both usage prices strictly greater than marginal costs.

References

Faulhaber, G.R., and Panzar, J.C. 1977. "Optimal Two-Part Tariffs with Self-Selection." Paper presented at the Fifth Annual Telecommunications Policy Research Conference, 1977. Bell Laboratories Economic Discussion Paper No. 74, January 1977.

Kahn, A.E., and Zielinski, C.A. 1976. "New Rate Structures in Communications." *Public Utilities Fortnightly* 97:19-24.

Leland, H. 1977. "Comments on Optimal Two-Part Tariffs with Self-Selection." Paper Presented at the Fifth Annual Telecommunications Policy Research Conference.

Littlechild, S.C. 1970. "Peak-Load Pricing of Telephone Calls." *Bell Journal of Economics and Management Science* 1, no. 2:171-210.

Mitchell, B.M. 1976 "Optimal Pricing of Local Telephone Service." Rand Report R-1962-MF, November.

Murphy, M.M. 1977. "Price Discrimination, Market Separation, and the Multipart Tariff." *Economic Inquiry* 15, no. 4:

Willig, R.D. 1978. "Pareto Superior Non-Linear Outlay Schedules." *Bell Journal of Economics* 9, no. 1:56-69.

33 Pricing Policies in Selected European Telephone Systems

Bridger M. Mitchell

This paper presents a brief survey of the principal policies for the pricing of telephone services in five European telephone systems: Norway, Sweden, Switzerland, the United Kingdom, and West Germany. After highlighting the important points at which European practices depart from those in use in the United States, I draw together evidence of consumer response to time-of-day rate structures for trunk calls and then consider the principal factors that should characterize an applied economic model of peak-load pricing for telephone services.

Pricing Policies

In nearly every European country telephone service has long been supplied by a public enterprise operating as a monopoly legally protected against competition. In the majority of countries postal as well as telegraph and telephone services are supplied by a single administration (PTT). In the others the telephone administration is operationally and financially separate from the postal system.

Generally the national public enterprise must, in financial terms, operate independently and cover its total costs. These costs may include the requirement that a specified surplus be paid into the national treasury. In most cases such transfers are in lieu of the value-added tax that applies to virtually all other goods and services. However, within the public enterprise there is considerable room for cross subsidies between services.

In many PTT-type enterprises revenues from the newer telecommunications services have been used to support postal deficits. (Such comparisons involve well-known difficulties of allocating common costs. In all cases the data are those reported by the public enterprises.) Within the telecommunications enterprise, whether it is self-standing or one component of a PTT, one typically finds an enduring pattern of cross subsidies. Telegraph service has been priced well below cost and telex service generally in excess of cost. Telephone services as a whole have at least covered their total costs (table 33-1).

I am happy to acknowledge the assistance of J. Schwalbach, the comments of J. Miller and P. Kleindorfer, and the cooperation of numerous officials of the several telephone administrators. The preparation of this paper has been supported by fellowships from the German Marshall Fund and the Alexander von Humboldt Foundation and by National Science Foundation grant No. APR77-16286 to the Rand Corporation.

Table 33-1
Degree of Cost Coverage, by Service
(*in percents*)

Service	*Switzerland 1976*	*Sweden 1973-1974*	*United Kingdom 1976*	*West Germany 1974*
Telephone services				113
Rentals	85-90	47		NA
Business	–	–	85	–
Residence	–	–	87	–
Calls			128	
Local	about 60	58		NA
Trunk	>100	245		NA
Private lines	206	NA	108	NA
Telegram	67	–	43	52
Telex	119	NA	84	120
Total telecommunications	108	–	104	112

Sources: Switzerland: PPT–*Geschäftsbericht* 1976; Sweden: Televerket-Författningssamling Serie A:29; United Kingdom: *Post Office Reports and Accounts*, 1976-1977; West Germany: Deutsche Bundespost, *Geschäftsbericht*, 1975.

Note: NA = not available.

Nearly all European telephone networks use a periodic-pulse method of metering and charging for telephone usage. At the beginning of each call an initial pulse is recorded by a meter on the subscriber's line and located in the central office. As long as the call is in progress, subsequent pulses are transmitted at regular intervals. The time interval between pulses generally depends on the distance called and frequently on the time of day as well.

Pulses are accumulated over the billing period (one month or one quarter), and the customer receives an invoice for the total number of telephone units (pulses) multiplied by the current price per unit. Thus there is no itemized billing of trunk calls (or of multiple-message units). The subscriber may, however, rent an optional meter for keeping track of his calling units. Such meters are commonly used with business phones.

General Price Structure

The charge for telephone service in Europe may be divided into four categories. In most countries, the same rates apply to both business and residential customers.

1. *Installation fee:* A one-time charge for installing a new main station, the required wiring inside the dwelling, and the drop line from the local network.
2. *Subscription fee:* Paid monthly, quarterly, or annually, for connection to the telephone network.

3. *Local usage charges:* Originally based on the number of calls made but increasingly on the length of call and time of day as well.
4. *Trunk usage charges:* Based on the distance, duration, and (generally) time of day of trunk calls.

In addition, there are separate charges for vertical services such as extension telephones and private exchange equipment (PBXs).

All European telephone administrations follow a general pattern of relatively uniform nationwide pricing. Thus connection fees are generally the same in both urban and rural areas, and per minute rates for trunk calls are equal for equivalent distances regardless of traffic density. However, the monthly subscription charges are frequently graduated with the size of the local community, reflecting the costs of supplying larger calling areas.

Within this overall structure, the relationship of prices to costs is not uniform. Furthermore, several telephone administrations are currently in the process of making major changes in their telephone tariffs.

In several countries the connection and subscription charges are significantly less than the average cost of adding a new subscriber to the network and probably also below incremental cost. Furthermore, charges for local calls do not fully recover costs. These deficits are recovered by surpluses earned from trunk call traffic.

For example, recent Swedish investigations find the following relationships between revenues and costs:

Service	*Average Revenue*	*Average Cost*
Subscription (monthly)	SKr. 11.67	SKr. 25.00
Local call	0.15	0.26
Trunk call	2.64	1.08

Figure 33-1 provides further insight into the relationship of trunk call revenues and costs in Sweden for an average trunk call as a function of distance. The revenue curve combines the calls of business and residential subscribers weighted by their respective call durations. The cost curve is based on the overall mean duration of 310 seconds and calculated from detailed studies of equipment usage priced at replacement cost for current technology, using distances between reference stations. Although costs do increase with the distance called, they climb much more slowly than the present schedule of rates for the various distance bands.

Detailed Structure of Telephone Rates

The principal charges for a connection to the network, for local calls and for domestic trunk calls are shown in tables 33-2, 33-3, and 33-4. There are significant

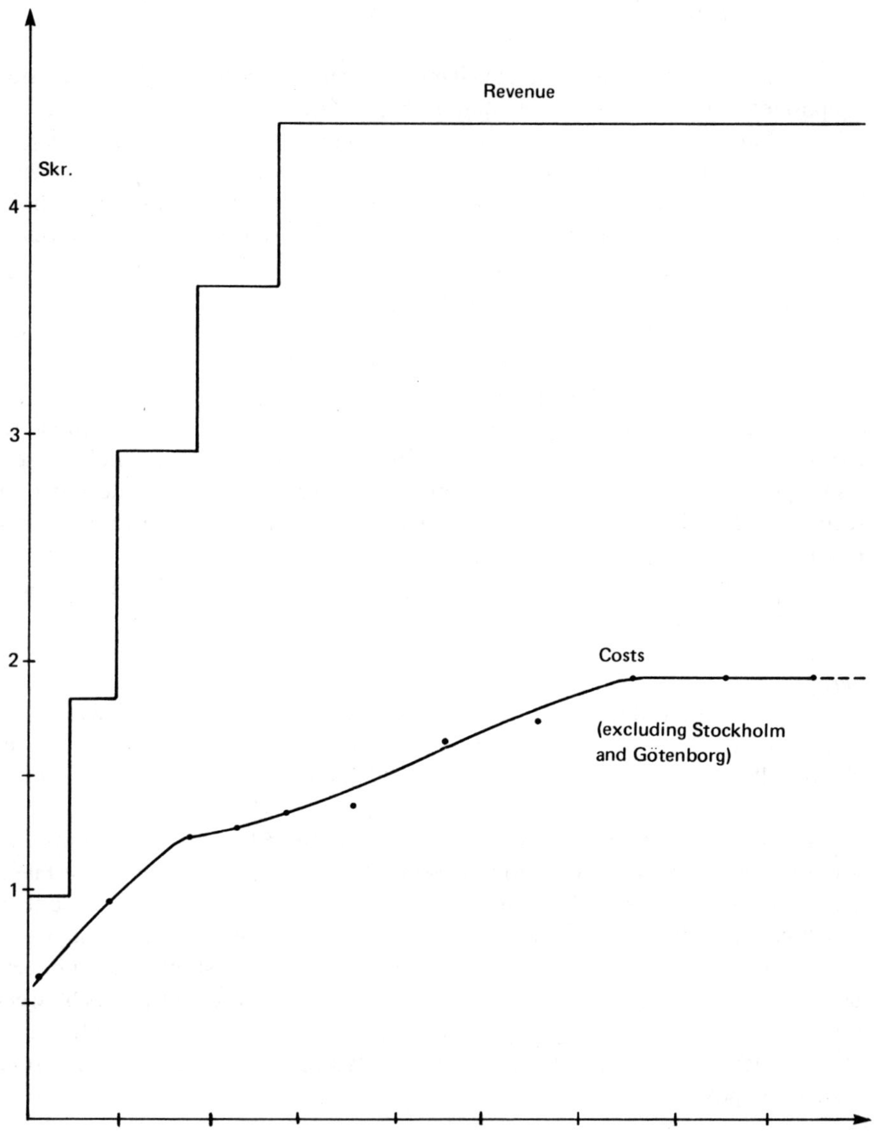

Figure 33-1. Average Trunk Call Revenues and Costs in Sweden

one-time charges for the installation of a main station in every country, and no distinction is made between business and residential subscribers. In some systems temporary "sales" (reduced installation fees) have been used to promote new connections in times of slack demand.

Subscription charges vary between about \$4 and \$16 per month in urban calling areas (table 33-2).[1] Only the British Post Office discriminates between classes of customers, charging residential users 50p less per month. The generally uniform treatment of all subscribers thus differs from the wide rate differentials for residential and business subscribers in the United States. Furthermore, customers have but a single rate open to them, whereas U.S. telephone companies frequently market a variety of service packages.

The structure of local call rates (table 33-3) is based on the periodic-pulse system. Here one sees one of the strongest contrasts with the practice of U.S. telephone companies, which in most areas use a flat-rate system for local calls. In 1978 three countries were charging for local calls according to duration. Moreover, West Germany will switch to an eight-minute pulse interval in July 1979, and the Swedish telephone administration has adopted a six-minute interval that will take effect in 1982. Off-peak rates will also be more widely used in the future. Both the West German and Swedish systems will have a twelve-minute interval during off-peak hours.

Rates for domestic trunk (long-distance) calls are based on a series of distance bands and on the time of day (table 33-4).[2] Time-differentiated rates are either two-level (peak, off-peak) or three-level (peak, shoulder, off-peak) and distinguish different periods on weekdays, Saturdays, and Sundays. Holidays are also charged at Sunday rates.

Table 33-2
Telephone Installation and Subscription Prices, 1978

Country		*Installation Charge*		*Monthly Subscription Charge*	
Norway	NKr	1,000[a]	(\$188.00	NKr 22-48	(\$4.14-9.02)
Sweden	SKr	300	(\$ 65.40)	SKr 32-40	(\$6.98-8.72)
Switzerland	SFr	100-150[b]	(\$ 54-81)	SFr 13-19	(\$6.98-10.20)
United Kingdom					
Business	£45		(\$ 84.15)	£3.25	\$6.08
Residential	£45		(\$ 84.15)	£2.75	\$5.14
West Germany	DM200		(\$ 99.40)	DM 20-32	\$15.90

Notes: Monthly subscription charges shown for the largest local calling areas. Prices in U.S. dollars at rates of exchange prevailing April 10, 1978.

[a]Plus a loan payment of NKr 2,000 which earns interest at 6.5 percent per year, returnable at the termination of subscription.

[b]Typical installation charges for private contractors.

Table 33-3
Tariff Structures for Local Calls, 1978

		Normal Rates		*Off-Peak Rates*	
Country	*Price per Unit*	*Pulse Interval*	*Hours*	*Pulse Interval*	*Hours*
Norway	67 øre (12.5¢)	3 min	08-17 Mon-Fri	None	Other Hours
Sweden	17 öre (3.7¢)	None	–	–	–
Switzerland	10 Rp (5.4¢)	4.8 min	All hours	–	–
United Kingdom	3 p (5.6¢)	2 min	09-13 Mon-Fri	3 min	08-09, 13-18 Mon-Fri
				12 min	18-08, Mon-Fri 00-24, Sat-Sun
West Germany	23 Pf (11.4¢)	None	–	–	–

Note: Prices in U.S. currency of rates of exchange prevailing April 10, 1978.

The price relationships between different distance bands and time periods are, to some extent, constrained by the specific pulse-generating technology that is in use. The pulse interval for a particular route is not infinitely variable; rather it is mechanically determined by dividing a fundamental time period (for example, 180 seconds) by an integer. Partly for this reason, the relative price structure has tended to remain fixed in many systems, with adjustments in the overall level of rates accomplished by changing the value of one unit (the pulse interval). Changes in the periods for time-of-day rates have been less frequent. Switzerland, however, is an exception–the unit price has been kept fixed at 10 Rp, and when revenue increases were necessary the pulse intervals have been shortened.

Consumer Response to Time-of-Day Rates

To employ peak-load pricing effectively in a telephone network, it is desirable to have reliable measurements of the parameters of consumer demand functions. However, the amount of data available for assessing demand elasticities is limited primarily to observations on the few occasions that tariff structures have been changed. For the most part telephone administrations have not had the flexibility to experiment with a variety of rate structures or to introduce a new rate on a trial basis for selected areas. (Norway is an exception.) Furthermore, although there has been some econometric modeling of aggregate traffic demands by telephone administrations in West Germany, the United Kingdom, and the United States, detailed traffic data have thus far not been available to outside researchers.

Table 33-4
Tariff Structures for Domestic Trunk Calls, 1978

Country	*Price per Unit*	*Distance Zone*	*Pulse Intervals (Seconds)* *Peak Hours*	*Shoulder Hours*	*Off-Peak Hours*
			08-20 Mon-Fri		20–08 Mon-Fri 00–24 Sat, Sun
Norway	67 øre	<50 km	36	–	48
	(12.5¢)	<50 km	20		26.7
			All hours		
Sweden	17 öre	<45 km	45		
	(3.7 ¢)	45-90 km	24	–	–
		90-180 km	15		
		180-270 km	12		
		>270 km	10		
Switzerland	10 Rp		08-17, 19-21	–	17-19, 21-08 Mon-Fri
	(5.4¢)		Mon-Fri		00-24 Sat-Sun
		10-20 km	36		36
		20-50 km	22.5		36
		50-100 km	16.3		25.7
		>100 km	12		20
United Kingdom			09-13 Mon-Fri	08-09, 13-18 Mon-Fri	18-08 Mon-Fri 00-24 Sat-Sun
	3p	<56 km	30	45	180
	(5.6¢)	>56 km	10	15	60
			06-18 Mon-Fri 06-14 Sat	18-22 Mon-Fri 14-22 Sat	22-06 Mon-Sat 00-24 Sun
West Germany	23 Pf	<25 km	45	67.5	67.5
	(11.4¢)	25-50 km	30	45	67.5
		50-100 km	15	22.5	67.5
		>100 km	12	18	67.5

Note: Prices in U.S. currency at rates of exchange prevailing April 10, 1978.

In this section, we bring together evidence of changes in traffic patterns that have been observed in the wake of rate structure changes in several countries. These data are highly aggregated and cover only selected aspects of the traffic patterns of economic interest. While they are probably too sketchy to permit any firm conclusions to be drawn, they do suggest a number of interesting hypotheses.

Norway

On January 1, 1976, the Norwegian Telecommunications Administration introduced its first off-peak trunk rates on an experimental basis in the county of Nordland. Rates were reduced by 40 percent from 8 p.m. to 8 a.m., Monday through Saturday, and at all hours on Sunday. This test of peak-load pricing in the trunk network formed the basis of the nationwide peak–off-peak rates introduced in January 1978 (see table 33-4).

A limited amount of data is available for the first months that the new rates were in effect. Measurements over this period permit a comparison of the traffic generated in Nordland after the rate change with traffic one year earlier. The changes shown in Table 33-5 were measured as a percentage of the peak-hour traffic in each month. These data show small reductions in traffic in the last hours before the discount rate goes into effect at 8 p.m. and a pronounced increase in calling during the first off-peak hour. The January 1976 traffic *to* Nordland, which remained subject to the standard rate at all hours, decreased over 1975 at both 7 p.m. and 8 p.m. (More recent data were not reported.)

Switzerland

During the 1920–1974 period, the Swiss PTT enjoyed an essentially constant level of telephone rates.[3] Then beginning in 1973 rapidly mounting costs led to pressure for increased revenue, and effective December 2, 1974, new rates were established for trunk and suburban calls. The old and new rates for the various distance zones are shown in table 33-6. When the distribution of the duration of calls is taken into account, these rate changes amount to about a 40 percent price increase for suburban calls and somewhat lesser amounts for trunk calls.

The December 1974 tariff revision also provided the PTT with the opportunity to adjust the structure of its peak and off-peak rates. By 1973 weekday trunk traffic was peaking in the early evening, shortly after the off-peak discount went into effect at 6 p.m. To discourage further growth in early evening traffic and to attempt to shift calls to hours when greater unused capacity existed, the PTT introduced peak rates during the 7 p.m. to 9 p.m. evening

Table 33-5
Traffic as a Percentage of Peak-Hour Traffic, Nordland, Norway

Weekday Hours	*January*			*March*		
	1975	*1976*	*Change*	*1975*	*1976*	*Change*
17-19[a]	61.2	58.4	− 2.8	63.2	62.8	− 0.4
19-20	52.0	51.1	− 0.9	59.0	57.3	− 1.7
20-21	55.0	65.0	+10.0	53.3	66.8	+13.5

Source: Bø, Ganstad, and Kosberg 1976.

[a]Per hour average of data spanning the two-hour period.

Table 33-6
Trunk Call Pulse Intervals Switzerland
(*in seconds*)

Zone	*January 1973-November 1974*		*December 1974-Present*	
	08-18 Mon-Sat	*18-08 Mon-Sat 00-24 Sun*	*08-17, 19-21 Mon-Fri*	*17-19, 21-08 Mon-Fri 00-24 Sat-Sun*
10-20 km	60	60	36	36
20-50 km	36	60	22.5	36
50-100 km	25.7	45	16.3	25.7
>100 km	18	30	12.8	20

Note: One pulse = 10 Rp (5.4¢).

period and simultaneously established off-peak rates for the preceding two hours, from 5 p.m. to 7 p.m.

In figures 33–2 and 33–3 the relative rates for the longest-calling-distance band at each hour are shown in the lower portion of a figure. Above them appear the graphs of trunk traffic before and after the change in rate structure.

The effect of the change in the Swiss rate structure, seen in the upper portion of figure 33–2, was extremely favorable.[4] The evening peak was reduced to just below the level of the midmorning peak, and traffic in the new late afternoon discount period grew appreciably. There were essentially no changes in the weekday traffic load curve at other hours.

The tariff revision also eliminated the use of peak prices on Saturdays, which until December 1974 had been in effect from 8 a.m. to 6 p.m. At that time peak traffic on Saturday was only about 65 percent of the weekday peak, and it also occurred in the early evening. Figure 33–3 suggests that in relative terms the change in the Saturday rate structure reduced evening calling and led to higher volume after about 3 p.m.

Apart from the shortened pulse interval for all off-peak calls, no change was made in the Sunday tariff, which had had off-peak rates at all hours. Over the 1973–1975 period small changes in the shape of the load curve can be seen (figure 33–4), and there is again a (somewhat smaller) reduction in the late evening peak load. The qualitative similarity of the daily and Sunday changes in load at the evening peak hours indicates the difficulties of inferring the effect of a single change in tariff structure. Is the relative reduction in 8 p.m. traffic on Sunday the result of an exogenous trend that would have manifested itself with no change in rates? Or is the Sunday evening reduction an echo of a tariff-induced change in weekday behavior that occurs either because subscribers establish the timing of their calling on a daily cycle or because they are ignorant of the rate differences on weekends?

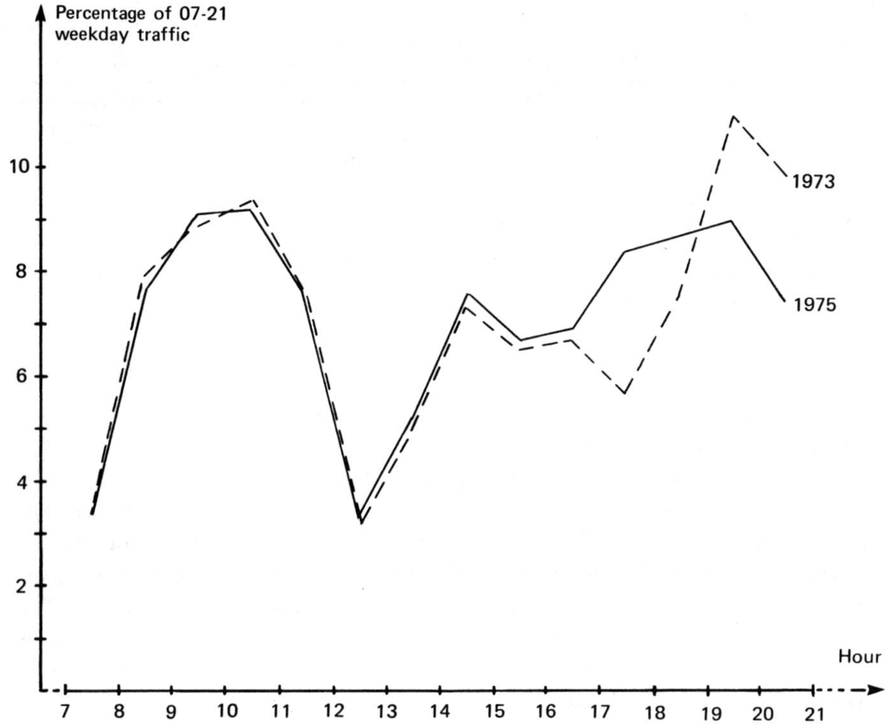

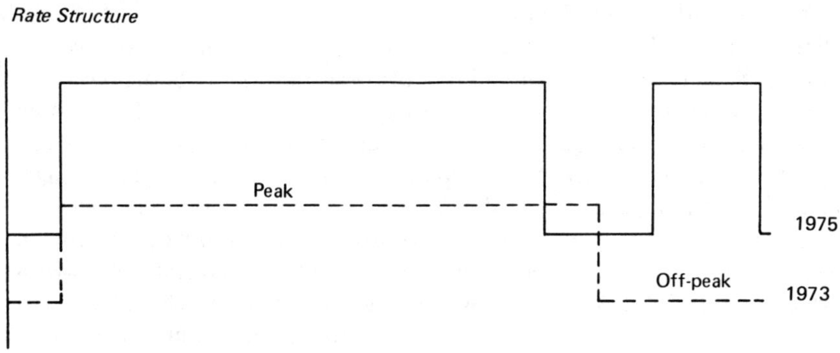

Figure 33-2. Weekday Trunk Traffic in Switzerland

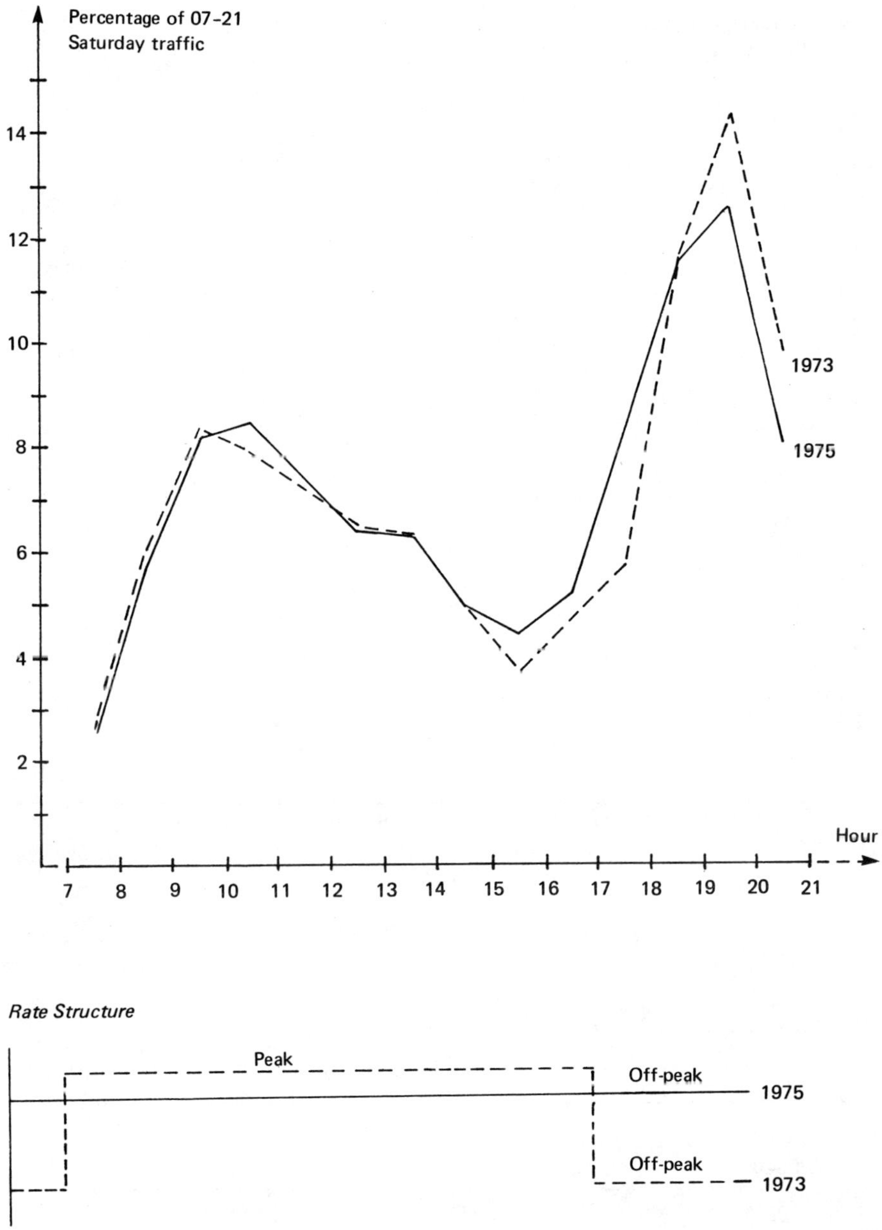

Figure 33-3. Saturday Trunk Traffic in Switzerland

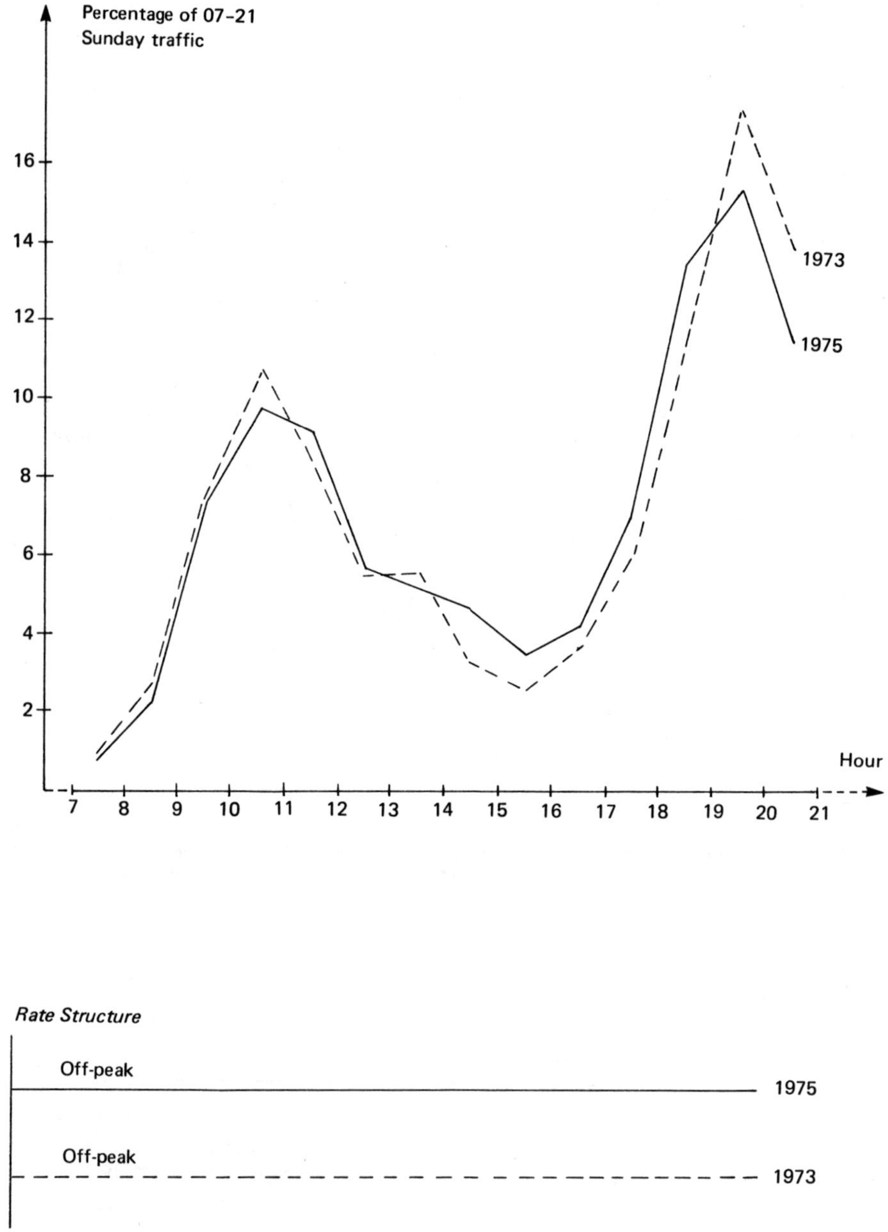

Figure 33-4. Sunday Trunk Traffic in Switzerland

United Kingdom

Before 1970 the rate structure for trunk calls in the United Kingdom had a standard charge from 8 a.m. to 6 p.m. on weekdays and a discount rate at other hours. To encourage subscribers to shift some calls away from the midmorning peak, the Post Office in July 1970 introduced a peak rate from 9 a.m. to 12 noon.

Since then there has been one minor revision in the rate structure: on April 28, 1975, the hour from noon to 1 p.m. was switched from the standard (shoulder) to the peak rate. The level of all telephone rates was raised later that year, on October 1, from 1.8p to 3p per telephone unit.

The small April 1975 rate structure adjustment provides only a limited opportunity to measure consumer responsiveness to midday tariff changes. Figure 33-5 compares the quarterly total number of trunk calls per half-hour before and after the noon–1 p.m. hour was repriced. In most cases a small shift of calls out of the peak period is observed.

The research staff at the British Post Office develops and maintains a series of econometric models to forecast telephone variables of major importance. Predictions for trunk calls are based on separate models for the weekday hours of 8 a.m. to 6 p.m. and for all other hours (when only off-peak rates apply).

In the full-rate model the quarterly number of weekday calls between 8 a.m. and 6 p.m. is explained as a function of an index of the peak and shoulder trunk rates, the number of business subscribers, the extent of automatic dialing, and the output level of the service sector (Cracknell, 1977). The model incorporates quarterly and annual moving-average disturbance parameters. Currently estimated for sixty-three observations, it yields a price elasticity of –0.10. The distribution of calls between the peak and shoulder periods is modeled in ratio form as a function of the ratio of the peak- to shoulder-period prices. This equation is estimated from monthly data beginning in July 1970 and includes a first-order moving-average residual.

For off-peak calls the cheap-rate model explains the quarterly number of off-peak calls as a function of the off-peak rate, the number of residential subscribers, the extent of automatic dialing, and aggregate consumer expenditure (Drew and Cracknell, 1976). Estimated with quarterly and annual moving-average error coefficients, the model produces a price elasticity of –0.08.

In their present form the models provide for tariff-induced shifts in traffic within business weekday hours but not otherwise. In later staff research these models were linearized to assess the effects of reducing peak rates so as to increase the use of current excess trunk call capacity. At the mean values of the volume of calls forecasted for 1977–1978, the own- and cross-price elasticities are

Variable	*Peak-Rate Elasticity*	*Shoulder-Rate Elasticity*
Peak-Rate Calls	–0.35	+0.18
Shoulder-Rate calls	+0.23	–0.26

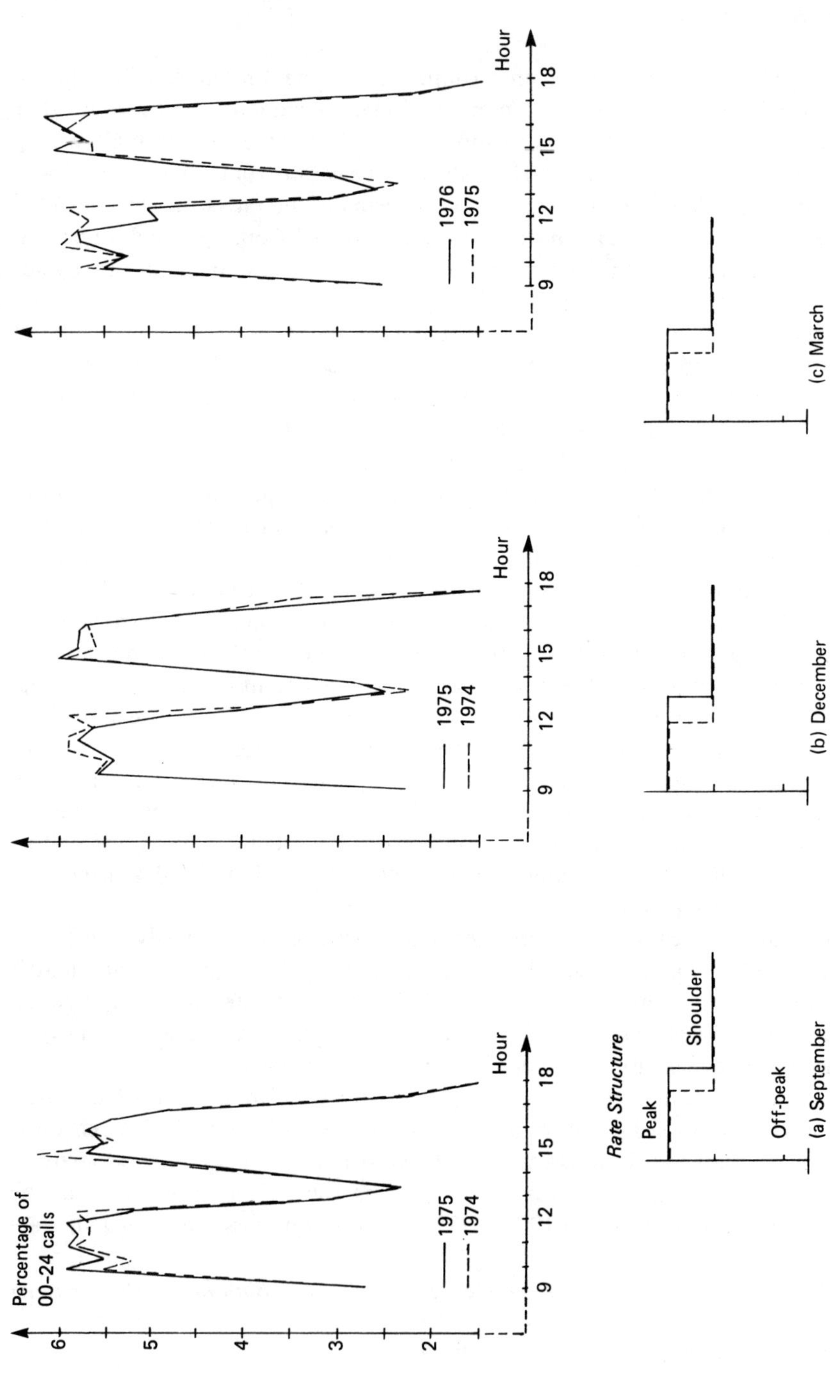

Figure 33–5. Quarterly Trunk Calls on Weekdays in the United Kingdom

Thus the econometric findings are that daytime calling patterns are susceptible to some degree of control through tariff policy.

West Germany

On July 1, 1974, the West German telephone administration made a major revision in its tariffs. Charges for virtually all services were increased, in many cases substantially. Monthly subscription rates were raised 25 percent and installation fees by 67 percent. The price of a telephone unit was, by comparison, increased by about 10 percent, from 21 Pf to 23 Pf.

In this climate the Bundespost was under political pressure to provide some indications of rate reductions. It used this occasion to introduce an important change in the rate structure for trunk calls (table 33-7). The so-called moonlight or off-peak rate, which has been designed to stimulate data traffic and had been in effect only between 1 a.m. and 6 a.m. every night, was reset to begin at 10 p.m. Monday through Saturday. The moonlight rate was also established for all hours on Sundays and federal holidays. Because no change was made in the pulse intervals (table 33-4), this liberalization of the moonlight hours provided significantly reduced rates, especially over longer distances. For calls beyond 100 km, subscribers could talk for 67.5 seconds for 23 Pf (11.4¢) rather than 18 seconds for 21 Pf (10.5¢), a price reduction of 60 percent.

The effects of the change in tariff structure can be seen by comparing the May 1974 and November 1976 weekday load curves in figure 33-6. At 10 p.m., when the off-peak rate now takes effect, there has been a very abrupt, threefold increase in measured network traffic. In fact, this needle peak indicates the existence of overload conditions in local segments of the network and therefore understates the actual demand for trunk calls at the moonlight rate. Over this two-and-a-half-year period the late-afternoon traffic has also increased, starting about one hour before the shoulder rate takes effect at 6 p.m. As this rate has

Table 33-7
Hours of Tariff Periods for Trunk Calls, West Germany

Period	*July 1971-June 1974*	*July 1974-June 1978*	*July 1978-Present*
Peak	06-18 Mon-Fri 06-14 Sat	06-18 Mon-Fri 06-14 Sat	06-18 Mon-Fri 06-14 Sat
Shoulder	18-01 Mon-Fri 14-01 Sat 06-01 Sun	18-22 Mon-Fri 14-22 Sat	18-22 Mon-Fri
Off-peak	01-06 Mon-Sun	22-06 Mon-Sat 00-24 Sun	22-06 Mon-Fri 14-24 Sat 00-24 Sun

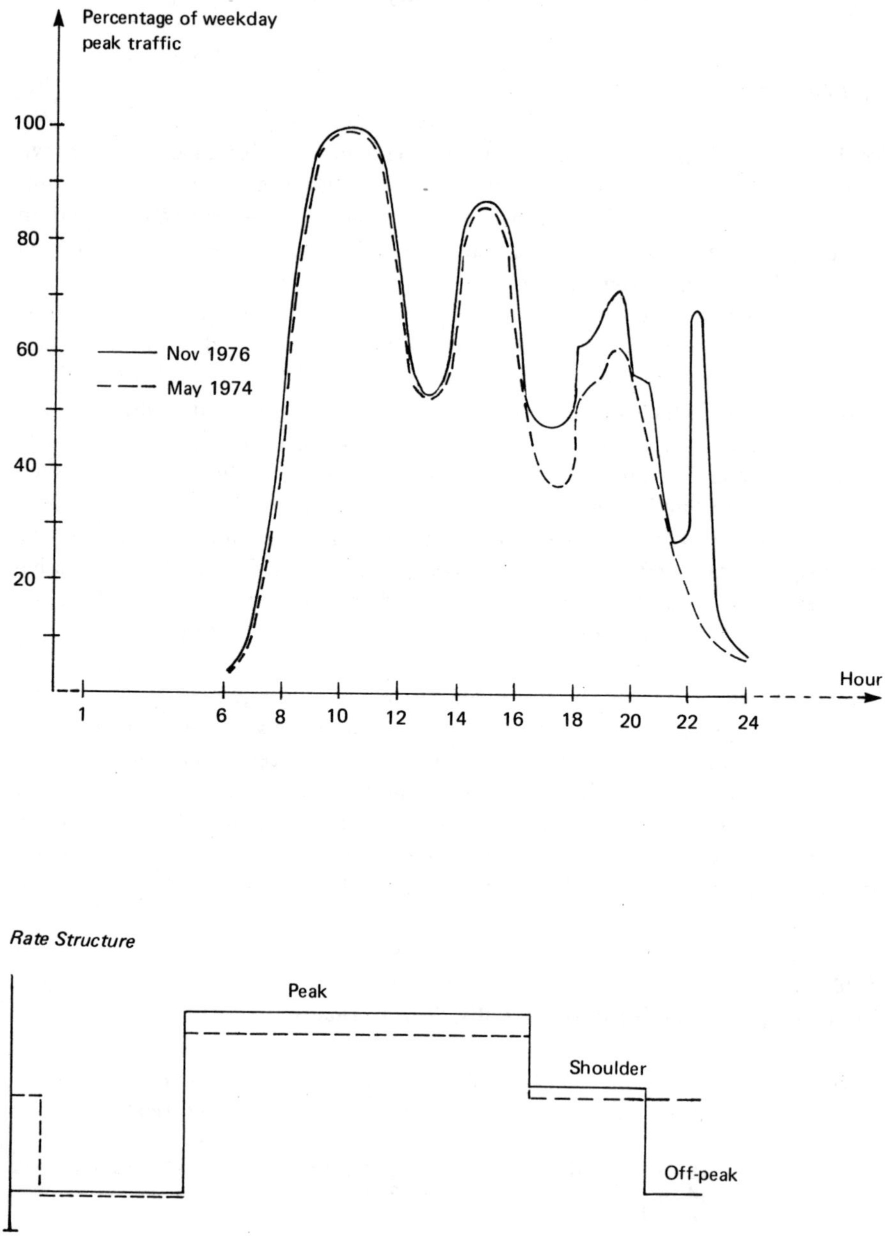

Figure 33-6. Weekday Trunk Traffic in West Germany

(in relative terms) not changed, the observed increase in shoulder-hour traffic may reflect a relatively more rapid secular growth in residential trunk calling that is concentrated outside business hours. During the same period the load curve during normal business hours has remained unchanged.

The mid-1974 tariff revision allowed the moon to shine for twenty-four hours every Sunday and holiday. Between May 1974 and November 1976 Sunday trunk traffic has almost tripled, and as shown in figure 33-7 the network peaks in both the morning and early evening now exceed the weekday peak. In these instances measured traffic again understates actual consumer demand, as the network is effectively blockaded in the peak hours. Figure 33-7 also shows that there is a sizable echo effect at 10 p.m., when a needle peak appears, even though the off-peak rate has been in effect the entire day. Both a combination of weekday habits and misinformation about the true rate structure may account for this behavior.

The Bundespost has committed itself publicly not to raise telephone rates before the early 1980s. Therefore, despite the overload conditions at some moonlight hours, the tariff revisions it has announced for July 1978 will leave the off-peak rate unchanged. However, on Saturday the off-peak moonlight rate will begin at 2 p.m. (rather than 10 p.m.), and during the weekday the shoulder-hour rate will be reduced some 20 percent to 25 percent. These adjustments are intended to ease the growing problem of a shifted peak—first, by encouraging Sunday calls to be made on Saturday afternoon and, second, by providing an incentive to move some late-evening weekday calls into the early evening period.

Aggregate traffic measurements combine call frequency and call duration into a single statistic. Data from a recent traffic sample of 861 households conducted for a two-week period by the Bundespost suggest that there is some price elasticity of demand with respect to both the number of calls as well as the mean duration. Figure 33-8 shows that the number of trunk calls as a proportion of total calls rises in the lower-priced periods for households headed by white-collar and blue-collar employees, and to a lesser extent for self-employed persons who presumably also use the telephone for business calling. Furthermore, as shown in figure 33-9, the duration of trunk calls is also inversely related to price. While suggestive, such patterns are doubtless also influenced by factors other than tariffs.

West Berlin

The economic problems of matching capacity to demand vary across different types of public utilities. Most utility services are provided through a supply network in which individual components must be sized according to local conditions. However, centralized capacity is of major importance in supplying services such as electric power, where every customer's demands adds to the generating

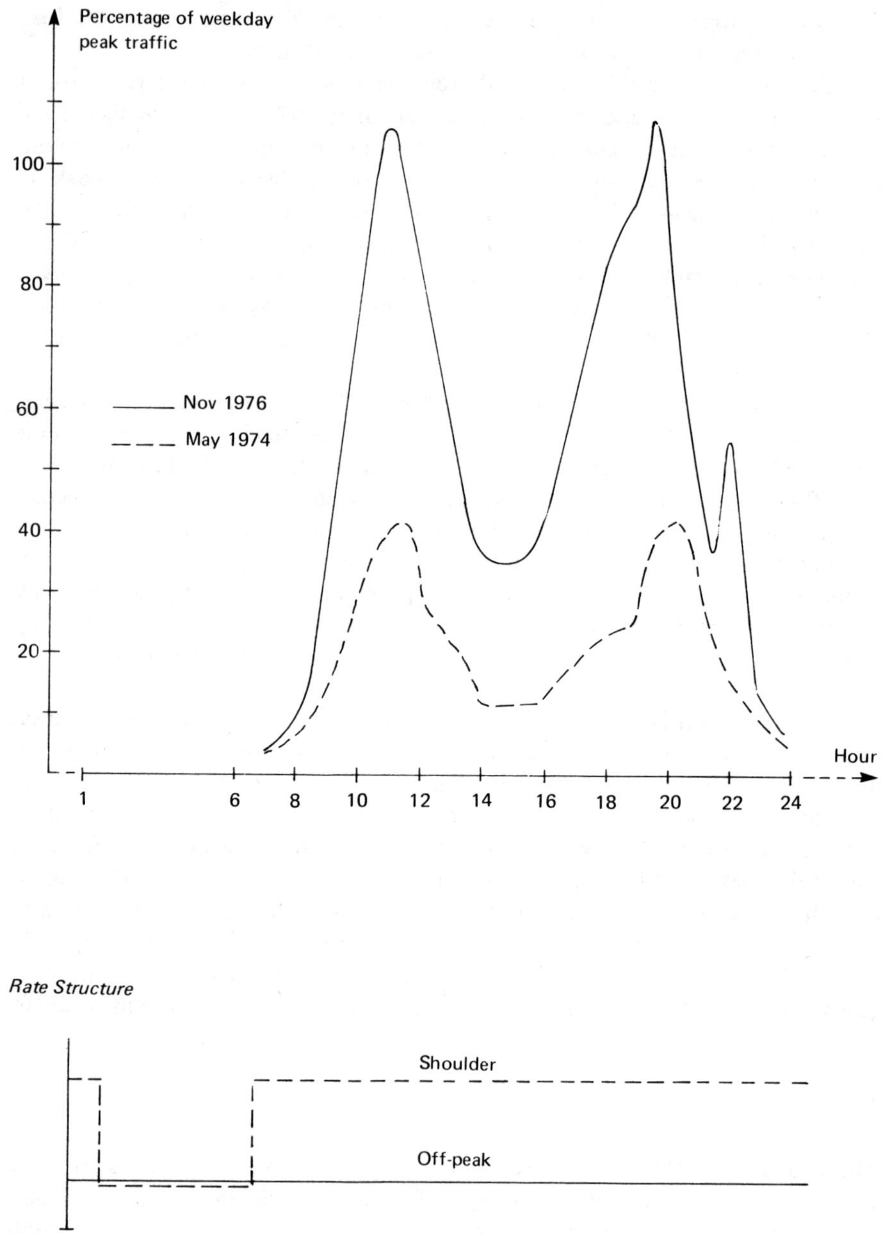

Figure 33-7. Sunday Trunk Traffic in West Germany

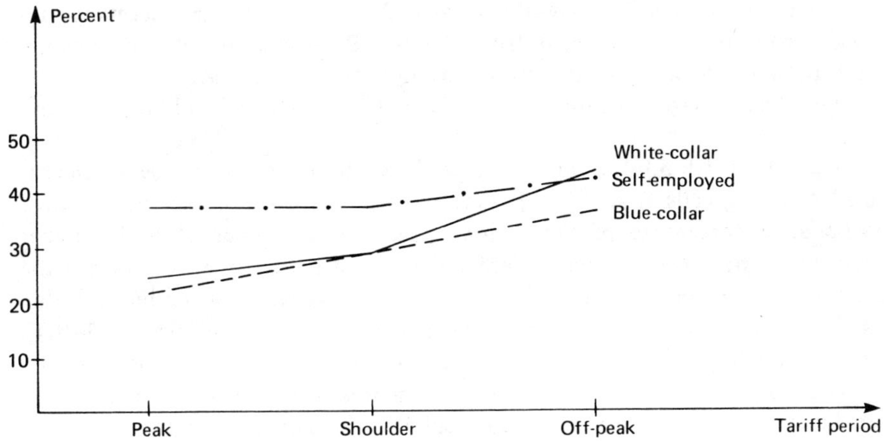

Figure 33-8. Trunk Calls as a Percentage of Total Calls for Residential Subscribers in West Germany

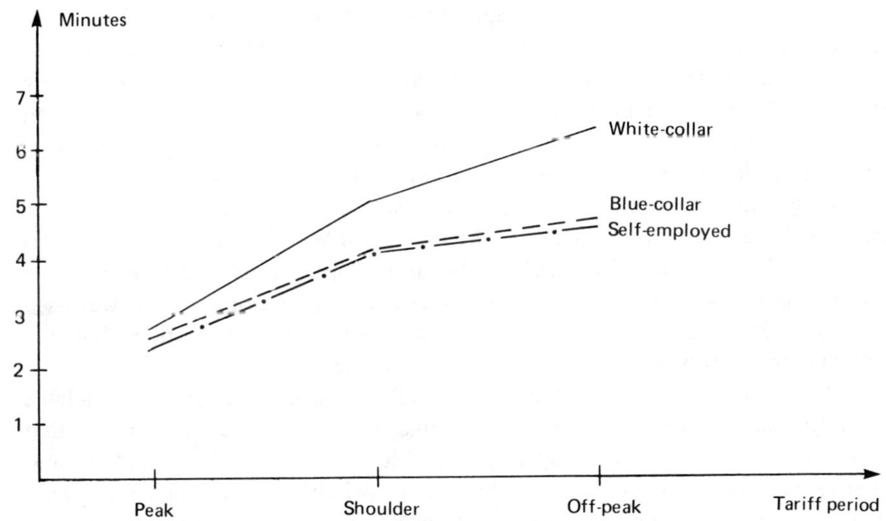

Figure 33-9. Average Duration of Residential Trunk Calls in West Germany

plant capacity that is needed. In contrast, the capacity required in a telephone system is decentralized among many local switching centers and individual routes.

The development of trunk traffic loads in West Berlin provides an especially interesting opportunity to examine one component of the aggregate West

German traffic curves. West Berlin, located 225 km inside East Germany, has a high proportion of long-haul trunk traffic. Because most of the telecommunication routes are provided by radio rather than microwave relay or cable, a longer-than-average amount of time is required to install additional trunk capacity.

Figures 33-10, 33-11, and 33-12 are based on quarter-hour measurements of all outgoing calls from West Berlin. In each figure the upper graphs show traffic as a percentage of the Sunday peak demand observed in December 1976, while the lower diagrams depict the rate structure in effect. Unlike the aggregate traffic in the entire West German network, West Berlin weekday traffic peaked in the early evening (7 p.m.) before July 1974 (figure 33-10a). The introduction of the off-peak moonlight rates at 10 p.m. caused an immediate surge in the late evening traffic from West Berlin. This demand quickly exceeded the available capacity and resulted in a blockaded network for the first half hour that the off-peak rate was in effect (figure 33-10b). And at the end of 1976, despite the addition of a significant amount of capacity, the growth in demand continued to overload the network at 10 p.m. (figure 33-10c).

The new off-peak hours also stimulated Saturday evening traffic after 10 p.m. (figure 33-11) and likewise resulted in excess demand by the end of 1976. In West Germany as a whole the greatest growth in aggregate traffic occurred on Sundays. In West Berlin the new Sunday off-peak rate had the particularly dramatic effect shown in figure 33-12. By 1975 the network was overloaded from 9:30 a.m. to noon and again at 8 p.m., and by the end of 1976 almost total overload conditions prevailed from 9 a.m. until after 10 p.m.

When a market is in disequilibrium and supply is insufficient, the measurement of demand poses problems. In the figures for West Berlin the traffic values that exceed the indicated capacity included attempted calls that were not successful due to overload conditions. The measurement equipment that was used to collect these data records a call attempt as soon as the subscriber dials the initial O digit that indicates that a trunk call is being placed.

The recorded amount of invisible traffic—traffic in excess of available capacity—may be greater than, about equal to, or less than the traffic that would exist were sufficient capacity available to connect all attempted calls. In overload periods subscribers are known to dial the same number repeatedly in an attempt to obtain a connection, thus increasing the number of attempts. On the other hand, the well-known difficulty of completing a call on Sunday may discourage calls that would otherwise be made. And in some cases local circuits feeding into the trunk center may also be full, so that no attempt is recorded. Increased understanding of consumer behavior in excess demand markets would be of considerable value in planning capacity expansion, as well as for analysis of market demand. However, at present gathering the requisite data requires the use of costly special-purpose equipment.

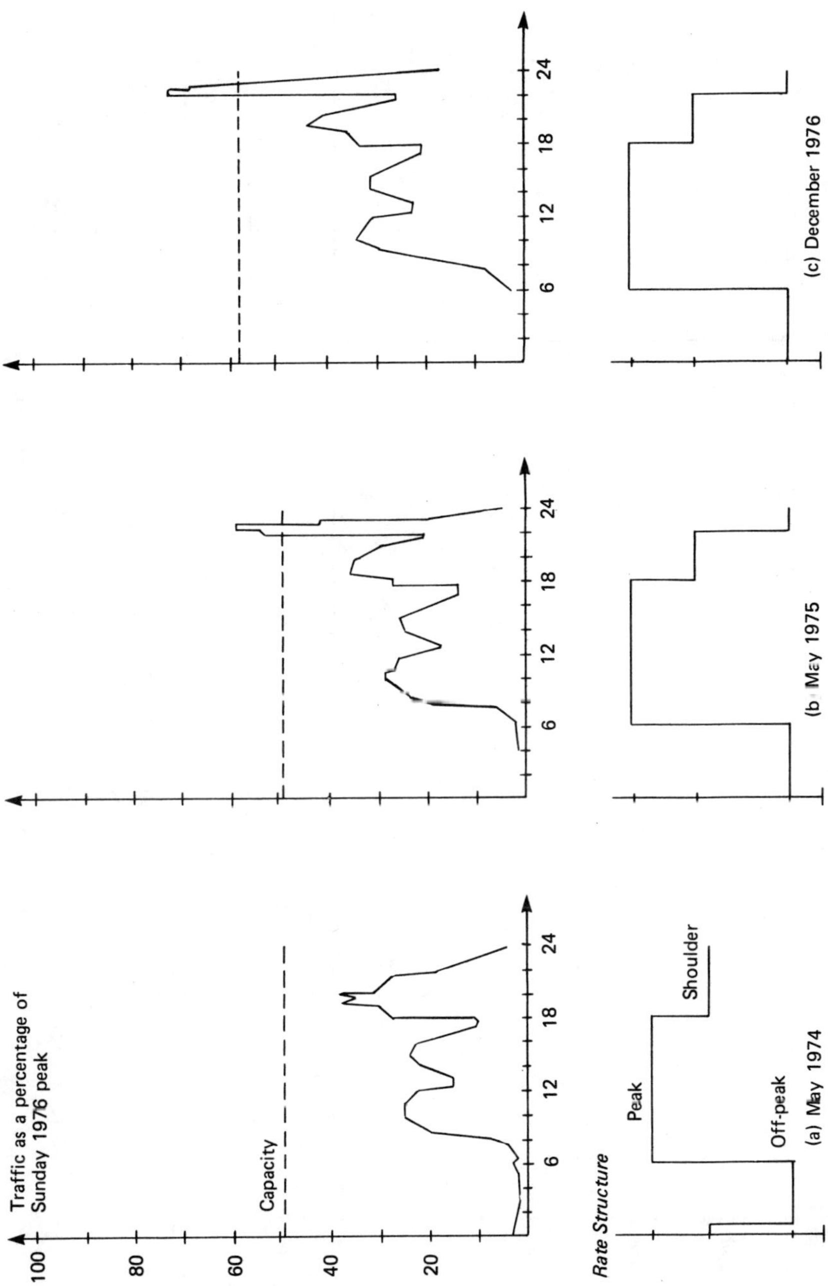

Figure 33-10. Weekday Trunk Traffic in West Berlin

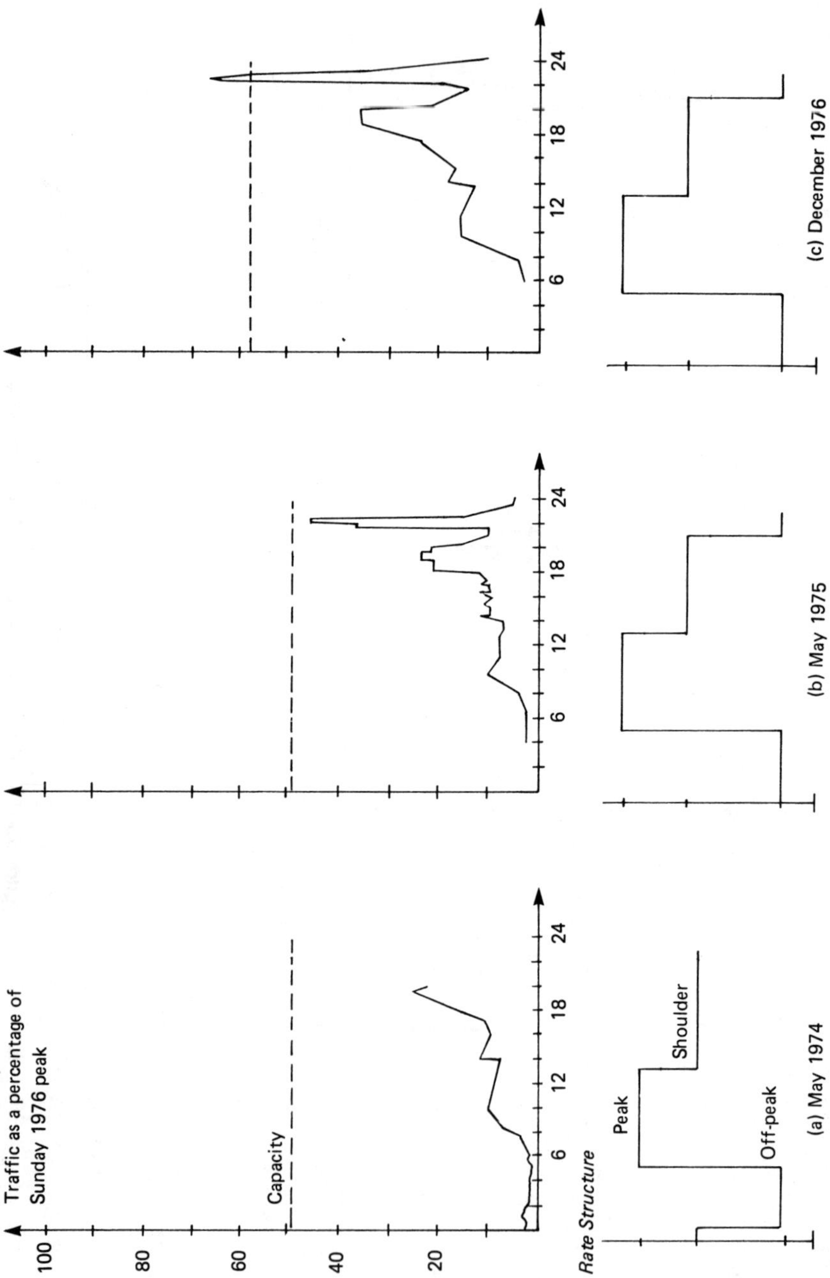

Figure 33-11. Saturday Trunk Traffic in West Berlin

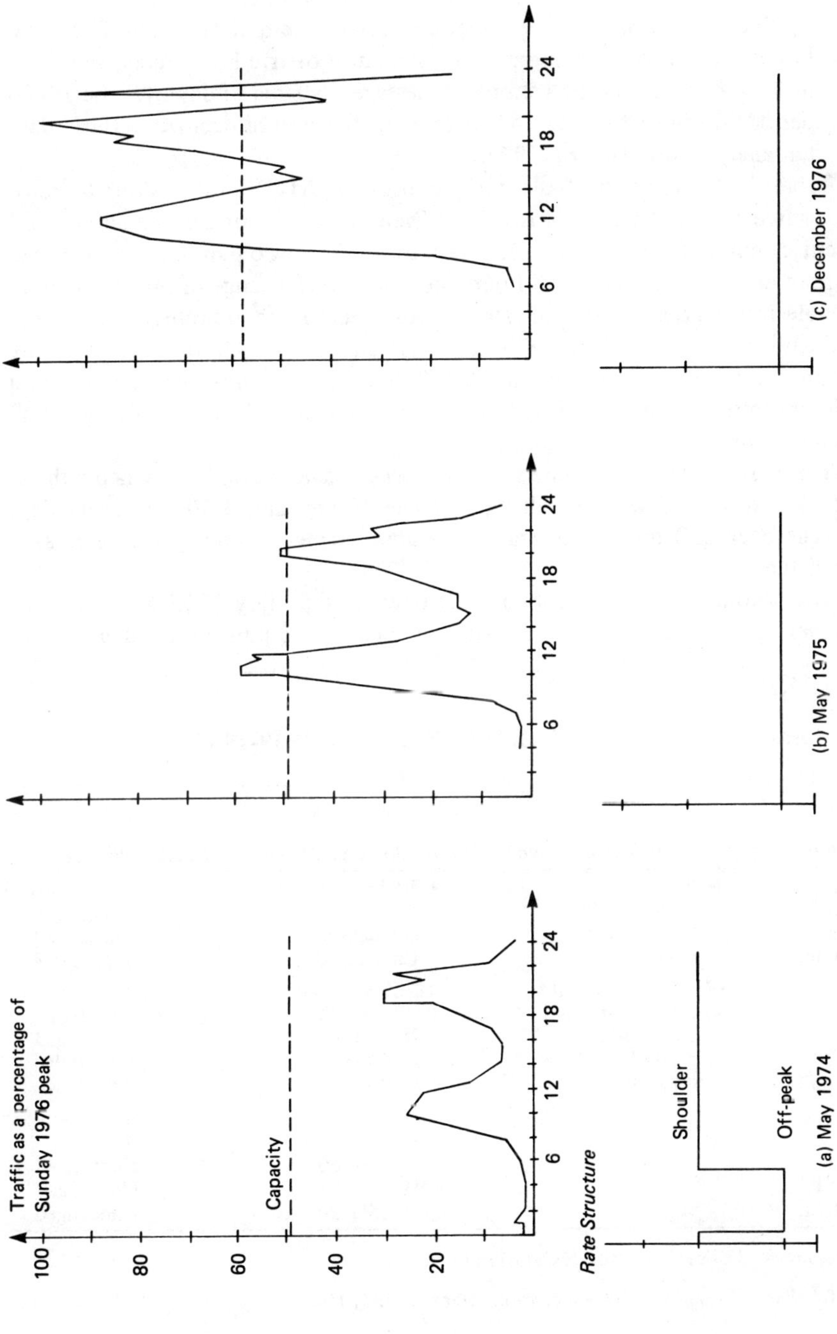

Figure 33-12. Sunday Trunk Traffic in West Berlin

United States

The long lines department of the American Telephone and Telegraph Company (AT&T) has made several changes in the structure of trunk call rates, and some data are now available on the volume of messages by time of day over the 1960–1967 period, during which two changes in rate structure occurred. These data were first analyzed by Auray (1977).

Table 33–8 shows the basic configuration of AT&T's rates. Prior to April 1963 a two-period structure was in use. Evening rates, representing about a 30 percent discount for calls over 100 miles, took effect at 6 p.m. Monday through Saturday and also applied at all hours on Sunday. The range of prices shown in the table reflects the tapering of rates with distance; in the full-rate schedule there are thirty-two separate distance bands between 25 and 3,000 miles. With the ticketing technology used by the Bell System trunk calls were billed for a minimum initial three-minute period, with lower marginal rates applying for all additional minutes.

The first change in the rate structure occurred in April 1963 when a third, off-peak rate period was introduced between 9 p.m. and 4:30 a.m. every day. For calls over 220 miles these rates provided further discounts that increased with distance.

The second change came two years later, in February 1965 when the off-peak period was enlarged. On weekdays it began at 8 p.m. rather than 9 p.m.

Table 33-8
Rate Structures for Interstate Trunk Calls, United States, 1959-1967

Rate Period and Mileage Bands	*Sept 19, 1959-Apr 3, 1963*	*Apr 4, 1963-Jan 31, 1965*	*Feb 1, 1965-Oct 31, 1967*
Peak	04:30-18 Mon-Sat	04:30-18 Mon-Sat	04:30-18 Mon-Fri
25-220 mi	$0.30-0.80	Unchanged	Unchanged
221-506 mi	$0.85-1.25	Unchanged	Unchanged
507-3,000 mi	$1.30-2.25	Unchanged	$1.30-2.00[a]
Shoulder	18-04:30 Mon-Sat 00-24 Sun	18-21 Mon-Sat 04:30-21 Sun	18-20 Mon-Fri 04:30-20 Sat
25-220 mi	$0.30-0.60	$0.30-0.60	Unchanged
221-506 mi	$0.65-0.95	$0.65-0.95	$0.65-0.90
507-3,000 mi	$1.00-1.75	$1.00-1.75	$0.90-1.50
Off-Peak		21-04:30 daily	20-04:30 Mon-Sat 00-24 Sun
25-220 mi	–	$0.30-0.60	Unchanged
221-506 mi	–	$0.60-0.65	Unchanged
507-3,000 mi	–	$0.65-1.00	Unchanged

Note: Prices are for initial three minutes, station to station.

[a]The rates for distances beyond 600 miles took effect April 1, 1965.

Furthermore, all hours on Sunday were now at off-peak rates. Small reductions of about 10 percent were made in the level of shoulder-period rates.[5]

Selected data on trunk message volume are available from measurements in October 1960, 1963, 1965, and 1966. Unfortunately, only partially aggregated weekday observations are now available, so that the analysis of demand responses that is possible is quite limited. In particular, because messages are aggregated into a single value for the 8 a.m.–5 p.m. period and into another value for the 5 p.m.–8 p.m. period, one cannot detect any shifting of demand that may have occurred near the beginning or end of a tariff period, nor can one establish the relationship between the actual daytime peak and the traffic levels in shoulder and off-peak periods. These data do, however, distinguish between business and residential customers.

Figure 33–13 shows the evolving pattern of evening trunk messages in the longest-haul markets, 507–3,000 miles, juxtaposed against the prevailing tariff structure. Beginning at 8 p.m. the number of hourly messages is shown as a percentage of the total number of messages in the 8 a.m.–5 p.m. period. However, because hourly data were not available, the dotted line shows only the mean volume per hour during the 5 p.m.–8 p.m. period. (The appearance of an abrupt increase in messages at 8 p.m. in 1960 is presumably an artifact produced by this deficiency).

The 1963 introduction of new off-peak rates at 9 p.m. had a dramatic effect, more than doubling the number of 9 p.m.–10 p.m. calls but also stimulating greater traffic later in the evening as well. Then, as can be seen in the third panel of the figure, the 1965 advancement to 8 p.m. of the starting hour for the off-peak rate led to an effective leveling of the 8 p.m.–10 p.m. demand. One year later, the demand during the first off-peak hour had again grown relatively faster and was determining the evening peak load.

This empirical evidence suggests several hypotheses about consumer behavior when telephone rate structures are changed and how such behavior might be included in an analytic model. The data themselves, however, should be regarded as only approximate indicators of the economically relevant measures of consumer demand.

For some purposes an aggregate measure of demand is both appropriate and necessary, as practical considerations require that a single, quite simple rate structure apply to all consumers. Nevertheless, a more detailed examination of consumer responses is desirable. There can be important differences between measures of the number of calls and total traffic. Similarly actual demand may not be accurately reflected by measured demand statistics at peak hours. In a network spanning several time zones the definition of common tariff periods may have varying implications for traffic loads in different regions. Finally, the variations across the network in the marginal costs of meeting peak loads should also be considered in developing an aggregate measure of trunk call demand. Continued interchange between economists and engineers should lead to increased understanding of these factors.

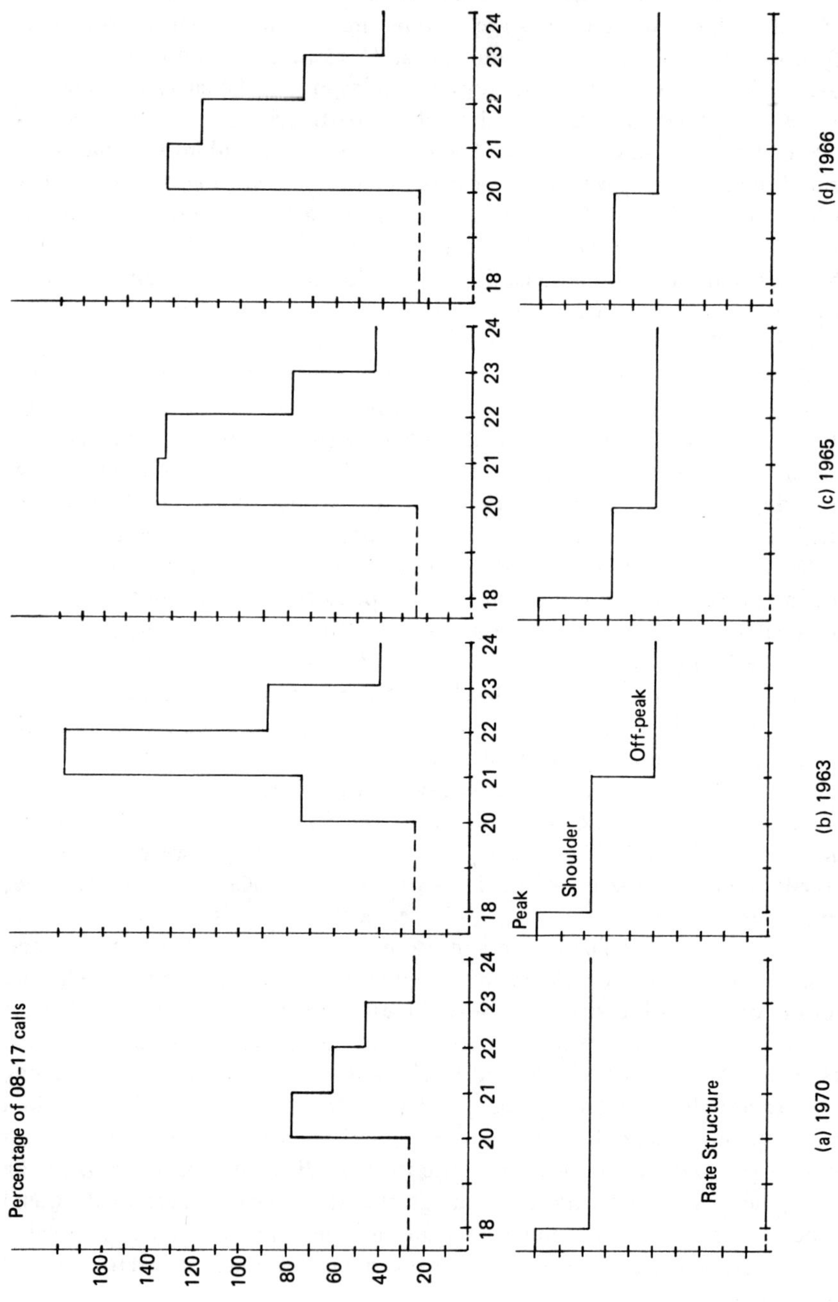

Figure 33-13. Distribution of Weekday Evening Calls in the United States, October 1960-1966 (507-3,000 miles)

Toward an Applied Model of Peak-Load Pricing

In the previous section we surveyed a variety of data on changes in traffic patterns that have resulted from revisions in trunk call rate structures. The combined experience accumulated by several telephone administrations and the detailed data they possess should enable a quantitative model of the demand for telephone calls under peak-load rates to be constructed and calibrated to national parameters. The development of such a model could be of value to the telephone administrations in their forecasting and revenue-planning activities, to national authorities and public utility regulators in assessing the performance of telephone enterprises, and to the general public who would benefit from the increase in efficiency of the telephone network that could be achieved under improved rate structures.

In this section I discuss some of the major elements that need to be incorporated into a model of peak-load pricing applicable to a telephone system. In order to be reliably used for quantitative forecasting and evaluation, such a model must incorporate the most important specific characteristics of telephone technology and call markets in addition to the general principles developed in the literature of optimal pricing. Nevertheless, some of the suggestions advanced here may be of use in other sectors, such as electric power, in which peak-load considerations are important.

Most of the discussion assumes that it is desirable to base prices on marginal costs, subject to a budget constraint that total costs be recovered, and that the level of service quality is exogenously given. The section concludes, however, with a short remark about the recognition of consumption externalities and consumer valuation of telephone service.

The Peak-Load Pricing Problem

In the classic peak-load pricing problem, first set out by Boiteux (1949) and Steiner (1957), one assumes that the quantities demanded are at uniform levels $Q_1 < Q_2$ in each of two equal-length time periods T_o (off-peak) and T_p (peak). To supply the service demanded requires resource expenditures of b (marginal operating costs) per unit consumed plus c (marginal capital costs) per unit of capacity.[6] Telephone service is perhaps the extreme form of this problem; almost all call-related costs are due to the provision of capacity, so that $b = 0$. Let $x(t)$ be the rate of calls demanded per unit time; then the cost of capacity per calling minute is

$$\beta = c \frac{\max x(t)}{\int_{T_p} x(t)\, dt} = c \frac{Q_2}{Q_2 \int_{T_p} dt} = \frac{c}{\text{Number of peak period minutes}}$$

Welfare-optimal pricing results if prices are set equal to the marginal costs of a call in each period: $p_1 = b = 0$ and $p_2 = b + \beta = \beta$. This result is correct provided that there is a firm peak, that is, that at these prices Q_1 is still less than Q_2. If, however, the peak shifts so that $Q_1 > Q_2$, then the optimal prices are those that lead to full utilization of capacity in each period and recover total costs, that is, $Q_1 = Q_2$ and $p_1 + p_2 = c$/total number of minutes.

The realistic peak-load pricing problem consists of a (large) number of periods n, each with different levels of demand, and the constraint that there can be at most $m < n$ distinct prices. Multiple periods are not, in themselves, a difficulty (Williamson, 1966) and the optimal pricing rules of the classical model are readily generalized to handle n periods. For the firm-peak case the price in the peak period is set equal to the per call cost of capacity, ansd the prices in off-peak periods are set to zero. If the peak period is shorter than in the two-period model, the per call peak price β must be higher to cover the same capacity cost. In the shifting peak case one must find those prices that equate demand in a set of periods and jointly sum to the capacity cost divided by the combined lengths of the peak periods.

However, the constraint that there be a limited number of distinct prices is the crux of the practical peak-load pricing problem. Craven (1971) and Rees (1976) provide the theoretical framework for calculating the optimal number of pricing periods. In practice the maximum number of price levels used for trunk telephone calls is three. In the pricing of electric power three is also the maximum number of distinct rates, although rate levels may be reset seasonally (Mitchell, Manning, and Acton, 1978). The fact that public utilities have so widely limited themselves to only two or three price levels is not accidental. It reflects their experience in dealing with millions of customers and their subscribers' interest and willingness to process price information. Even with three price levels consumer surveys report a significant degree of ignorance and actual misinformation about the periods at which discount rates apply. In economic terms consumers' transactions costs of processing detailed price information climb very rapidly and presumably exceed the efficiency gains that might otherwise be achieved by a more complex rate structure.[7]

The immediate implication of the pricing constraint is that there will be nonuniform levels of demand within a pricing period and that the peak-load pricing problem becomes one of simultaneously finding prices and pricing methods that satisfy an optimality criterion.

If a narrowly defined peak period is chosen, the per call price must be very high to recover capacity costs over a small number of calls. The high price, however, increases the likelihood that, because of the own-price elasticity of demand, the peak will shift. Moreover, a narrow period means that through the cross-price elasticity of demand a relatively high proportion of calls initially within the period could be shifted to adjacent periods of time, thereby increasing the likelihood of a shifting peak. On the other hand, if a wide peak-price period is selected, then charging the peak price for calls during subperiods when demand is less than capacity results in unnecessary rationing, and at the same time there is no price incentive to shift calls from the subperiod in which the peak occurs to the adjacent subperiods where excess capacity is available.

It should be clear from the evidence on consumer responses to rate structure changes that the shifting peak case is more than a theoretical curiosity. Shifting peaks for trunk traffic, both intraday and to the weekend, have occurred at off-peak prices well above zero.[8] If one considers the peak-load problem to originate from the weekday peak, the requirement that weekend price levels be chosen from those used during the twenty-four-hour weekday pricing periods can be an important constraint. In this case the higher own-price elasticity of demand from residential customers during weekend hours reduces the possibilities for leveling weekday loads.

Modeling the Trunk Call Market

The task of constructing a realistic model for applying peak-load pricing to telephone systems can be divided into two parts: (1) the measurement of demand functions by time-of-day and day of week and (2) the solution of an optimization problem that incorporates constraints on the number of prices to be charged and the requirement that revenues cover total costs.

Two separate problems in evaluating the development of telephone traffic may be distinguished: forecasting under constant rate structure and evaluating alternative rate structures.

Several telephone administrations have developed extensive data bases and econometric models to forecast revenue growth and to guide investment planning. In such models total traffic or messages are disaggregated by consumer class or sector of economic activity and in some cases by region or by call distance. The principal explanatory variables are the number of active main stations an PBX trunk lines, the national or regional level of economic activity, the rate of growth of key sectors (services), disposable income, and a trunk call price index. For examples of this type of model see Auray (1977), Cracknell (1977), Davis, Caccappolo, and Chaudry (1973), Drew and Cracknell (1976), and Pousette (1976).

In contrast, little research to date has been directed toward modeling the effect of changes in the peak-load structure of telephone rates.[9] Changes in rate structure occur relatively infrequently. In the past such innovations have resulted primarily from a perceptible imbalance between demand and capacity or a political need to offer a lower price in some part of the rate structure, rather than from a deliberate attempt to improve the efficiency of the network.

The major questions to be answered in considering a change in rate structure concern the alterations in the level and timing of peak loads and the effects on total revenues. To a first apprximation, these effects will primarily modify the ongoing growth trends that are central to conventional traffic and revenue forecasting models. A useful model for analyzing rate structures should, by the principle of parsimony, emphasize a rather different set of major variables than are employed in a forecasting model. Such an applied model of peak-load pricing will give primary attention to the detailing of own- and cross-price elasticities of demand.

How best to solve the model, once constructed, will presumably depend to some extent on its particular mathematical structure and will require numerical search or iterative procedures. The model can incorporate several alternative objective functions and be used, for example, to maximize profits, social welfare, or benefits to particular subscriber groups.

The major problem is to simultaneously determine the optimal prices *and* pricing periods. Initial feasible solutions can readily be calculated from an ad hoc establishment of the pricing periods. One can then test for local optimality by perturbing the boundaries of the periods, calculating new optimal prices, and comparing values of the maximands.

Although the properties of such an iterative procedure will need to be investigated to determine whether they produce a global optimum, iterative methods will be of practical importance in indicating the direction of improvement of proposed changes in rate structures and enabling the user to evaluate the changes in traffic flows that would result.

Own-Price Elasticities

It is a priori clear that, just as business and residential subscribers have widely been found to have different total traffic demand elasticities, they are certain to have different time-of-day characteristics. It would, therefore, be desirable to have separate hourly (or more frequent) measurements of trunk traffic for each customer class, but such data are unlikely to be available in time-series form except when special traffic or marketing surveys have been conducted. Nevertheless, when it is necessary to work with aggregate load curves, one should keep in mind the changing composition of total traffic. For example, in systems that have not yet reached a high level of main station penetration, one should allow for a relative increase in the number of residential subscribers.

Separate own-price elasticities will be needed for almost all hours of the typical weekday. While some adjacent hours can probably be combined into subperiods with a single average elasticity, at the outset it is not obvious which periods should be grouped together, except, perhaps, for the early morning hours. One possibly interesting technique would be to fit the hourly elasticities in terms of a smoothing function, such as a cubic spline, using a relatively small number of parameters. Because there will be less time differentiation of rates on weekends, it should suffice to work with only a few elasticities for multihour weekend periods or to use a parsimonious smoothing function.

Cross-Elasticity Effects

The evidence of consumer responses to traffic changes strongly indicates that at least within a twenty-four-hour period, the introduction of time-of-day rates has a highly nonuniform effect on traffic at different hours. As a generalization, it appears that the greatest load changes occur during the minutes just preceding and following the time that the new price takes effect, smaller changes are found at greater distances in time, and little or no effect is discernible more than a few hours before or after the price-switching time.

To focus most clearly on the interhour cross-elasticity effect, it is convenient to assume that all own-price elasticities are zero, so that price changes merely redistribute a fixed total amount of traffic during the twenty-four-hour day. Further, for clarity, assume that the initial load curve $x(t)$ which prevails under a uniform price $p(t) = p_0$ is the simple two-level curve shown in figure 33–14a. In the conventional peak-load pricing model the introduction of a peak-load price, $p(t) = p_1$ for the period $T_p \in (t_1, t_2)$, would uniformly depress the curve in the peak period and uniformly raise it at other times (figure 33–14b). For telephone traffic, however, the result is something like that shown in figure 33–14c. The greatest load changes occur in the minutes just after the peak price takes effect and just before the lower price again becomes effective. Most of this load is shifted just outside the peak period, and the added load diminishes rapidly as the distance in time from the peak-pricing period increases.

This nonuniform cross-elasticity effect might be modeled from a number of perspectives. One approach would begin with a consumer whose preferences for a particular pattern of calling over the day can be summarized in a utility function. A call placed at a time other than the preferred moment is of lesser utility, and this disutility increases with the magnitude of the call's displacement in time. The consumer maximizes his utility function subject to the budget constraint, trading off disutility of call displacement for the savings on peak-period charges. Such an approach would represent an extension of a model introduced by Koenker (1977). It would then be necessary to aggregate over individual consumers. But unlike the approach used by Mitchell (1978), this would probably require explicitly including stochastic factors, since the average consumer makes only about four calls in a twenty-four-hour period.

An alternative approach is to impose a suitable structure directly on the market demand functions, sacrificing the elegance of deriving demand relationships from formal consumer theory but shortcutting some formidable aggregation problems. Because the most readily available data for empirical estimation

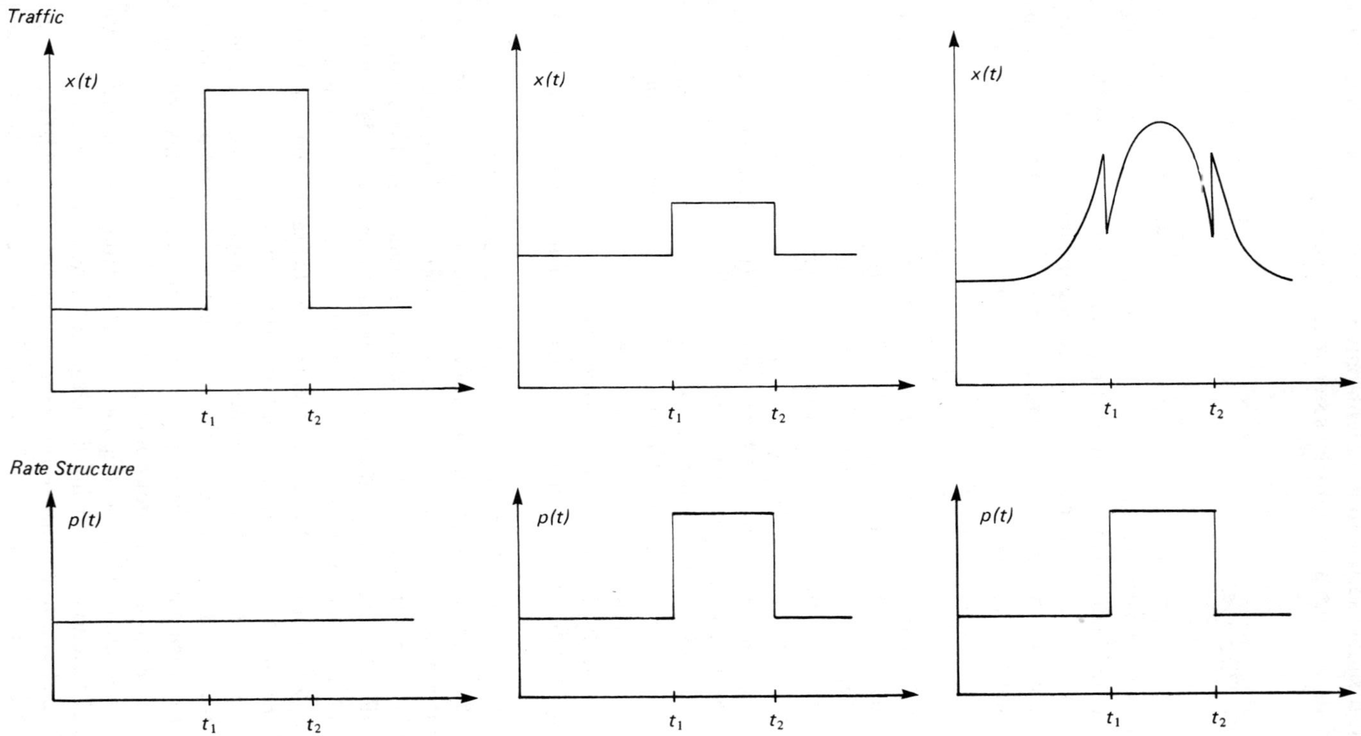

(a) Load Curve under Uniform Pricing

(b) Uniform Cross-Elasticity Effect under Peak-Load Pricing

(c) Nonuniform Cross-Elasticity Effect under Peak-Load Pricing

Figure 33–14. Theoretical Effects of Peak-Load Pricing

or calibration of a model are aggregate observations on calls or traffic, this may initially be the more promising approach.

How might nonuniform cross-elasticity effects be introduced in a peak-load pricing model? Consider a simple example, using again the initial load curve $x(t)$ of figure 33–14a. Let the amount of load shifted (figure 33–15) from a moment within the peak period, at time $t + \tau$, to the moment t be $z(t,\tau)$. We assume that this amount is equal to a cross-elasticity coefficient $\in(\tau)$ times the relative increase in price multiplied by the initial volume of calling at time $t + \tau$.

$$z(t,\tau) = \in(\tau)\frac{p(t+\tau) - p(t)}{p(t)}x(t+\tau)$$

The coefficient $\in(\tau)$ decreases (absolutely) with the length of time τ by which the calling is displaced. An exponential $\epsilon(t) = \epsilon e^{-\lambda t}$ or some other simple function will probably be satisfactory for empirical work.

The displacement $\Delta x(t)$ of load at a time t in the off-peak period is the total of such amounts integrated over the period of price change:

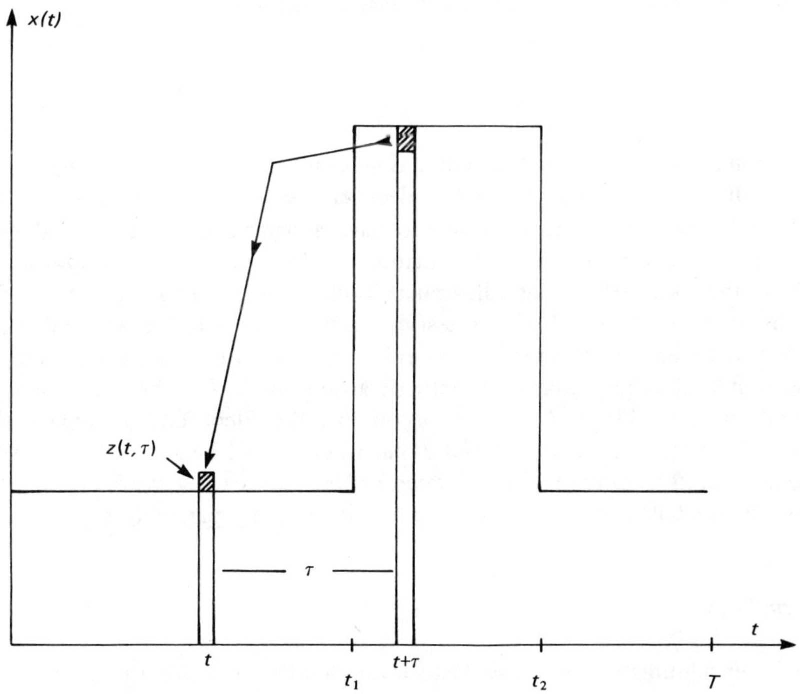

Figure 33–15. Displacement of Peak Traffic to Off-Peak Period

$$\Delta x(t) = \int_{t_1-t}^{t_2-t} z(t,\theta)d\theta, \quad t \in (t_1, t_2)$$

During the peak period the displacement is

$$\Delta x(t) = \int_{0}^{t_1} z(t,\tau)dt + \int_{t_2}^{T} z(t,\tau)dt, \; \tau \in (t_1, t_2)$$

The formulation of interday cross-elasticity effects can be more straightforward. From the empirical evidence available to date, it seems reasonable to assume that there are no significant demand shifts between days, except perhaps between Saturday and Sunday.[10] For this case, a uniform cross elasticity from each Saturday hour to the same hour on Sunday may be satisfactory. However, in empirically establishing the model's parameters, one should also check carefully for the existence of weekend echos at the hours that lower rates take effect on weekdays. However irrational, such an increase in demand represents a particular form of cross elasticity that, if it does not fade away shortly after the new tariff takes effect, could result in capacity shortages.

Trends

A new long-run equilibrium load curve is no doubt reached more rapidly in telephone call markets than, say, in the electricity sector, where changes in long-lived appliance stocks can continue to have a significant effect several years after a new rate structure has been introduced. Nevertheless, it is important to establish the speed of market adjustment, both in order to use data in the first months following a rate change to test model predictions against actual developments and to be able to separate the effects of the structural change from ongoing trends in other variables affecting traffic growth. For these purposes the methods used by Pioger (1977) to decompose the Electricité de France load curve and estimate the annual trend at each hour could be adapted to monthly observations. By comparing trend rates before and after a tariff change, this approach could also be used to estimate hourly elasticity parameters.

Externalities

Telephone administrations have historically tended to price the subscription service below its costs. They and the parliamentary bodies that approve their rates have justified this policy in terms of making telephone service widely available, insuring that it is affordable by economically disadvantaged groups, and so on. Moreover, when rate increases are necessary there may be greater public resistance to increasing the fixed monthly charge than the unit charge.

A telephone provides valuable service only if there is someone at the other end of the line to talk to. This simple observation suggests the general proposition that the value of an individual telephone subscription increases with the total number of subscribers or with the number of subscribers who belong to an individual interest group. If the costs of supplying telephone service consist simply of a fixed cost per subscriber plus a constant marginal cost per call, it is welfare-optimal to recognize this externality in individual consumption functions and to set the subscription fee below marginal cost in order to enlarge the number of telephone subscribers (Littlechild, 1976). This cross-subsidy pattern is reinforced when the heterogeneity of consumer tastes is considered and a subscriber's willingness to pay depends on the number of calls made (Mitchell, 1978).

Some telephone administration studies have taken account of the role of the number of telephone connections in stimulating demand for subscriptions as well as increasing total traffic (for example, Böhm and Wacker, 1977). However, to date empirical studies have not been able to quantify the extent of the externality or to calculate what degree of subsidy to the subscription price would optimize the development of the network over time.

Once they have been measured, it is straightforward to incorporate externality effects into the applied model for peak-load pricing. The revenue constraint is modified to include a subsidy to cover a specified amount of shortfall from subscription fees. Other constraints—for example, a required amount of "separations" payment to state jurisdictions—can as easily be added to the costs that must be covered.

Congestion and Service Quality

In the short run the capacity of the telephone system is fixed, and the allocation of scarce telephone resources is optimally guided by pricing at the opportunity costs of those resources. As demand approaches the technical capacity of the network, the stochastic process of arrival and duration of calls will result in a rapidly increasing proportion of attempts and calls that are incomplete. Because such attempts consume dialing and waiting time of subscribers, they impose a congestion cost like that created by the extra car entering the freeway at rush hour.

In this situation, it would be optimal to set the price of a call equal to the marginal congestion cost that it caused (Pyatt, 1972). Note that such a pricing rule is obtained solely by considering valuations derived from consumers' demands and is not directly related to the marginal capacity of running costs of supplying additional calls. However, the optimal investment policy requires that the capacity of the telephone network be expanded wherever marginal investment costs are not greater than the marginal congestion costs (which have been set equal to the optimal prices). As a result, this pricing rule is similar to pricing at long-run marginal cost.

There is, however, one important difference between congestion pricing and marginal-cost pricing. In the congestion pricing model the quality of service is endogenous, and routes that have high marginal expansion costs will tend to have lower service quality (Kay, 1968). For this reason, it would appear that service quality would be lowest on long-distance routes. However, the declining importance of distance in the total cost of supplying trunk calls may diminish the practical significance of this factor.

An important and unanswered empirical question is whether residential consumers value service quality less highly than business subscribers. If they do, it is indeed desirable to ration capacity in off-peak hours by allowing some increase in congestion to occur,[11] and the practice of dimensioning the trunk network to meet peak demands during weekday business hours is consistent with this view. Nevertheless, households are, no doubt, willing to pay somewhat higher rates to reduce congestion, and the persistently blockaded network that exists on Sundays in regions of West Germany is unlikely to be optimal.

Notes

1. International comparisons of rate levels are problematic, and prices at U.S. dollar exchange rates provide only a rough indication of the relative charges for different services.

2. The Swedish telephone administration will introduce off-peak rates in 1982.

3. Trunk call rates were effectively reduced by 20 percent when the network was converted from manual to automatic operation in 1963.

4. Traffic measurements are for the Friborg, Olten, Sion and Thun areas.

5. In addition, consolidation of the 601-675 and 676-800 mile bands resulted in reductions in peak-period rates for all higher distances, as of April 1965.

6. In this discussion we leave aside the effect of economies of scale (Mohring, 1970).

7. It may be that a more detailed rate structure will be of interest to subscribers with specialized telecommunications needs, such as data transmission. In such cases optional rates could be designed to cater to their individual load characteristics, much as interruptible power tariffs are offered to selected large industrial users of electricity.

If consumers could be provided with price quotations at the moment they demand telephone service, a dynamic rate structure could be used, with price levels varying to reflect the level of capacity actually available when the call is placed. Proposals for dynamic pricing have met with widespread skepticism, but with fully electronic networks such an idea would be cost-effective. As a two-way medium, telephone service is well suited to such a fundamental innovation.

8. Evidence of shifting peaks under time-of-day rate structures is also found in some European electric utilities (Mitchell, Manning, and Acton, 1978).

9. Littlechild (1970) represents a first step in this direction. Willig and Bailey (1977) consider repricing day, evening, and night calls by assuming that demand functions are independent; their major interest is the structure of call prices as it relates to distance.

10. This simplification is, of course, also justified by the fact that traffic peaks are sufficiently similar Monday through Friday that the same rate structure is almost certain to apply on every weekday.

11. For distributive reasons, political decision making may also prefer this outcome to the alternative of higher off-peak prices.

References

Auray, R. 1977. "Customer Response to Message Toll Telephone Rate Changes." Paper presented at the Ninth Annual Conference, Institute of Public Utilities, Williamsburg, Va., December 12–14.

Bailey, Elizabeth E., and White, Lawrence J. 1974. "Reversals in Peak and Off-Peak Prices." *Bell Journal of Economics and Management Science* 5:75–92.

Boiteux, Marcel. 1949. "La Tarification des Démandes en Pointe." *Revue Générale de l'Electricité* 58:321–340. Translated as "Peak-Load Pricing." *Journal of Business* 33 (1960):157–179.

Böhm, F., and Wacker, W. 1977. "Analytisches Prognosemodell fur Fernsprechhauptanschlüsse." *Der Fermelde-Ingenieur* 31, No. 3: March 15. 1977.

Bø, K., Ganstad, O., and Kosberg, J.–E. 1976. "Some Traffic Characteristics of Subscriber Categories and the Influence from Tariff Changes." Paper presented at the Eighth International Teletraffic Congress, Melbourne, Australia, November.

Cracknell, D.R. 1977. "New Econometric Models for Full Rate Trunk Calls." Report No. 89, Statistics and Business Research Department, British Post Office, April.

Craven, J. 1971, "On the Choice of Optimal Time Periods for a Surplus Maximizing Utility Subject to Fluctuating Demand." *Bell Journal of Economics and Management Science.*

Dansby, Robert E. 1975. "Welfare Optimal Peak-Load Pricing and Capacity Decisions with Intraperiod Time Varying Demand." Bell Laboratories, Economic Discussion Paper No. 39, November.

Davis, B.E., Caccappolo, G.J. and Chaudry, M.A. 1973. "An Econometric Planning Model for American Telephone and Telegraph Company." *Bell Journal of Economics and Management Science* 4, no. 1:29-56.

Deutsche Bundespost, 1976. *Geschäftsbericht 1975*, Bundesministerium für das Post-und Fernmeldewesen.

Drew, J.M., and Cracknell, D.R. 1976. "An Econometric Model of Cheap Rate Trunk Calls," Report No. 73, Statistics and Business Research Department, British Post Office, November.

Hirshleifer, J. 1958. "Peak Loads and Efficient Pricing: Comment." *Quarterly Journal of Economics* 72:451-452.

Kay, J.A. 1968. "Pricing and Investment Criteria in Public Utilities: The Case of Telephone Service." M.A. dissertation, University of Edinburgh. 1968.

Koenker, Roger. 1977. "Optimal Peak-Load Pricing with Time-Additive Consumer Preferences." In *Forecasting and Modeling Time-of-Day and Seasonal Electricity Demands,* ed. A.C. Lawrence. Electric Power Research Institute, Palo Alto, Calif., December.

Krapf, J. 1977. "Auswirkungen der Tariferhönungen und der Änderung der Tarifzeiten vom 2. Dezember 1974 auf den Telefonverkehr." *PTT Technische Mitteilungen* 55, no. 2:62-70.

Littlechild, Stephen C. 1970a. "Marginal Cost Pricing with Joint Costs." *Economic Journal* 80:323-325.

——. 1970b. "Peak-Load Pricing of Telephone Calls." *Bell Journal of Economics and Management Science.* 1, no. 2:171-210.

——. 1975a. "Two-Part Tariffs and Consumption Externalities." *Bell Journal of Economics and Management Science* (Autumn):661-670.

Littlechild, Stephen C., and Rousseau, J.J. 1975. "Pricing Policy of a U.S. Telephone Company." *Journal of Public Economics* 4:35-56.

Marchand, M.G. 1973. "The Economic Principles of Telephone Rates under a Budgetary Constraint." *Review of Economic Studies* 40, no. 4:507-515.

Mitchell, B.M. 1978. "Optimal Pricing of Local Telephone Service." *American Economic Review* (September).

Mitchell, B.M., Manning, W.G., Jr., and Acton, J.P. 1978. *Peak-Load Pricing: European Lessons for U.S. Energy Policy.* Cambridge, Mass.: Ballinger Publishing Co.

Mohring, H. 1970. "The Peak-Load Problem with Increasing Returns and Pricing Constraints." *American Economic Review* 60:693-705.

Mossler, Elisabeth. 1977. "Peak-Load Pricing for Trunk Calls," *Tele,* English edition, No. 1:49-55.

Panzar, J.C. 1976. "A Neoclassical Approach to Peak-Load Pricing." *Bell Journal of Economics* 7.

Pioger, Y. 1977. "Eléments de Reflexion sur la Déformation des Courbes de Charge Globales et Sectorielles: Evaluation de l'Effet Induit par la Tarification." Electricité de France, 8 July.

Pousette, Tomas. 1976. "The Demand for Telephones and Telephone Services in Sweden." Presented at the European Meetings of the Econometric Society, Helsinki, August.

——. 1978. "Samhällsekonomiska Aspekter på Televerkets Prissättning." Industriens Utredningsinstitut, Stockholm, March.

Praz, L. 1971. "Eine Methode zur Ermittlung der Kosten von Telefonverbindungen, *PTT Technische Mitteilungen* 55, no. 2:52-61.

Pyatt, G. 1972. "Some Economies of a Public Utility." Discussion Paper No. 28, University of Warwick, Coventry, September.

Rees, Ray. 1976. *Public Enterprise Economics.* London: Weidenfeld and Nicolson.

Socher, J. 1972. "Die Elastizität des offentlichen Fernsprechverkehrs gegenüber Gebührenänderungen (unter besonderer Berücksichtigung des künftigen Nahdienstes)." In *Jahrbuch des elektrischen Fernmeldewesens.*

Steiner, Peter O. 1957. "Peak Loads and Efficient Pricing." *Quarterly Journal of Economics* 71:585-610.

Televerkets Författningssamling. 1975. *Enhetstaxa för Kortväga Telefonsamtal samt Nedsatt Kvälls och Veckoslutstaxa,* Serie A:29, Stockholm.

Televerket. 1976. *Tele,* English edition, Stockholm, vol. 27, 1976.

Willig, R.D., and Bailey, E.E. 1977. "Ramsey-Optimal Pricing of Long Distance Telephone Services." In *Pricing in Regulated Industries: Theory and Application,* ed. J.T. Wenders. Denver, Colo: Mountain States Telephone and Telegraph Co.

Williamson, O.E. 1966. "Peak Load Pricing and Optimal Capacity under Indivisibility Constraints." *American Economic Review.*

About the Contributors

Richard P. Adler is with the Aspen Institute for Humanistic Studies in Palo Alto, California, and is a former director of the Aspen Workshop on Television.

Charles Atkin is associate professor in the Department of Communication of Michigan State University in Lansing.

Walter Baer is a senior scientist at the Rand Corporation in Santa Monica, California, and has written extensively on the effects of government regulation and technological change in telecommunications.

Yale M. Braunstein is assistant professor of economics at Brandeis University and has published in the areas of regulation and public policy in telecommunications and the science information industries.

Muriel G. Cantor is professor and chairperson of the Department of Sociology at American University in Washington, D.C.

David Caulkins is the coordinator of the Personal Computer NETwork (PCNET) Committee, a nonprofit volunteer organization. He is also chief engineer of Cabledata Associates.

George A. Codding, Jr., is professor of political science at the University of Colorado. He has worked in the Secretariat of the International Telecommunications Union and acted as a consultant to one of its secretary generals.

Robert Faber is a PH.D. candidate in the School of Journalism and Mass Communication at the University of Wisconsin, Madison.

Donna J. Frese is presently a graduate student in the Department of Speech Pathology and Audiology of Wichita State University.

Bradley Greenberg is professor in the Department of Communication at Michigan State University.

Emilie Griffin is vice-president and creative supervisor at Compton Advertising and was the first director of the Children's Advertising Review Unit of the National Advertising Division of the Council of Better Business Bureaus.

Joseph Grundfest, a graduate of Stanford Law School and Stanford's Department of Economics, is currently a research fellow at the Brookings Institution in Washington, D.C.

Felix Gutiérrez is associate professor of journalism at California State University, Northridge, where he also teaches classes in the Department of Chicano Studies.

Allen S. Hammond IV is program manager of the Minority Telecommunications Development Program at the National Telecommunications and Information Agency. He has served as an assistant professor of Communications Regulation at the University of Maryland.

William E. Hanks is assistant professor of mass communication at the University of Pittsburgh.

Charles L. Jackson is on the staff of the House Subcommittee on Communications as staff engineer. He has been an engineering assistant to FCC commissioner Glen Robinson and special assistant to the chief of the Common Carrier Bureau of the FCC.

Suzanne Jeffries-Fox is a researcher with Yankelovitch, Skelly, and White in Stamford, Connecticut.

Norge W. Jerome, a nutrition anthropologist, is a professor in the Department of Community Health, School of Medicine, at the University of Kansas Medical Center. She is a founding member and member of the board of directors of the Council on Children, Media and Merchandising, and has served as a member of the National Advisory Council of the Children's Television Workshop's series "Feeling Good."

Robert Johansen is senior research fellow at the Institute for the Future in Menlo Park, California. He is coauthor of *Electronic Meetings: Technical Alternatives and Social Choice* (Addison-Wesley, 1979) and numerous articles in the telecommunications field.

Jeffrey Krauss is assistant chief of the Office of Plans and Policy of the Federal Communications Commission. He was previously with the American Satellite Corporation, where he was concerned with regulatory matters.

Lester B. Lave is professor and head of the Department of Economics with a joint appointment in the School of Urban and Public Affairs at Carnegie-Mellon University. He is also adjunct professor in the Graduate School of Public Health of the University of Pittsburgh.

Samuel Leinhardt is professor of sociology in the School of Urban and Public Affairs, Carnegie-Mellon University, Pittsburgh. His primary research interests are the development of models for social networks and their analysis in substantive contexts. He is currently editor of the American Sociological Association's yearbook, *Sociological Methodology*.

John F. Lyons, Jr., is with the Office of Policy Analysis and Development, National Telecommunications and Information Agency, in Boulder, Colorado, where he is involved in policy research and analysis in all areas of communications law.

Richard K. Manoff is president of Manoff International and since 1956 has been active in nutrition, family planning, and health education communications programs in the United States and in some twenty-five developing countries. His firm, which bears his name, is a marketing and advertising firm representing many major American firms.

Steven McDermott is assistant professor of speech communication at San Jose State University. He was a graduate assistant at Michigan State University at the time of the study reported in this volume.

Bridger M. Mitchell is senior staff economist at the Rand Corporation and research fellow at the International Institute of Management, Berlin He has taught at Stanford University and at the University of California, Los Angeles, and has served as a Brookings economic fellow in the Department of Health, Education, and Welfare.

Donald L. Nielson is director of the SRI Telecommunications Sciences Center, a member of the U.S. Commissions C and E of URSI, and a registered professional engineer in the state of California.

John C. Panzar is a member of the Economics Research Department at the Bell Laboratories in Murray Hill, New Jersey. His area of research is applied microeconomic theory with a focus on pricing in regulated markets.

Terry A. Pickett is assistant professor of Communication and Human Relations at the University of Pittsburgh with research interests in telecommunications, citizen participation, decision making, and systems analysis of communications organizations.

Artur Raviv is professor of economics at the Graduate School of Industrial Administration at Carnegie-Mellon University.

Byron Reeves is assistant professor in the School of Journalism and Mass Communication at the University of Wisconsin at Madison. His publications and research have been concerned with children and television, specifically children's perceptions of television characters and their perceived reality of television.

John O. Robinson is a member of the Office of Plans and Policy at the Federal Communications Commission. He holds degrees in aeronautical engineering and communications and is a senior member of the Institute of Electrical and Electronics Engineers.

Walter J. Schaar is general manager of station KSPS-TV Channel 7 in Spokane, Washington. He is an assistant professor at Gonzaga University and president of the Washington State Division of the Western Educational Network.

Nancy Signorielli is research coordinator at the Annenberg School of Communications at the University of Pennsylvania. She has conducted research for the Civil Rights Commission, the Corporation for Public Broadcasting, and the Foundation for Child Development.

Robert D. Smith is vice-president and general manager of the Greater Toledo Educational Television Foundation, licensee and operator of public broadcasting stations WGTE-TV and WGTE-FM, in Toledo, Ohio.

Paul J. Steen is general manager of the KPBS stations in San Diego, California. He is also acting director of University Affairs, San Diego State University, and professor in the Department of Telecommunications and Film.

R. Michael Tyler is managing director of Communications Studies and Planning, London. Communications Studies and Planning has played a significant role in the planning of the British teletext demonstrations, including Prestel.

Albert J. Walderhaug is chief of the Research and Analysis Branch of the Interindustry Economics Division, a part of the Bureau of Economic Analysis in the U.S. Department of Commerce. He participated in the formulation and performance of the 1958 input-output study of the U.S. economy and has been associated with the development of the input-output program in government since then.

Ellen Wartella is assistant professor in the Department of Communication at Ohio State University. Her major research interests are the developmental aspects of children's understanding of television messages. She is also editor of the eighth volume of the Sage Annual Review of Mass Communications research, *Children Communicating.*

Frederick Williams is dean of the Annenberg School of Communications of the University of Southern California and president of the International Communications Association.

About the Editor

Herbert S. Dordick is visiting professor and director of the Center for Communications Policy Research at the Annenberg School of Communications at the University of Southern California. He is currently on leave as a visiting scholar at the Center for International Studies at the Massachusetts Institute of Technology.